CPSS Power Electronics Series

This series comprises advanced textbooks, research monographs, professional books, and reference works covering different aspects of power electronics, such as Variable Frequency Power Supply, DC Power Supply, Magnetic Technology, New Energy Power Conversion, Electromagnetic Compatibility as well as Wireless Power Transfer Technology and Equipment. The series features leading Chinese scholars and researchers and publishes authored books as well as edited compilations. It aims to provide critical reviews of important subjects in the field, publish new discoveries and significant progress that has been made in development of applications and the advancement of principles, theories and designs, and report cutting-edge research and relevant technologies. The CPSS Power Electronics series has an editorial board with members from the China Power Supply Society and a consulting editor from Springer.

Readership: Research scientists in universities, research institutions and the industry, graduate students, and senior undergraduates.

Indexed by EI Compendex

Deshang Sha · Peisong Ma · Jiankun Zhang

High Frequency Isolated Semi Dual Active Bridge DC-DC and AC-DC Converter

Springer

Deshang Sha
School of Automation
Beijing Institute of Technology
Beijing, China

Peisong Ma
School of Automation
Beijing Institute of Technology
Beijing, China

Jiankun Zhang
School of Automation
Beijing Institute of Technology
Beijing, China

ISSN 2520-8853 ISSN 2520-8861 (electronic)
CPSS Power Electronics Series
ISBN 978-981-95-1640-7 ISBN 978-981-95-1641-4 (eBook)
https://doi.org/10.1007/978-981-95-1641-4

This Springer imprint is published by the registered company Springer Nature Singapore Pte Ltd.
The registered company address is: 152 Beach Road, #21-01/04 Gateway East, Singapore 189721, Singapore

Preface

Dual Active Bridge (DAB) converters are widely used for energy storage applications needing bidirectional power flow capability. For some unidirectional power flow applications, two controllable switches on the output side can be replaced by two diodes. In this way, shoot-through can be avoided and this simplifies the power circuit and the driver circuit design. It is easy to achieve ZVS on for all power switches and ZCS off for rectifying diodes under most operating conditions with proper design.

This book presents different semi dual active bridge (semi-DAB) converter topologies and their closed loop control strategies to achieve better power conversion performances for applications requiring only unidirectional power flow and galvanic isolation. The semi-DAB DC-DC converter can be used for electrical vehicles, more electrical aircrafts, and power conditioning system for renewable power sources such as PV panels, fuel cell, etc. The semi-DAB AC-DC converter can be a good solution for on-board charger for EVs or adapters for consumer electronics such as laptop computers.

The converter topologies, different operating modes within entire operating range, closed loop implementation, ZVS achievement condition and conversion efficiency improvement are analyzed in this book. Based on that, the control strategies are applied to achieve better performance aimed at different control goals. This book will benefit the researchers and engineers in the field of topology isolated DC-DC and isolated AC-DC converters. The main objectives are as follows,

1. To review and summarize the existing researches on semi-DAB DC-DC converter which has many distinctive features. The key challenges and issues are also presented and the possible single-stage AC-DC power conversion based on semi-DAB DC-DC converters.
2. For voltage fed semi-DAB DC-DC converter, different control strategies relying on PWM together with phase shift control can be used to reduce the nonactive power, the current stress, and improve the overall conversion efficiency within entire operating range. Operating modes varies as different load requirement and seamless transition can be obtained among different varying modes.

3. To reduce input DC side current ripple, current fed semi-DAB DC-DC topology is proposed. Different control strategies are used to reduce the nonactive power, current stress and conduction loss, improve the full load range ZVS achievement, operating stability and conversion efficiency.
4. To achieve single-stage isolated AC-DC converter, semi-DAB AC-DC topologies with relevant control are proposed. In order to improve the conversion efficiency within full load range, different modulations are used and seamless transition can be obtained. To minimize the total THD of the AC side current due to the lack of inner current tracking loops, decoupled inductor modulation strategy is presented. To achieve full load ZVS range, an improved modulation and variable switching frequency is proposed.
5. To reduce the turn off current of power switches, a resonant semi-DAB AC-DC topology is proposed with fixed switching frequency.

This book consists of 11 Chapters based on our several research projects, and covering the aspects of converter topologies and power control strategies for semi-DAB DC-DC and AC-DC converters. Like most books, this book starts with an introduction in Chap. 1 to present a brief introduction focusing on the applications and derivation of semi-DAB DC-DC and AC-DC converters, review of the control methods for dual active bridge converters, and key issues and challenges of this kind of converters.

For the rest contents, from the power circuit point of view, this book can be divided into two parts. The first part is focused on semi-DAB DC-DC converter supplied by DC power source which is from Chaps. 2–7, and the second part discusses semi-DAB AC-DC topologies and controls supplied by AC power sources, which are included from Chaps. 8–11. The detail organization of the book content is summarized in Chap. 1.

The Authors would be very grateful for some research programs granted by Nature Science Found of China(51577012, 5197701), Beijing Nature Science Foundation of China(3192033) and the Technology Innovation Talent Plan of BIT (2021CX01014). Those projects inspire us pursue our interests with great enthusiasm in this field.

The Authors would like to express their sincere thanks to Dr. Deliang Chen, Mr. Fulin You and Mr Yichao Fu for their contributions on semi-DAB DC-DC converters. Of course, many APCC members have done excellent work related with the semi-DAB power converters.

The Authors would also like to thank their families who have given tremendous support all the time. Finally, the authors are extremely grateful to Springer and the editorial staff for the opportunity to publish this book and help in all possible manners.

Beijng, China

Deshang Sha
Peisong Ma
Jiankun Zhang

Competing Interests The authors have no competing interests to declare that are relevant to the content of this manuscript.

Contents

List of Figures

Chapter 1
Introduction

Abstract This chapter presents a brief introduction focusing on the applications and classifications of unidirectional high frequency isolated semi-dual active bridge (semi-DAB) DC-DC converters which are widely used in many applications such as energy power conversions for renewable energies for PV or fuel cells, power supply for all electrical aircrafts and electrical vehicles, etc. High frequency isolated semi-DAB AC-DC converters can be used to charge EV batteries. To achieve high power conversion efficiency and high power density, the topology candidates and the corresponding power control methods has been a challenge in both industry and academy research fields. This chapter provides a basic foundation for the whole work, and it gives the goal of this book which is to provide valuable knowledge, advanced control methods and practical design guides for the high frequency isolated unidirectional semi dual active bridge DC-DC converters with wide voltage gain.

Keywords Unidirectional high frequency isolated converters · Semi dual active bridge (semi-DAB) · Energy power conversions · Renewable energy · Electrical vehicles · DC-DC converters · AC-DC converters

1.1 Applications of High Frequency Isolated DC-DC Converters and AC-DC Converters with Unidirectional Power Flow Capability

The fact that batteries can store electricity has been well known for most people all over the world, because they are widely used in our daily appliances such as cameras, shavers, laptops and so on. On a larger scale, batteries are also used to store energy for the electrical grid, electric vehicles and DC microgrid. The energy storage of these batteries plays an important role for the system stability.

High frequency isolated DC-DC converters are widely used in many applications. The input is a DC while the output is a DC, the transformer can achieve galvanic isolation for safety considerations and the voltage conversion gain can be adjusted

D. Sha et al., *High Frequency Isolated Semi Dual Active Bridge DC-DC and AC-DC Converter*, CPSS Power Electronics Series,
https://doi.org/10.1007/978-981-95-1641-4_1

very flexibly by changing the transformer turs-ratio. The isolated DC-DC can be used in many applications.

It has the following values:

(a) Energy storage can be used as a flexible resource for the power grid.
(b) Achieve the peak shaving function for the grid.
(c) Frequency regulation, transmission capacity relief and VAR supporting.
(d) Reduce the energy-related emissions.
(e) Increase the efficiency and reliability for microgrid.

1.1.1 DC-DC Converters Supplied by Renewable Energy Sources, DC Batteries, Fuel Cells

Figure 1.1 shows the basic architecture of a new energy microgrid system. It can be seen that the DC power generated by photovoltaic panels needs to be connected to the DC bus through the DC-DC converter, the AC power obtained by wind turbines needs to be connected to the DC bus through the AC-DC converter, the AC power grid also needs to be connected to the DC bus through the isolated AC-DC converter. The energy storage system like battery needs to be connected to the DC bus through the DC-DC converter. When the power generated by the renewables exceeds the demanded power, the bidirectional DC-DC converter will charge the battery, thus storing these energies for backup. On the other hand, once the power is not enough to support the facilities, the bidirectional DC-DC converter would be used to discharge the battery. The DC bus provides power to the DC load through the DC-DC converter and to the AC load through the DC-AC converter. The performance of the DC-DC and AC-DC converter will have impact on the system conversion efficiency, cost, dynamic response and reliability.

Figure 1.2 gives the power circuit topology for the PV microinverter, where two single-ended flyback DC-DC converters are interleaved to improve the power handling capability and minimize the input current ripple [1]. The low voltage PV DC voltage is boosted to a relatively high DC voltage working as the input for the inverter stage composed of thyristor switched as low frequency. The switching frequency is fixed and soft switching are difficult to be achieved for the DC-DC power conversion stage. DC power supplies can be obtained not only from DC sources such as PV panels, batteries, fuel cells, etc. In many cases, DC voltages are obtained by rectifying AC voltage form the grid or generators. A non-isolated AC-DC rectifier is used generate DC source. This will be discussed later.

The push–pull DC-DC converter shown in Fig. 1.3 is used to boost low DC voltage to a relatively high DC voltage. Switches normally work at hard switching states due to the fixed switching frequency using PWM modulation technique. The reverse recovery issue of rectifying diodes is a concern and it causes not only additional losses but also the output noises.

A widely used DC-DC topology is phase shifted full bridge (PS-FB) DC-DC converter, which is shown in Fig. 1.4 [2, 3]. It can be used to supply power for

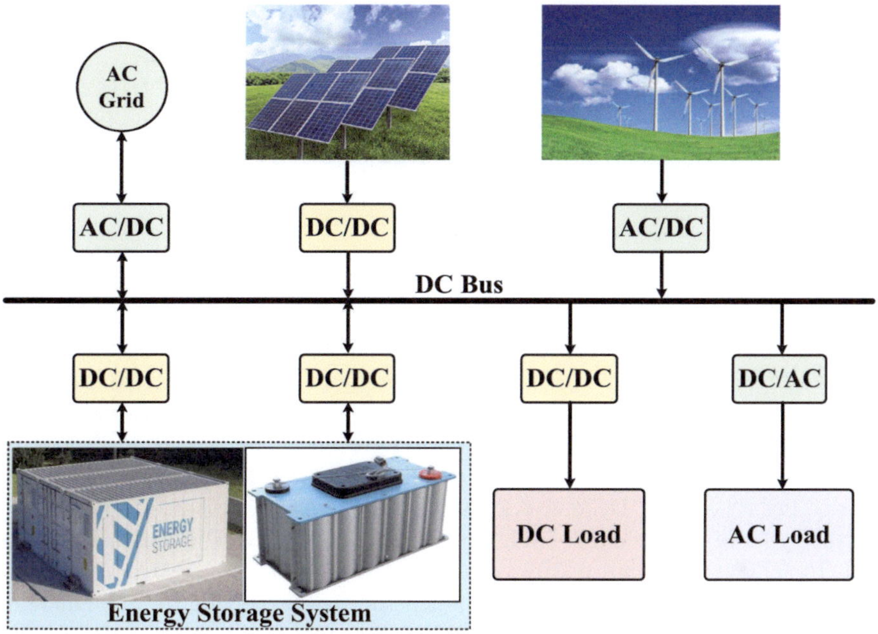

Fig. 1.1 Renewable energy based microgrid system architecture

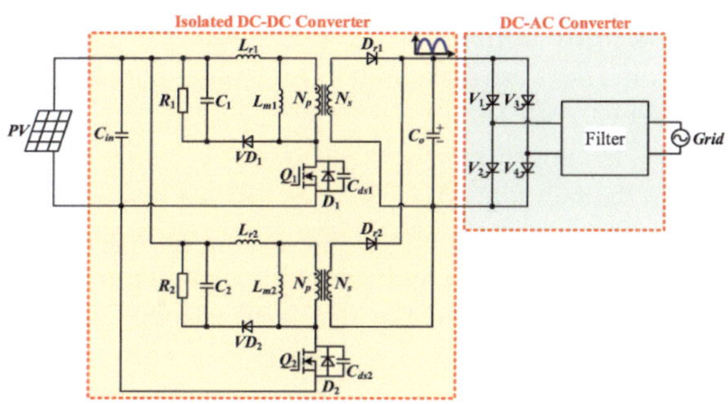

Fig. 1.2 Flyback-based microinverter topology

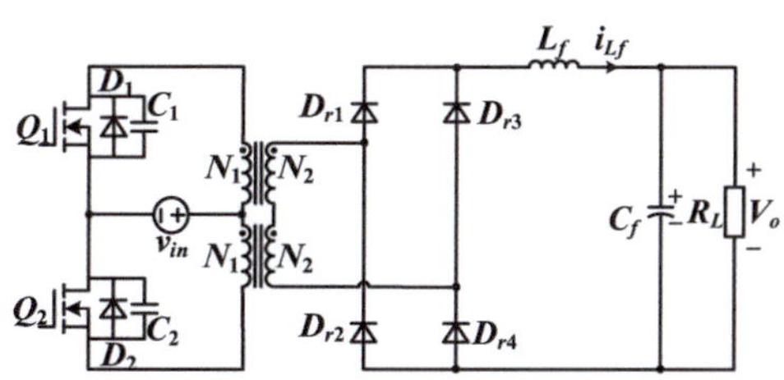

Fig. 1.3 Push–pull high frequency isolated DC-DC converter

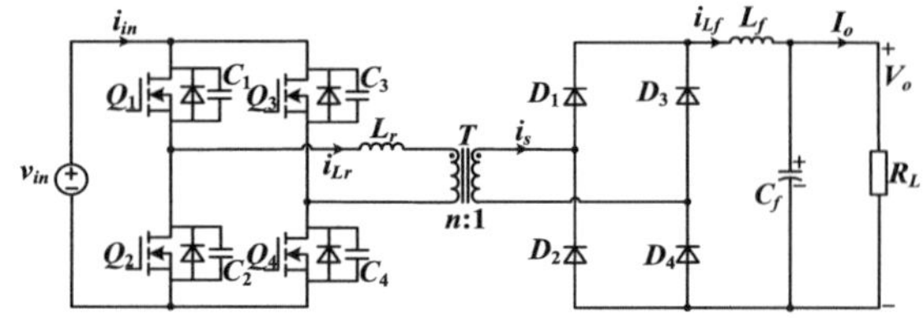

Fig. 1.4 Phase shift full bridge high frequency isolated DC-DC converter

telecommunication equipment, welding machines and cutting machines et al. But the conventional PS-FB converter cannot achieve high efficiency over the CC–CV charging profile due to its unique drawbacks such as large circulating current and narrow zero-voltage-switching (ZVS) range for lagging leg switches. To extend the ZVS range for lagging leg switches and minimize the conduction loss, the lagging leg can be used to form a half bridge LLC [4, 5]. However, an additional transformer has to be added in the converter, which reduces the conversion efficiency and the power density. For PS-FB DC-DC converter, snubber circuits are required to suppress the voltage spike of the secondary side output diodes [2–5], which causes additional loss. To achieve ZCS turn-off for output rectifying diodes, a diode in series with an IGBT can work as a controllable switch for the output side by implementing phase shift control among the primary and secondary side [6]. But this brings more conduction loss. The additional snubber circuits and diodes limit the power density and conversion efficiency under light load condition.

It is worth noting that the push–pull and the phase shift full bridge isolated DC-DC converters are classified as an isolated buck type DC-DC converter. It is difficult to achieve ZVS for all the switches within the full load of range. For the rectifying diodes, to suppress the voltage spike over diodes, snubber circuits are normally used due to the fact that its ZCS off is lost for diodes.

To achieve ZVS for switches within the wide load range and minimize the voltage stress over secondary rectifying diodes without the need to add snubber circuits, LLC converter as shown in Fig. 1.5 with variable switching frequency is a good option [7, 8]. The LLC converter can achieve very high conversion efficiency especially as its switching frequency approaches the resonant frequency. However, since a battery charger featuring wide output voltage range, the switching frequency range is wide. When the switching frequency stays far away from the resonant frequency, the significant circulating loss will reduce the conversion efficiency. Besides, under light load conditions, the current in the resonant inductor keeps continuous. This causes significant conductions loss and high current stress. Thus, the light load conversion efficiency for a LLC converter used for a battery charger is not high. For an LLC converter, its voltage conversion gain is determined by the switching frequency and also sensitive to power circuit parameters such as the transformer leakage inductance, magnetizing inductance and even transformer junction capacitors.

The frequency range of a series resonant converters can be narrowed by using delay time control on the secondary side composed of full bridge MOSFETs [9]. However, four full controllable devices have to be used to replace the four rectifying

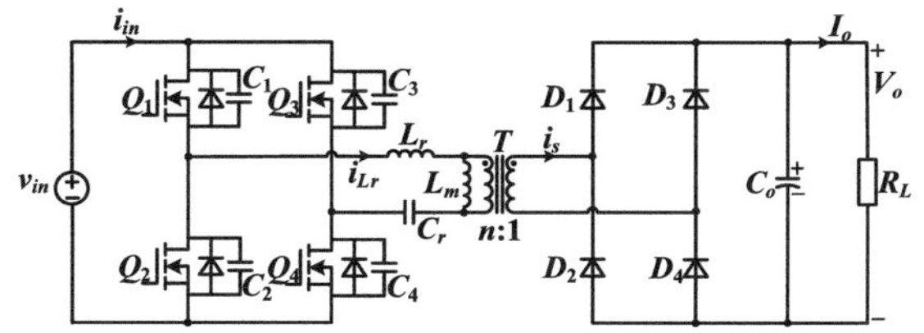

Fig. 1.5 LLC high frequency isolated DC-DC converter

diodes. Under light load conditions, the maximum switching frequency has to be increased to decrease the voltage conversion gain especially when the output voltage is low. This increases the switching loss, thus limiting the light load conversion efficiency.

A variable bus control can be employed for the LLC converter to narrow the switching frequency operating range, thus improving the conversion efficiency [10]. But it complicates the control implementation and normally needs additional power stage. Even at light load conditions, the operating switching frequency is pretty close to resonant frequency, the driving loss, turn-off loss and non-active power at high frequency reduces the conversion efficiency. For the LLC DC-DC converter, once the switching frequency is above than the resonant one, ZCS off for rectifying diodes is lost.

1.1.2 Battery Charger Supplied by AC Grid

Grid connected ac-dc converters with power factor correction (PFC) are widely used in many applications such as EV charging, telecom power supply and future data centers [11–13]. Figure 1.6 gives the electric vehicle charging system architecture. To achieve unity power factor and high frequency isolation at the same time, two stage power conversion is normally used as shown in Fig. 1.7.

For the high frequency isolated DC-DC converter, the input voltage of the DC-DC converter is regulated to be fixed and the voltage conversion gain of the DC-DC is relatively narrow. This lowers the burden of the regulation of the DC-DC stage. For the two stages, if the PFC stage works in the continuous conduction mode (CCM), an inner current loop is needed to achieve the tracking voltage regulator output amplitude and utility phase angle [11–13]. However, this causes the complexity of the control system since each stage needs separated control independently. Current sensors are needed to sense the input AC side grid current. If the PFC stage works in the boundary conduction mode [14–16], a tight current tracking loop is unnecessary. The inductor current zero crossing point needs to be sensed. For the PFC stage working in the discontinuous conduction mode (DCM) [17], the inductor current does not need to be sensed anymore. But for the PFC stage in the two-stage power conversion, regardless of CCM, BCM or DCM, a capacitor with large capacitance is required to smooth the DC bus voltage. In order to prevent the overcurrent, some relays and

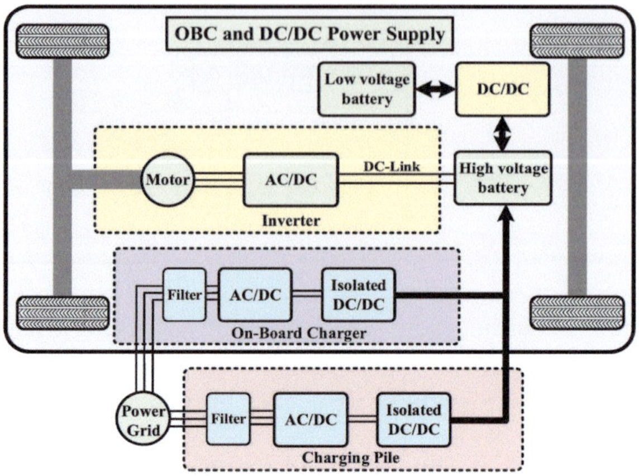

Fig. 1.6 Electric vehicle charging system architecture

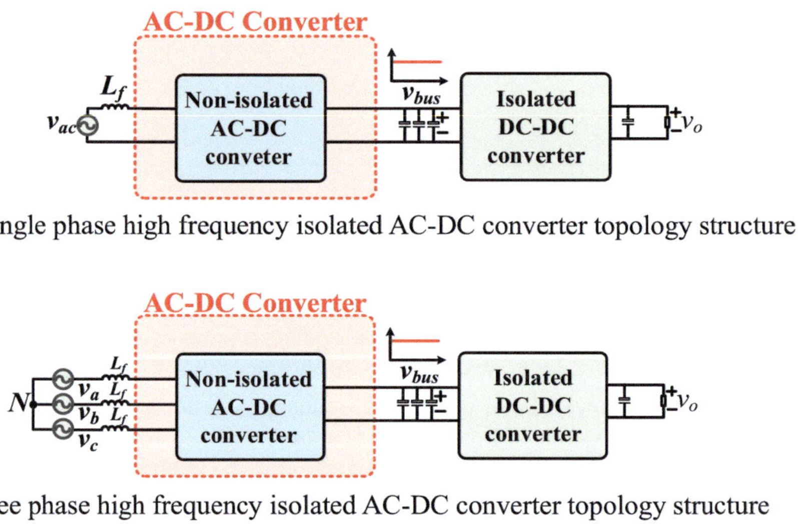

Fig. 1.7 High frequency isolated AC-DC converter topology structure

resistors are used to suppress the inrush current during the start up process. The two stage power circuit configuration can be composed of a grid connected series resonant AC-AC converter followed by a buck/boost converter [18]. It should be noted that the conversion efficiency is low due to the two stage power conversion [11–18].

To improve the conversion efficiency, single stage power conversion with high frequency isolation interfaced with utility offers a better technical solution. Meanwhile, for the high AC input voltage, low output voltage power conversion, single stage power conversion does not need a high voltage rating electrolytic capacitor with

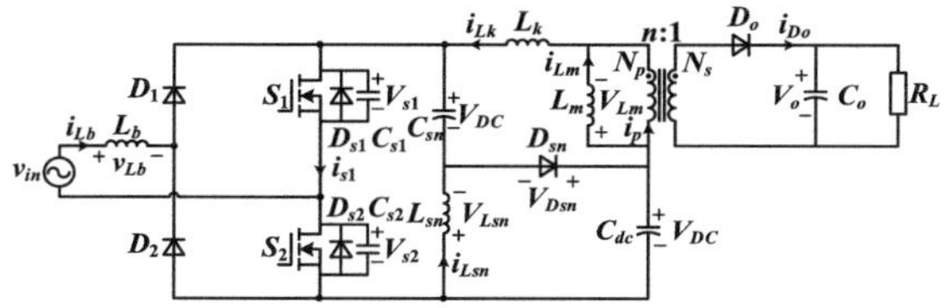

Fig. 1.8 Equivalent circuit diagram of the proposed converter

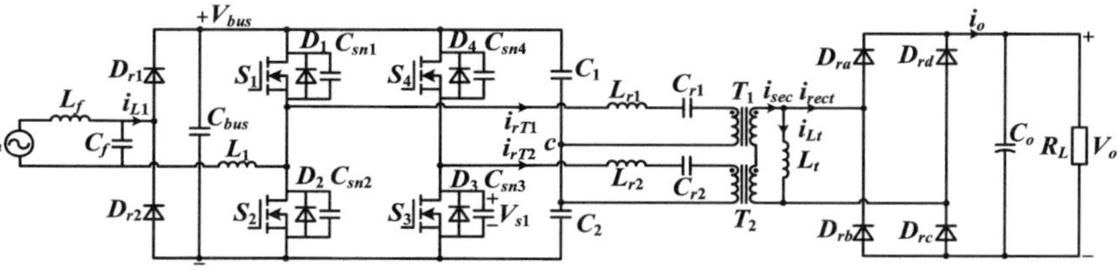

Fig. 1.9 HF isolated dual-tank LCL-type series resonant ac-dc converter

large capacitance values in the DC bus but a low capacitance value film capacitor instead. The high voltage electrolytic capacitor size is big and its lifetime is short. Thus, single stage conversion is useful to improve the power density and improve the reliability. The bridgeless PFC stage and the isolated DC-DC converter can share the same power switches as shown in Fig. 1.8, this simplifies both the power circuit and the control circuit [19]. A single stage isolated AC-DC converter combines a diode rectifier, a boost converter, and two identical half-bridge DC–DC LCL resonant converters. Some power switches are shared both by the PFC stage and the DC-DC conversion stage as shown in Fig. 1.9. Thus the number of switches can be reduced and the control design is simplified [20].

1.2 Concept of Dual Active Bridge DC-DC Converter

1.2.1 Voltage-Fed Dual Active Bridge DC-DC Converter and Its Control

Dual active bridge DC-DC converter as shown in Fig. 1.10 is widely used for bidirectional power industrial applications. The dual active bridge consists of two active full bridges. The leakage inductance of the transformer L_k works as a power link between the input side and output side. Through controlling the voltage over the leakage inductance, the transferred power value and the power direction can be regulated. DAB converter has many advantages: wide ZVS range for all the power switches, naturally bidirectional power flow control and symmetrical architecture in topology.

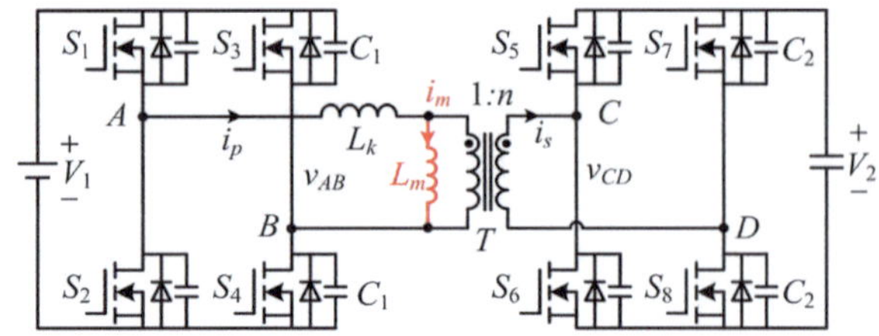

Fig. 1.10 Bidirectional isolated dual active bridge converter

Galvanic isolation of the converter is actually required for many applications due to some safety reasons. In addition, in some applications where the converter conversion ratio is very high, the transformer can be used to adjust the gain of the converter to extend to optimal operating range.

Many control strategies have been presented in the literature. Single phase shifted (SPS) control is very simple and widely used for DAB converters [21–24]. ZVS can be achieved easily, but the non-active power is significant. As the voltage conversion ratio varies, the ZVS range is limited. Besides, high non-active power causes higher peak current and additional conduction loss and thus the conversion efficiency is reduced [25, 26]. To reduce the non-active power, many improved control strategies have been proposed.

The extended phase shifted control [27–31] is a typical improvement of SPS to extend the ZVS operation range and also reduce non-active power. However, non-active power is still not minimized.

To reduce non-active power further, both the two bridges are PWM modulated while phase shifted control is made between the two active bridges [26, 32–35]. Since there are three control variables with many possible combinations to choose, the instantaneous close-loop control is really a challenge. A fast controller is required to handle the real-time computation.

The dual PWM control plus phase shifted control, also called triangular modulation, causes the leakage inductance current to operate in a discontinuous mode [36, 37]. With this strategy, the non-active power can be minimized, but ZVS is lost for some switches. Besides, the current stress over transformer windings and switches is relatively high due to the discontinuous operating mode.

A boundary control and a unified triple-phase-shift control for DAB were proposed [38, 39]. With this strategy, the leakage inductor current operates in the continuous mode. Both non-active power and current stress are minimized. However, the magnetizing current in the transformer is used to compensate the bias current generated by the dual PWM control to achieve ZVS for secondary side switches. This causes additional conduction loss and core loss.

To further reduce the turn off current and turn off loss of switches, resonant DAB topology structure and a series of modulation and control strategy are proposed [40–45], which could reduce the turn off loss, EMI and improve the conversion efficiency of the converter.

1.2.2 *Current-Fed Dual Active Bridge DC-DC Converter and Its Control*

In view of the type of sources at the two sides of the DAB converter, it can also be divided into two parts: voltage fed dual active bridge and current fed dual active bridge. The converter belongs to the category of 'voltage fed', if the input and output are voltage sources and is directly connected to the switch nodes. Otherwise, it can be regarded as a current fed DAB converter. For instance, the DAB converter previously shown in Fig. 1.10 is a typical voltage fed converter, because both the input and output are directed connected to a voltage source (or capacitor). There are also some other voltage fed DAB converters that are similar to this converter, such as the DAB converter based on two half bridges [46], the DAB converter based on a full bridge and a half bridge [47] and the DAB converter based on the voltage-doubler rectifier [48].

Figure 1.11 shows an example of a current fed dual active bridge converter [49]. In this topology, the DC bus is connected in series with an inductor, and then connected to the full bridge DAB converter. For this connection, due to the fact that the inductor current cannot be interrupted as an open-loop, the PWM signals for the two switches in the same bridge should be overlapped to provide the circulating current path. As pointed out in [49], Zero Current Switching (ZCS) for the primary side switches can be achieved.

In high voltage and low current applications, actually ZVS matters more in the aspect of switching loss. To active wide ZVS range for the power switches, Fig. 1.12 shows another example of a current fed dual active bridge converter [50]. As seen, the voltage source at the low voltage side is connected to the switch node through an inductor. In the middle part of the converter, it is actually a DAB converter with two half bridges. By combing the boost converter and the half bridge converter together, the LVS side has double functions which serves as:

(1) A boost converter to step up the voltage to adjust the clamping voltage of the low voltage side.
(2) An inverter to generate the high frequency AC voltage to pass through the transformer for isolation.

To reduce the current ripple of the low voltage side, the interleaved current fed DAB topology [51] can be employed as shown in Fig. 1.13. The topology at the low

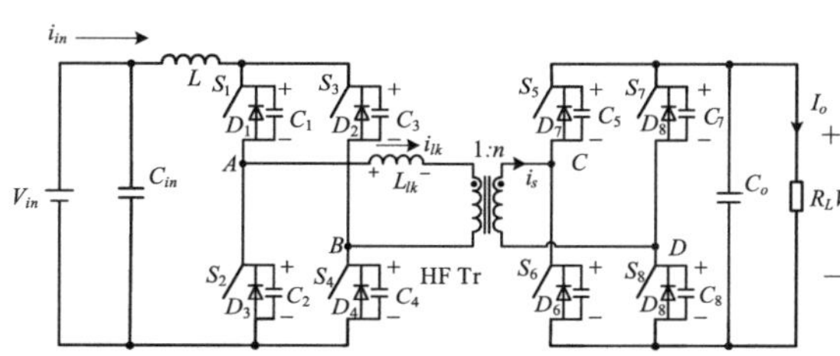

Fig. 1.11 ZCS current fed dual active bridge

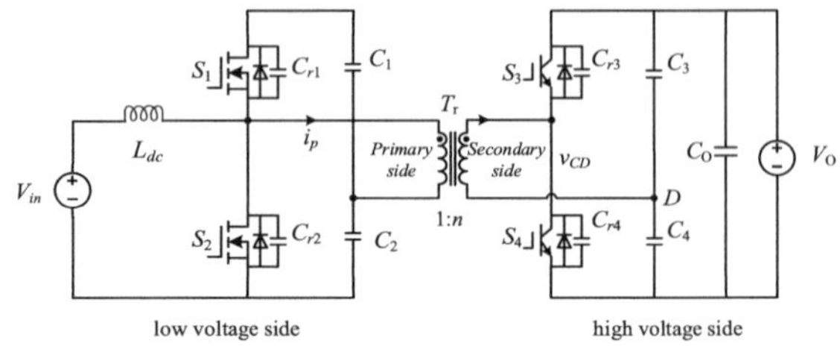

Fig. 1.12 Current fed dual active bridge converter

voltage side consisted of two boost converters with input parallel and output parallel connections. The PWM signals for the two boost converters are 180° phase-shifting with each other. This topology can achieve a low current ripple at the low voltage side and also wide ZVS ranges for all the power switches, which makes it popular in the low voltage to high voltage conversion applications.

In order to reduce the peak current and circulating current of current-fed full-bridge and half-bridge DC-DC converter, the same duty cycle control for the current-fed full-bridge converter as shown in Fig. 1.14 is proposed to reduce the peak current. In order to improve the dynamic performance, the feedforward control with an optimal feedforward coefficient is proposed [52]. In order to realize ZVS soft-switching within the full load range under wide gain, a new idea of fixed duty cycle compensation is proposed [53]. To further improve the performance of this converter, the duty cycle of the bottom switch at the low voltage side is controlled within 0.5 for the first time, and the proposed active mismatch control strategy can realize ZVS soft-switching of all switches within the full load range, further reduce the circulating current and improve the conversion efficiency. The ZVS on within full gain and full load range is realized [54].

A high-frequency isolated bidirectional three-level DC–DC converter employing diode clamped and flying capacitor as shown in Fig. 1.15 is proposed for battery charging/discharging with a high-voltage interface [55]. Voltage over all the power switches in the three level topology at turn off transients can be clamped well. For the battery side, a full-bridge converter by sharing a common capacitor has been used to reduce the clamp capacitor voltage ripple and avoid the magnetic bias issue. The DPDPS control strategy using identical duty cycles employing the magnetizing

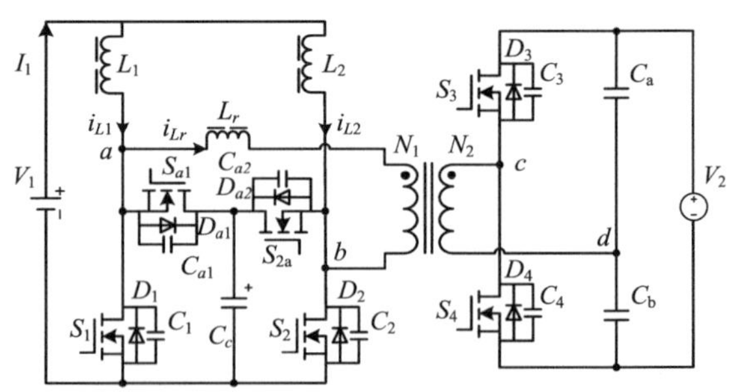

Fig. 1.13 Interleaved current fed dual active bridge converter

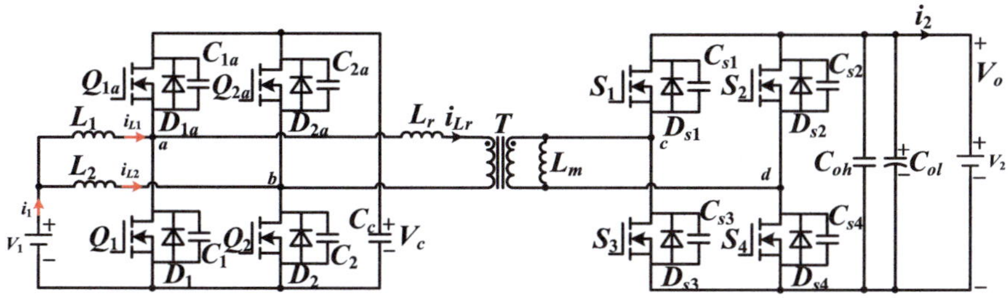

Fig. 1.14 Circuit of the current-fed bidirectional DC-DC converter

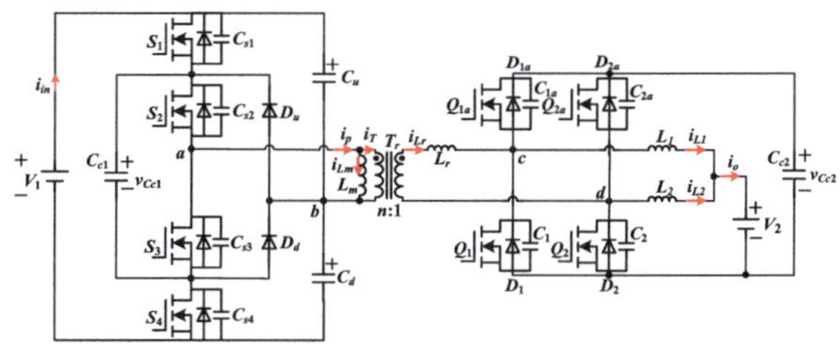

Fig. 1.15 Topology of the bidirectional CF DC-DC three-level converter

inductance design is proposed to significantly reduce the circulation loss. ZVS for all the power switches can be achieved throughout full-load range with the help of the magnetizing current of the transformer.

1.2.3 Three Phase Dual Active Bridge DC-DC Converter

The current stress is very high when the conversion power is high. Therefore, 1p-DAB is not suitable for high power application due to the high current stress in the switches. Therefore, 1p-DAB can be extended to 3p-DAB as shown in Fig. 1.16 and the current stress can be minimized. Compared to 1p-DAB, multiphase converters have smaller voltage and current ripple, smaller filter size, and higher power capacity [56–60].

An enhanced modulation by combining the triangular, trapezoidal, and SPS modulations is raised in [61] to improve the conversion efficiency of the 3p-DAB DC–DC converter. Two phases of the converter are paralleled to improve the light load conditions. But the control strategy is very complicated and hard to be implemented. A 3p CF-DAB as shown in Fig. 1.17 using double duty cycle control on both sides is proposed in [62] to achieve wide ZVS variations. But it is an awfully big price to achieve ZVS and its power density is affected by a large number of magnetic components. A simultaneous PWM control for 3p-DAB is given in [63] to handle several operating modes. However, the optimal mode is not discussed.

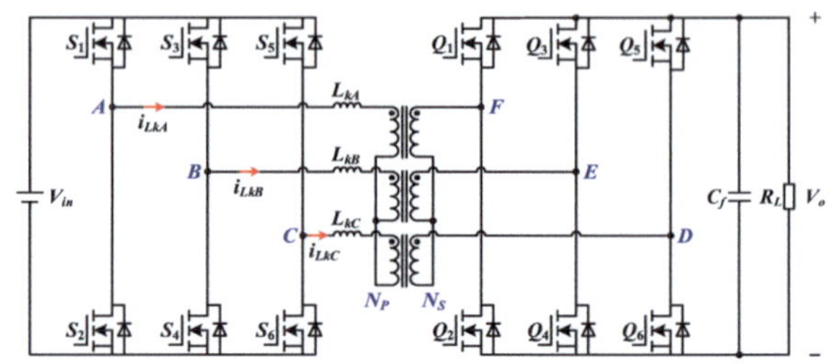

Fig. 1.16 Three phase voltage-fed dual active bridge DC-DC converter

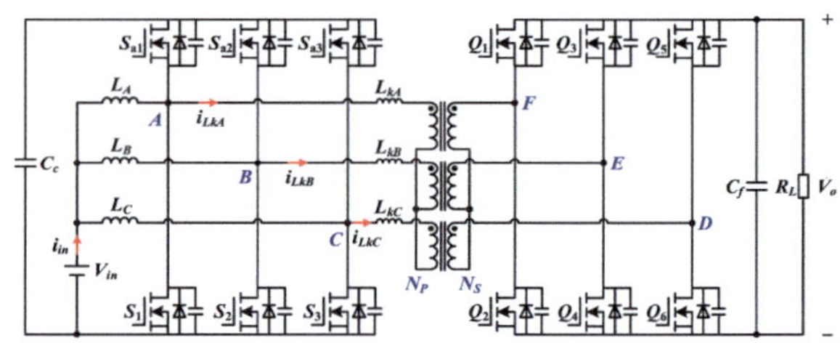

Fig. 1.17 Three phase current-fed dual active bridge DC-DC converter

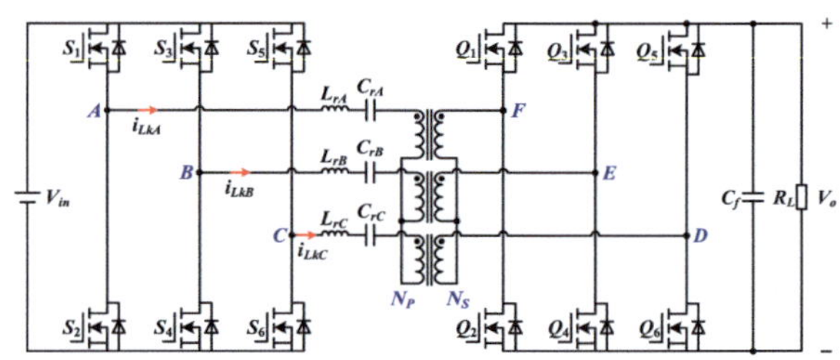

Fig. 1.18 Three phase resonant dual active bridge DC-DC converter

Similar with the single phase DAB DC-DC converter, three phase resonant DAB DC-DC converter as shown in Fig. 1.18 is proposed to reduce the turn-off currents of switches and achieve ZVS within extremely wide range, which could further improve the conversion efficiency [64, 65].

1.3 Single-Stage DAB-Based AC-DC Converters

1.3.1 Classification of Power Circuit

The single-stage DAB-based isolated AC-DC converter topology can be divided into two categories, including no-resonant and resonant topology structures.

The typical no-resonant DAB-based isolated AC-DC converter topology is shown in Figs. 1.19 and 1.20, which are composed of a diode rectifier or a synchronous rectifier followed by a DAB converter. A series of modulation and control strategy are proposed to improve the performance of this converter [66–70]. Another novel no-resonant topology structure of the single-stage AC-DC converter are shown in Figs. 1.21 and 1.22, which does not need a line frequency rectifier but a cycloconverter whose switch is made of anti-series connected switches. Similarly, a series of modulation and control strategy are proposed [71–76].

For resonant DAB-based isolated AC-DC converters, the typical topology structures are shown in Figs. 1.23 and 1.24, which are composed of a synchronous rectifier followed by a resonant-DAB converter, and the resonant tank has different types such as LC, LLC and LCL, etc. A series of modulation and control strategies are proposed to improve the performance of this converter [77–79]. Another topology structure of the single-stage resonant-DAB-based AC-DC converter is similar with Fig. 1.21,

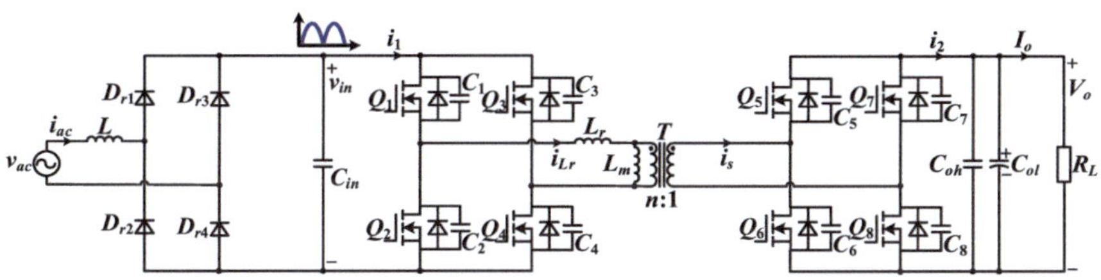

Fig. 1.19 DAB-based single-stage isolated unidirectional AC-DC converter

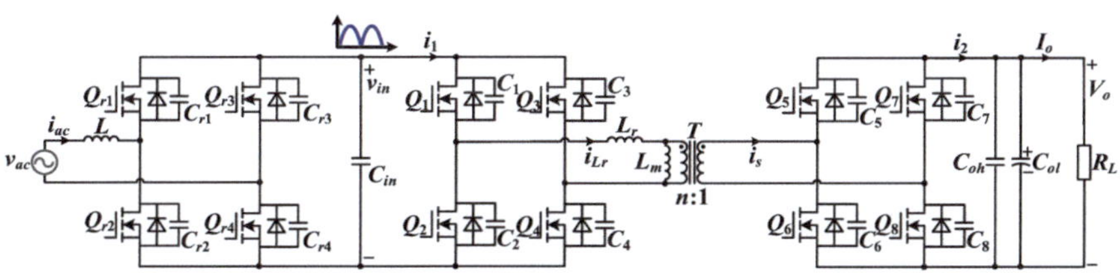

Fig. 1.20 DAB-based single-stage isolated bidirectional AC-DC converter

Fig. 1.21 Half bridge matrix single-stage isolated AC-DC converter

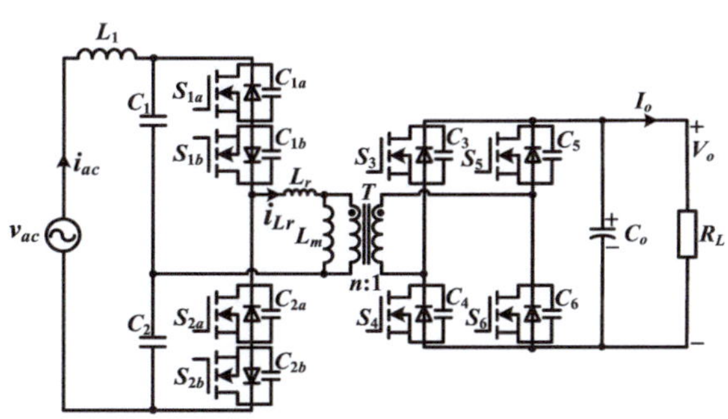

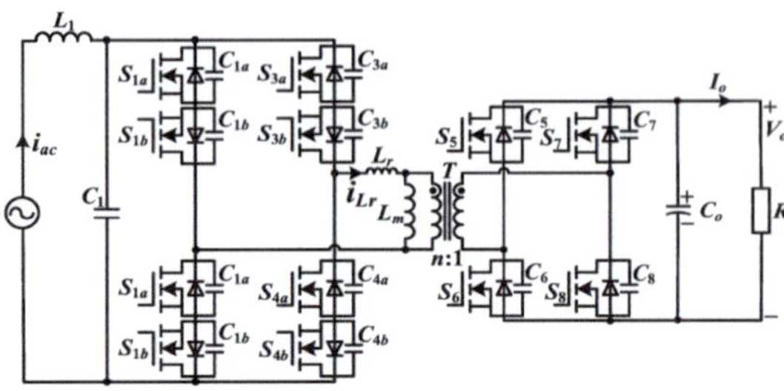

Fig. 1.22 Full bridge matrix single-stage isolated AC-DC converter

which is shown in Fig. 1.25, the only difference is that the middle inductor is replaced by a resonant tank [80].

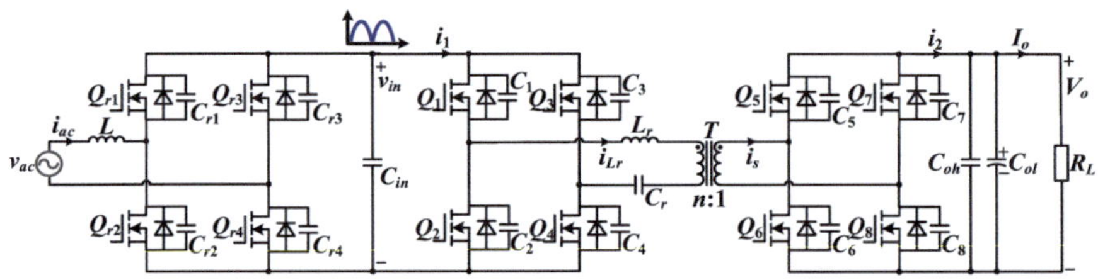

Fig. 1.23 Series resonant DAB-based single-stage isolated AC-DC converter

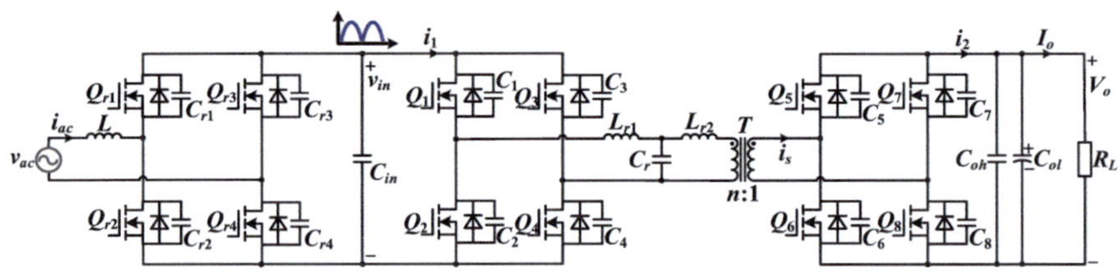

Fig. 1.24 LCL resonant DAB-based single-stage isolated AC-DC converter

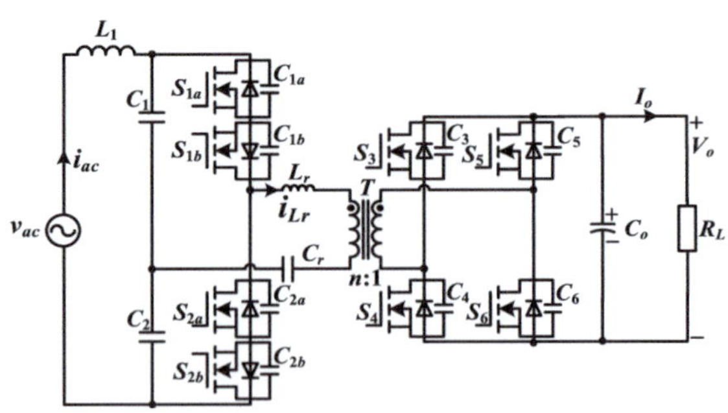

Fig. 1.25 Half bridge matrix resonant single-stage isolated AC-DC converter

1.3.2 Existing Control for Single-Stage DAB AC-DC Converter

For no-resonant DAB-based single-stage AC-DC converter topology structure, the modulation and control strategies have been studied widely. A novel modulation and control strategy of the single-stage DAB AC-DC converter is proposed [66]. The system control block diagram is shown in Fig. 1.26. It consists of a phase shift control loop and the modulation scheme. The phase shift control loop is used to control the transmitted power by stabilizing the output voltage. The modulation scheme is based on the input instantaneous voltage of the DAB, the fundamental frequency and the phase of the input AC voltage to generate duty cycles D_1 and D_2. $G_v(s)$ is the phase shift control compensator composed of the proportional and integral regulation. $H_o(s)$ is a notch filter to reduce double line frequency appearing in the output voltage, which is generated by the AC voltage. Triple-phase-shift (TPS) modulation scheme is used and the inner current tracking loop and lookup tables is not used, but the soft-switching range of the secondary side switches will be narrowed as the load increases under the constant switching frequency control strategy, and the only one mode cannot optimize the performance of the converter within the entire ac line period.

In order to extend soft switching range and optimize the performance of the converter, a multi-mode control strategy with seamless transition is presented [67]. The operating modes under different voltage conversion gains are shown in Fig. 1.27, including the situations of $m < 1$ and $m > 1$. The schematic of the control strategy for the single-phase single-stage isolated AC-DC converter is shown in Fig. 1.28. The DC side voltage v_{dc2} is controlled by the voltage loop, which outputs the P.U. value of the input reference current i^*_{ref}. This value together with i^*_{ZVS} and m are fed into the control variable calculation unit by which the control variables d_1, d_2, σ and φ are calculated. The control variable calculation unit selects the operating mode by the mode selection unit using the value of i^*_{ref}, i^*_{ZVS} and m. Based on the selected mode, the control variables is calculated and fed to the PWM generation unit to generate PWM signals for switches. As seen, the control variables are obtained by the ZVS condition and the input average current of DAB equal to the reference ac current, implying that the leakage inductor peak or RMS value of the transformer current is not optimal. In addition, the full operating range ZVS can not be realized under this constant switching frequency control strategy.

A single-stage isolated AC-DC converter composed of a synchronous rectifier followed by DAB is proposed in [68, 69], more accurate ZVS condition (current-dependent charge-based) is presented, this control strategy requires lookup tables for

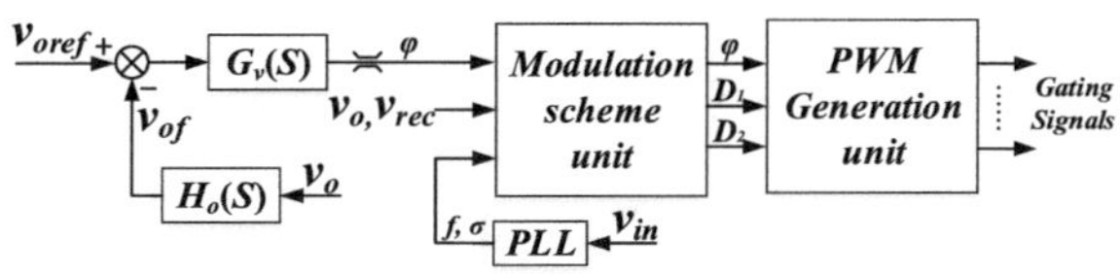

Fig. 1.26 The system control block diagram

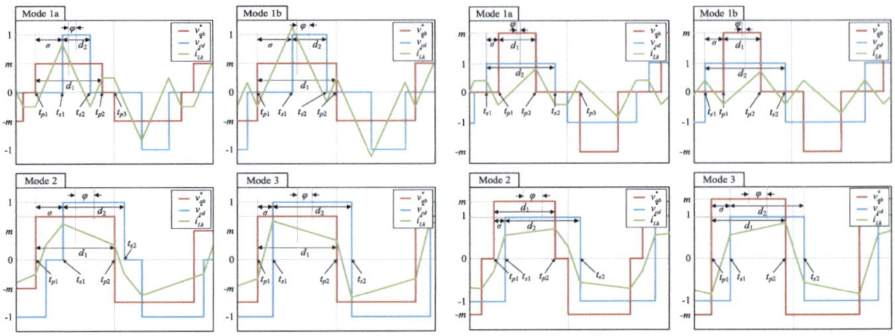

(a)m<1 (b)m>1

Fig. 1.27 The operating modes under different voltage conversion gains

Fig. 1.28 The schematic of the system control strategy

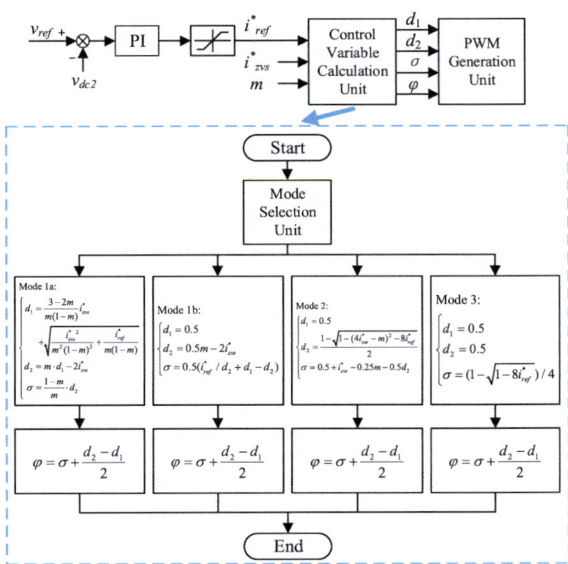

real-time closed-loop control and extra inductor current will cause more magnetic loss. In addition, this modulation and control strategy is only suitable for the situation of $m > 1$. A improved hybrid current modulation scheme is proposed in [81], which solves the current distortion problem without increasing the root mean square (RMS) value of current, while the large switching loss reduces the efficiency. The nonlinear parasitic output capacitance is analyzed for the AC-DC converter [82]. The nonidealities that existed in the circuit will cause the error of theoretical analyses. A modeling approach of the DAB AC-DC converter is presented in [83] to reduce the modeling error. In [84], an accurate nonlinear model during the commutation process between two switches in each bridge is considered, but ZVS is still lost in

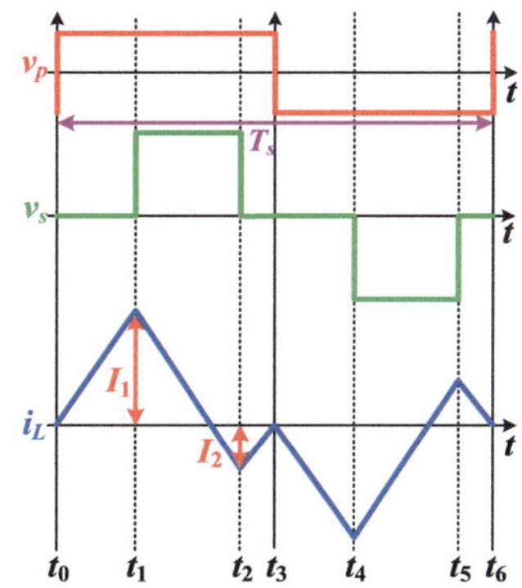

Fig. 1.29 The operating mode

some operating regions especially around the zero-crossing-point of grid voltage under light load conditions.

A single-stage AC-DC converter based on the full-bridge matrix structure is shown in [73], the operating mode as shown in Fig. 1.29 will cause a large peak current, and some switches can not achieve ZVS.

A single-stage AC-DC converter based half-bridge structure is shown in [74, 75], which adopt two operating modes within the entire ac line period as shown in Fig. 1.30, while ZVS can not be achieved for all operating modes under the constant switching frequency control strategy. In order to increase the efficiency, the phase shift plus frequency modulation is proposed in [76], which could achieve ZVS in entire a line frequency period. Furthermore, an optimal control strategy is proposed, while the control is very complex by relying on lookup tables.

For the resonant single-stage AC-DC converter topology, LLC-based single-stage AC-DC converter are researched extensively. Full bridge resonant AC-DC converter using T-type voltage doubler is proposed [85], which could minimize switching loss, while only the DCM mode is used, and the voltage gain is always larger than 1, causing higher peak current of the transformer. The similar topologies are given [86], and the same problems still exist due to only one operating mode. A series of quasi-single-stage bridgeless electrolytic capacitor-free AC-DC converter for supplying multiple light-emitting diode strings in parallel are proposed [87–89], which is the integration of boost PFC and half-bridge LLC, reducing device numbers and cost, while the grid side current quality is relatively poor. A simple synchronous rectification strategy is proposed [90], but the proposed SR driving strategy needs accurate time-domain analysis, thus lookup tables are needed and the control is complex. A power boundary control strategy is proposed for the single-stage isolated AC-DC converter based half-bridge LLC [91], the parameters are designed optimally. Although the rms current of the resonant tank is be achieved, but the grid side current quality is relatively poor. An interleaved totem-pole LLC single-stage AC-DC converter with a novel hybrid control strategy is proposed [92], utilizes PWM to shape the input current, and PFM to regulate the output voltage. LLC-based single stage AC-DC converters have less freedom of control.

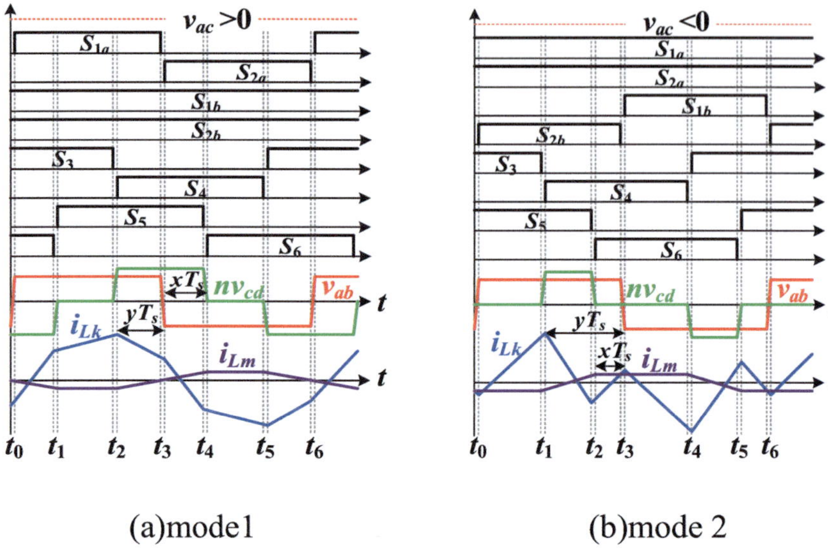

Fig. 1.30 The operating modes

So in order to increase the flexibility of control, resonant DAB-based single-stage AC-DC converters have drawn some attention. A variable switching frequency control strategy employing EPS modulation is proposed [77], which could reduce the turn-off current of switches and obtain high efficiency, while only one mode exists within the entire ac line period, which is hard to optimize the performance in facing varying grid side voltages, so the variable frequency range is relatively wide. A single-stage resonant-DAB AC-DC converter with a novel modulation and constant frequency control is presented, which can obtain ZVS during most of the grid period, but the current loop is necessary and the ZVS around the zero-crossing-point of AC voltage and light load are not given [78].

1.4 Existing Research for Semi-DAB DC-DC Converters

For unidirectional power flow applications, two MOSFETs in the secondary side of the DAB converter can be replaced by two diodes to reduce the cost and the driver circuit numbers. This topology is called semi dual active bridge (Semi-DAB) [93], which is shown in the Fig. 1.31. With single phase shift (SPS) modulation strategy, all the switches can achieve ZVS as the input voltage matches with the output voltage. But as the output voltage varies, ZVS for some switches may be lost under light load conditions, which causes high current spike in switches, external inductors and transformer windings. This lowers the conversion efficiency and may bring reliability issues.

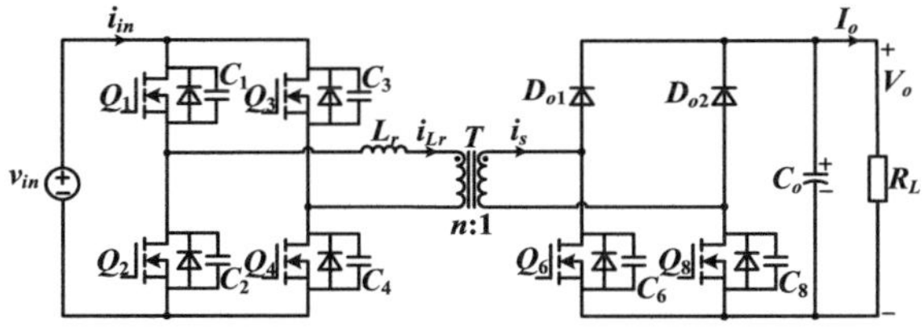

Fig. 1.31 Semi-DAB DC-DC converter

To minimize the circulating current, reduce the non-active power and the RMS value of the leakage inductor current, a BCM control strategy was proposed for Semi-DAB converters [94]. However, the effectiveness of this strategy under light load condition is not verified. Theoretically, under light load condition, the output voltage cannot be regulated to track the reference value when the load drops under a certain value. It means that the BCM control strategy is not effective under very light load conditions.

To extend the ZVS range, phase shift plus PWM modulation and control strategy [95] is proposed. To optimize the conversion efficiency within full load range, a multi-mode control [96] is used, three working modes work together and seamless transition can be obtained among them once the output power requires. With the proposed control strategy, the semi-DAB DC-DC converter can obtain high conversion efficiency for full load range and wide voltage range operations, while the switching frequency is not fixed throughout the entire load. The peak current of the transformer can be minimized with optimal operating modes [97], and the conversion efficiency is high throughout full load range compared with traditional modulation strategy. In [98], the semi-DAB converter only operates in the discontinuous current mode (DCM) and boundary current mode (BCM) in all operating conditions, which has the excellent soft switching performance, a low circulation energy and high efficiency. To adapt the wider voltage gain range application, a hybrid three-level semi-DAB converter has been proposed [99], while for the output side, it has the risk of shoot through the problem and the drive for the top switch needs to be isolated from the bottom one. Another method to ensure ZVS condition within the wide voltage range is to use a current-fed semi-DAB topology [100, 101], achieving ZVS for all MOSFETs and ZCS for all rectifying diodes. However, the leakage inductor current operates in the discontinuous mode, thus increasing the conduction loss. Meanwhile, the current stress over switches, especially the bottom switches at the low voltage side, is relatively high. Besides, two additional DC inductors are required, thus reducing the system power density and the conversion efficiency. On the other hand, this topology is not suitable for a relatively high input voltage since the clamp voltage will be very high, which requires high voltage rating switches. A semi-DAB converter based LLC resonant tank has been proposed [102], while which operates in DCM, the peak current is high. In order to reduce the circulating current and the non-active power, a novel modulation strategy of zero backflow power based resonant semi-DAB converter has been proposed, while the switching frequency is variable when the load power is less than a certain critical value [103]. In order to handle higher power level, three-phase voltage-fed semi-DAB and three phase current-fed semi-DAB DC-DC converter are proposed in [104, 105], but each phase is coupled.

1.5 Key Issues of Semi-DAB DC-DC and AC-DC Converters

1.5.1 Full Load Range Operating Stability and ZVS Range in the Entire Operating Range

For the semi-DAB DC-DC converter, full load range stable operation within the wide voltage gain ranges especially in the light load even with no load is an important issue. While most of the papers do not show the stable operation and ZVS on situations under light load or even no load conditions.

In addition, similar to the DAB DC-DC converter, semi-DAB DC-DC converter with the single-phase-shift (SPS) modulation strategy can achieve ZVS of all switches in the full-load range when the voltage conversion gain is 1, while the ZVS range is narrow as the gain deviates from 1, and which will also cause high current spike over switches, external inductors and transformer windings. This will lower the conversion efficiency and may bring reliability issues. In order to extend ZVS range of all switches, lower the peak value and RMS value of the current in switches, inductor and transformer, extend-phase-shift (EPS), triple-phase-shift (TPS) modulation and various control strategies are proposed [94–99]. Another method to improve the ZVS performance is to use the current-fed semi-DAB topology structure, while the external DC inductor will increase the loss and volume, lower the power density and conversion efficiency [100, 101]. More important is that it remains a huge challenge when faced in extremely wide voltage conversion gain range and extremely light load even with no load.

1.5.2 Conversion Efficiency Improve Within Entire Operating Range

Similar with the traditional DAB DC-DC converter, semi-DAB DC-DC converter with SPS modulation can achieve full-load range ZVS and best performance when the voltage gain is 1, while the ZVS range is narrow as the gain deviates from 1, and which causes large non-active power and circulating current. The non-active power and circulating current [106, 107] causes higher peak current and additional conduction loss for the semi-DAB DC-DC converter. Non-active power loss is defined in [94] and used as the optimization object to reduce the conduction loss with TPS control. To reduce the non-active power and the circulating current, phase shift plus PWM control [95] is proposed, while the non-active power and the circulating current are still not minimized. So, an improved boundary operation is proposed [94]. The leakage inductance current operates in boundary mode in most operating conditions. The current stress in switches and transformer windings can be reduced. With the propose control, ZVS can be obtained for all power switches by proper design of the duty cycles for the both sides. ZCS can be ensured for output diodes in most conditions. Besides, the non-active power and circulating current can be reduced under most operating conditions. The conduction loss of the output MOSFETs can be reduced dramatically too. The closed loop control is easy to be implemented since there is only one independent variable. While the proposed BCM control strategy is not effective under very light load conditions. One way to extend the output power range of the conventional control strategies is to increase the switching frequency under light load conditions. The increased switching frequency can extend the ZVS range, but switching loss and driving loss will increase as well. Compared with fixed frequency control, this strategy can improve the light load operation performance. However, ZVS is still lost under very light load conditions. The increasing driving loss and switching loss lowers the conversion efficiency. So a multi-mode control strategy is proposed to achieve the full load operating range for semi-DAB [96], the conversion efficiency under this control strategy can be improved compared with the conventional modulation and control within the full load range.

Another method to reduce the non-active power and circulating current is to use a current-fed semi-DAB topology [100, 101]. ZVS can be achieved for all MOSFETs and ZCS can be obtained for rectifying diodes. The non-active power can be suppressed dramatically. However, the leakage inductor current operates in the discontinuous mode. The current stress through switches especially the bottom switches at the low voltage side is relatively high. Besides, two additional DC inductors are required, so the system power density is not competitive and the conversion efficiency may be affected.

1.5.3 Concept of Single-Stage Semi-DAB AC-DC Converter and Possible Challenges

No doubt, the DAB DC-DC converter can be used in single-stage AC-DC converters where the power flow is unidirectional. Actually, DAB converter has the bidirectional power flow capability feature. However, for unidirectional power flow applications, there are many switches and this causes more gating signals and drivers and it is not cost effective. Compared with traditional DAB employing eight switches, semi-DAB is simplified by replacing some switches with diodes. Thus the power circuit and the driver circuits can be simplified, it becomes cost effective and more reliable by using diodes. Is it possible for semi-DAB DC-DC to be used as a part of the single-stage AC-DC converter as shown in Fig. 1.32? The answer is yes. The folded AC voltage varying from zero to peak value works as the input voltage of the semi-DAB DC-DC converter. The voltage conversion gain of the single-stage semi-DAB AC-DC

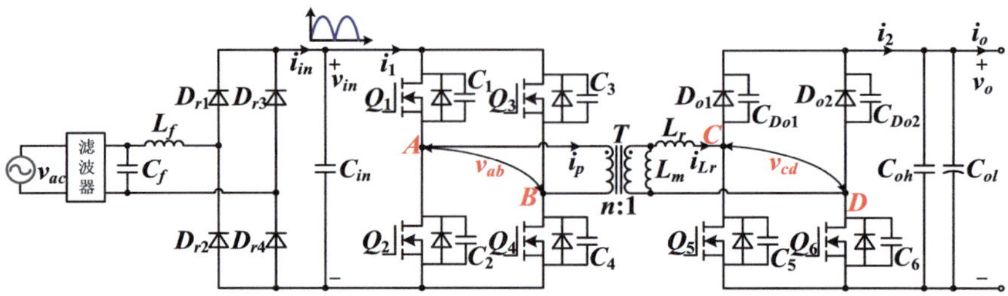

Fig. 1.32 Semi-DAB-based single-stage high frequency isolated AC-DC converter

converter varies extremely very wide, if the switching frequency is fixed, as the grid voltage value is small especially as it approaches zero, ZVS is difficult to obtain. How to achieve ZVS within the entire operating range (the entire ac line period and full load range) should be a big challenge.

For the DAB-based single-stage high frequency isolated AC-DC converter, which could achieve a wider output voltage range and high power factor without mandatory current tracking loop and look-up tables. The input current of semi-DAB DC-DC converter can be sinusoidal controlled by controlling the leakage current. While the effect of most modulation strategies on THD of i_{ac} and the mechanism are seldom considered in current works [67–70, 108]. How to minimize the grid side AC current distortion and achieve a very high power factor should be concern. A novel decoupled inductor current modulation strategy for a single-stage AC-DC converter based on the semi-dual-active-bridge (semi-DAB) is proposed [109], which helps to minimize the total harmonics distortion (THD) and improve the conversion efficiency.

For DAB-based single-stage high frequency isolated AC-DC converter using PWM modulation, some switches are turned off at peak current, which causes very high turn off loss. To minimize the turn off current, resonant power circuit with varying switching frequency control can reduce the turn off current [110], while the variable switching frequency will complicate the transformer design. So a resonant-DAB-based single-stage AC-DC converter with a novel symmetrical modulation and fixed switching frequency control is proposed [78]. The research on its modulation strategy is still in its infancy, most current studies only use one modulation mode in the entire ac line frequency cycle. While the only one modulation mode is not conducive to optimize the converter performance within the entire ac line frequency cycle. So it is necessary to study the multi-mode modulation and control strategy within the entire ac line frequency cycle, the time domain analysis for different modulation modes and the transition points among these modes are hard to calculate, which is a big challenge.

1.6 Organization of the Book

From the topology point of view, this book can be divided into two parts. The first part is focused on topology on semi-DAB DC-DC converters which are from Chaps. 2 to 7, and the second part presents the topologies and control strategies for semi-DAB AC-DC converters which are included from Chaps. 8 to 11. The content for each chapter is organized as follows.

Chapter 2 proposes an improved boundary operation for the voltage fed semi-DAB converter. The leakage inductor current works in boundary mode under most operating conditions. The operating performances and that with conventional single-phase-shift modulation are compared. The non-active power can be suppressed and the conduction loss for output side can be reduced significantly. In Chap. 3, a multi-mode control strategy is proposed. Operating modes vary as the load varies. Thus the conversion efficiency within full load range can be improved. Seamless transition can be obtained among different operating modes. Light load conversion efficiency can be improved greatly under the proposed multi-mode control.

In Chaps. 4 and 5, current-fed semi-DAB topologies and relevant controls are studied for wide voltage gain applications. In Chap. 4, a symmetrical dual PWM and phase shifted control is proposed. The clamp voltage over the capacitor from the input side is kept fixed in spite of the input voltage variation. All the power switches can achieve ZVS and output rectifying diodes can achieve ZCS. To further improve the conversion efficiency of the current-fed converter, a hybrid control featured with varying clamp voltage control strategy is proposed in Chap. 5. For the low-voltage-side (LVS), the duty cycle of the bottom switches will be changed with the load variation. An instantaneous turn-off current tracking loop is designed to achieve duty cycle control and ZVS, which helps to reduce the circulating current turn off current of the switch.

To handle high power applications, Chaps. 6 and 7 study voltage fed and current fed three phase semi-DAB DC-DC converters respectively. In Chap. 6, a varying operating mode control is proposed for voltage fed three phase semi-DAB DC-DC converters and full load operating stability especially at light load condition can be achieved. In Chap. 7, a current fed three phase semi-DAB DC-DC converter is proposed to reduce the input current ripple. A hybrid operating mode control is proposed to handle the operating stability throughout the full load range especially the light load conditions. Fixed clamping voltage and varying clamping voltage are used depending on the load conditions.

A semi-DAB-based single-stage high frequency isolated AC-DC converter suitable for unidirectional power flow applications、modulation and control strategy is proposed in Chap. 8. Two optimal peak current modulation modes are proposed within entire an ac line cycle, a smooth transition strategy between two modes is also presented. And the detailed transition process between the two modes is analyzed. The proposed circuit topology and the control strategy can achieve high AC power quality and conversion efficiency. Chapter 9 reveals the influence mechanism of different modulation modes on THD of the ac input current and the influence mechanism of different modulation modes on various losses of the converter, and introduces a decoupled inductor current modulation strategy for this semi-DAB AC-DC circuit topology. The THD of the ac input current can be reduced and the waveform quality of the current can be improved. The modulation modes can also greatly reduce the switching loss and the transformer core loss. The turns ratio of the high frequency transformer is optimized, which is helpful to improve the conversion efficiency. Chapter 10 proposed a basic variable frequency control strategy to extend the ZVS range, which has some shortcomings. So an improved modulation and variable switching frequency control strategy is proposed, which can reduce the envelope of leakage inductor peak current and the primary side peak current; all switches can achieve ZVS on within the full operating range; the control strategy can minimize the sensitivity of the switching frequency to the load power; the design of the magnetic components are not affected despite the switching frequency becoming a variable; In addition, more accurate ZVS condition considering the magnetizing inductor effect is analyzed. In order to further reduce the turn-off current of switches, a resonant semi-DAB-based single-stage high frequency isolated AC-DC converter is proposed in Chap. 11. And a decoupled resonant inductor current multi-mode modulation strategy and a novel closed loop control mechanism among these modes is proposed. The smooth transition can be achieved among different modes. The proposed modulation and control strategy can reduce the turn-off current of the corresponding switches, the peak value and the root mean square value of the resonant inductor current and the primary side current. The resonant tank components have been optimally designed, which is helpful to improve the conversion efficiency.

It should be noted that the effectiveness of all the converter topologies and control methods are verified with experimental results.

References

1. Tamyurek B, Kirimer B (2015) An interleaved high-power flyback inverter for photovoltaic applications. IEEE Trans Power Electron 30(6):3228–3241
2. Sha D, Deng K, Liao X (2012) Duty cycle exchanging control for input-series-output-series connected two PS-FB DC-DC converters. IEEE Trans Power Electron 27(3):1490–1501
3. Lo Y, Lin C, Hsieh M, Lin C (2011) Phase-shifted full-bridge series-resonant DC-DC converters for wide load variations. IEEE Trans Ind Electron 58(6):2572–2575
4. Gu B, Lai J, Kees N, Zheng C (2013) Hybrid-switching full-bridge DC-DC converter with minimal voltage stress of bridge rectifier, reduced circulating losses, and filter requirement for electric vehicle battery chargers. IEEE Trans Power Electron 28(3):1132–1144
5. Lee I (2016) Hybrid PWM-resonant converter for electric vehicle on-board battery chargers. IEEE Trans Power Electron 31(5):3639–3649
6. Mishima T, Akamatsu K, Nakaoka M (2013) A high frequency-link secondary-side phase-shifted full-range soft-switching PWM DC–DC converter with ZCS active rectifier for EV battery chargers. IEEE Trans Power Electron 28(12):5758–5773
7. Musavi F, Craciun M, Gautam DS, Eberle W, Dunford WG (2013) An LLC resonant DC–DC converter for wide output voltage range battery charging applications. IEEE Trans Power Electron 28(12):5437–5445

8. Kim D, Kim M, Lee B (2017) An integrated battery charger with high power density and efficiency for electric vehicles. IEEE Trans Power Electron 32(6):4553–4565
9. Liu G, Jang Y, Jovanović MM, Zhang JQ (2017) Implementation of a 3.3-kW DC–DC converter for EV on-board charger employing the series-resonant converter with reduced-frequency-range control. IEEE Trans Power Electron 32(6):4168–4184
10. Wang X, Jiang C, Lei B, Teng H, Bai HK, Kirtley JL (2016) Power-loss analysis and efficiency maximization of a silicon-carbide MOSFET-based three-phase 10-kW bidirectional EV charger using variable-DC-bus control. IEEE J Emerg Sel Top Power Electron 4(3):880–892
11. Whitaker B et al (2014) A high-density, high-efficiency, isolated on-board vehicle battery charger utilizing silicon carbide power devices. IEEE Trans Power Electron 29(5):2606–2617
12. Kasper M, Bortis D, Deboy G, Kolar JW (2017) Design of a highly efficient (97.7%) and very compact (2.2 kW/dm^3) isolated AC–DC telecom power supply module based on the multicell ISOP converter approach. IEEE Trans Power Electron 32(10):7750–7769
13. Rothmund D, Guillod T, Bortis D, Kolar JW (2019) 99% efficient 10 kV SiC-based 7 kV/400 V DC transformer for future data centers. IEEE J Emerg Sel Top Power Electron 7(2):753–767
14. Kim J, Moon G (2016) Minimizing effect of input filter capacitor in a digital boundary conduction mode power factor corrector based on time-domain analysis. IEEE Trans Power Electron 31(5):3827–3836
15. Wu X, Yang J, Zhang J, Qian Z (2012) Variable on-time (VOT)-controlled critical conduction mode buck PFC converter for high-input AC/DC HB-LED lighting applications. IEEE Trans Power Electron 27(11):4530–4539
16. Choi H (2013) Interleaved boundary conduction mode (BCM) buck power factor correction (PFC) converter. IEEE Trans Power Electron 28(6):2629–2634
17. Lee J, Chae H (2014) 6.6-kW onboard charger design using DCM PFC converter with harmonic modulation technique and two-stage DC/DC converter. IEEE Trans Ind Electron 61(3):1243–1252
18. Huber JE, Rothmund D, Wang L, Kolar JW (2016) Full-ZVS modulation for all-SiC ISOP-type isolated front end (IFE) solid-state transformer. In: IEEE energy conversion congress and exposition, pp 1–8
19. Lee S, Do H (2016) Single-stage bridgeless AC–DC PFC converter using a lossless passive snubber and valley switching. IEEE Trans Ind Electron 63(10):6055–6063
20. Du Y, Bhat AKS (2016) Analysis and design of a high-frequency isolated dual-tank LCL resonant AC–DC converter. IEEE Trans Ind Appl 52(2):1566–1576
21. Kheraluwala MN, Gascoigne RW, Divan DM, Baumann ED (1992) Performance characterization of a high-power dual active bridge DC-to-DC converter. IEEE Trans Ind Appl 28(6):1294–1301
22. Inoue S, Akagi H (2007) A bidirectional DC-DC converter for an energy storage system with galvanic isolation. IEEE Trans Power Electron 22(6):2299–2306
23. Bai H, Mi CC, Gargies S (2008) The short-time-scale transient processes in high-voltage and high-power isolated bidirectional DC-DC converters. IEEE Trans Power Electron 23(6):2648–2656
24. Costinett D, Maksimovic D, Zane R (2013) Design and control for high efficiency in high step-down dual active bridge converters operating at high switching frequency. IEEE Trans Power Electron 28(8):3931–3940
25. Oggier G, García GO, Oliva AR (2011) Modulation strategy to operate the dual active bridge DC-DC converter under soft switching in the whole operating range. IEEE Trans Power Electron 26(4):1228–1236
26. Bai H, Mi C (2008) Eliminate reactive power and increase system efficiency of isolated bidirectional dual-active-bridge dc-dc converters using novel dual-phase-shift control. IEEE Trans Power Electron 23(6):2905–2914
27. Demetriades GD, Nee HP (2008) Characterization of the dual-active bridge topology for high-power applications employing a duty-cycle modulation. In: IEEE power electronics specialists conference, pp 2791–2798

28. Oggier GG, García GO, Oliva AR (2009) Switching control strategy to minimize dual active bridge converter losses. IEEE Trans Power Electron 24(7):1826–1838
29. Zhao B, Yu Q, Sun W (2012) Extended-phase-shift control of isolated bidirectional DC-DC converter for power distribution in microgrid. IEEE Trans Power Electron 27(11):4667–4680
30. Choi W, Rho KM, Cho BH (2016) Fundamental duty modulation of dual-active-bridge converter for wide-range operation. IEEE Trans Power Electron 31(6):4048–4064
31. Zhao B, Song Q, Liu W (2012) Power characterization of isolated bidirectional dual-active-bridge DC-DC converter with dual-phase-shift control. IEEE Trans Power Electron 27(9):4172–4176
32. Bai H, Nie Z, Mi CC (2010) Experimental comparison of traditional phase-shift, dual-phase-shift, and model-based control of isolated bidirectional DC-DC converters. IEEE Trans Power Electron 25(6):1444–1449
33. Krismer F, Kolar JW (2009) Accurate small-signal model for the digital control of an automotive bidirectional dual active bridge. IEEE Trans Power Electron 24(12):2756–2768
34. Krismer F, Kolar JW (2012) Efficiency-optimized high-current dual active bridge converter for automotive applications. IEEE Trans Ind Electron 59(7):2745–2760
35. Krismer F, Round S, Kolar JW (2006) Performance optimization of a high current dual active bridge with a wide operating voltage range. In: 37th IEEE power electronics specialists conference (PESC), pp 1–7
36. Hirose T, Takasaki M, Ishizuka Y (2014) A power efficiency improvement technique for a bidirectional dual active bridge DC-DC converter at light load. IEEE Trans Ind Appl 50(6):4047–4055
37. Zhou H, Khambadkone AM (2009) Hybrid modulation for dual-active-bridge bidirectional converter with extended power range for ultracapacitor application. IEEE Trans Ind Appl 45(4):1434–1442
38. Xu G, Sha D, Zhang J, Liao X (2017) Unified boundary trapezoidal modulation control utilizing fixed duty cycle compensation and magnetizing current design for dual active bridge DC-DC converter. IEEE Trans Power Electron 32(3):2243–2252
39. Huang J, Wang Y, Li Z, Lei W (2016) Unified triple-phase-shift control to minimize current stress and achieve full soft-switching of isolated bidirectional DC-DC converter. IEEE Trans Ind Electron 63(7):4169–4179
40. Li X, Bhat AKS (2010) Analysis and design of high-frequency isolated dual-bridge series resonant DC/DC converter. IEEE Trans Power Electron 25(4):850–862
41. Hu S, Li X, Bhat AKS (2019) Operation of a bidirectional series-resonant converter with minimized tank current and wide ZVS range. IEEE Trans Power Electron 34(1):904–915
42. Hu S, Li X, Zheng Q-F (2021) A dual-bridge DC/DC resonant converter using extended PWM and phase-shift control. IEEE Trans on Ind Appl 57(4):4009–4020
43. Song W, Zhong M, Deng Y, Yin S, Yu B (2022) Model predictive power control for bidirectional series resonant isolated DC-DC converters with steady-state and dynamic performance optimization. IEEE J Emerg Sel Top Ind Electron 33(3):604–615
44. Song W, Yin S, Deng Y, Zhong M (2023) Steady-state performance optimization of dual-bridge series resonant DC/DC converters. IEEE J Emerg Sel Top Power Electron 11(2):1767–1777
45. Deng Y, Song W, Yin S, Zhong M, Chen J, Feng X (2023) Zero-backflow power control scheme of dual bridge series resonant DC–DC converters with high-accuracy time domain modeling. IEEE Trans Power Electron 38(9):10985–10996
46. Xu X, Khambadkone AM, Oruganti R (2007) A soft-switched back-to-back bi-directional DC/DC converter with a FPGA based digital control for automotive applications. In: 33rd annual conference of the IEEE industrial electronics society, pp 262–267
47. Morrison R, Egan M (1998) A new single transformer, power factor corrected UPS design. IEEE Trans Ind Appl 36(1):237–243
48. Zhan X, Wu H, Xing Y, Ge H (2016) A high step-up bidirectional isolated dual-active-bridge converter with three-level voltage-doubler rectifier for energy storage applications. In: IEEE applied power electronics conference and exposition, pp 1424–1429

49. Xuewei P, Rathore AK (2013) Novel bidirectional snubberless soft-switching naturally clamped zero current commutated current-fed dual active bridge (CFDAB) converter for fuel cell vehicles. In: Energy conversion congress and exposition, pp 1894–1901
50. Peng F, Li H, Su G, Lawler J (2004) A new ZVS bidirectional DC-DC converter for fuel cell and battery application. IEEE Trans Power Electron 19(1):54–65
51. Xiao H, Xie S (2008) A ZVS bidirectional DC–DC converter with phase-shift plus PWM control scheme. IEEE Trans Power Electron 23(2):813–823
52. Sha D, Zhang J, Wang X, Yuan W (2017) Dynamic response improvements of parallel-connected bidirectional DC-DC converters for electrical drive powered by low-voltage battery employing optimized feedforward control. IEEE Trans Power Electron 32(10):7783–7794
53. Sha D, Wang X, Chen D (2018) High-efficiency current-fed dual active bridge DC-DC converter with ZVS achievement throughout full range of load using optimized switching patterns. IEEE Trans Power Electron 33(2):1347–1357
54. Sha D, Wang X, Liu K, Chen C (2019) A current-fed dual-active-bridge DC-DC converter using extended duty cycle control and magnetic-integrated inductors with optimized voltage mismatching control. IEEE Trans Power Electron 34(1):462–473
55. Sha D, Chen D, Zhang J (2018) A bidirectional three-level DC-DC converter with reduced circulating loss and fully ZVS achievement for battery charging-discharging. IEEE J Emerg Sel Top Power Electron 6(2):993–1003
56. Inoue S, Akagi H (2007) A bidirectional DC–DC converter for an energy storage system with galvanic isolation. IEEE Trans Power Electron 22(6):2299–2306
57. Liu C, Johnson A, Lai J-S (2005) A novel three-phase high-power soft switched DC/DC converter for low-voltage fuel cell applications. IEEE Trans Ind Appl 41(6):1691–1697
58. Segaran D, Holmes DG, McGrath BP (2008) Comparative analysis of single and three-phase dual active bridge bidirectional DC–DC converters. In: 2008 Australasian universities power engineering conference, pp 1–6
59. Wang Z, Li H (2010) Three-phase bidirectional DC–DC converter with enhanced current sharing capability. In: 2010 IEEE energy conversion congress and exposition, pp 1116–1122
60. van Hoek H, Neubert M, Kroeber A, de Doncker R (2012) Comparison of a single-phase and a three-phase dual active bridge with low-voltage high-current output. In: 2012 International conference on renewable energy research and applications (ICRERA), pp 1–6
61. Hoek H, Neubert M, de Doncker RW (2013) Enhanced modulation strategy for a three-phase dual active bridge–boosting efficiency of an electric vehicle converter. IEEE Trans Power Electron 28(12):5499–5507
62. Wang Z, Li H (2012) A soft switching three-phase current-fed bidirectional DC–DC converter with high efficiency over a wide input voltage range. IEEE Trans Power Electron 27(2):669–684
63. Huang J, Wang Y, Li Z, Jiang Y, Lei W (2015) Simultaneous PWM control to operate the three-phase dual active bridge converter under soft switching in the whole load range. In: 2015 IEEE applied power electronics conference and exposition (APEC), pp 2885–2891
64. Dung NA, Chiu H, Lin J, Hsieh Y, Chen H, Zeng B (2019) Novel modulation of isolated bidirectional DC–DC converter for energy storage systems. IEEE Trans Power Electron 34(2):1266–1275
65. Jia P, Xie J, Liu M, Zhang Y, Xing Y, Qiu K (2024) A parameter design method and a fixed-frequency control strategy for the three-phase DAB-SRC based on the time domain analytical model. IEEE J Emerg Sel Top Power Electron (Early Access)
66. Sha D, Wang S (2021) A single-stage natural power factor corrector based on dual active Bridge DC-DC converter without inner current tracking loop. IEEE Trans Power Electron 36(1):342–352
67. Zhang J, Sha D, Ma P (2022) A dual active bridge DC-DC based single stage AC-DC converter with seamless mode transition and high power factor. IEEE Trans Ind Electron 69(2):1411–1421
68. Everts J, Krismer F, Van den Keybus J, Driesen J, Kolar JW (2014) Optimal ZVS modulation of single-phase single-stage bidirectional DAB AC-DC converters. IEEE Trans Power Electron 29(8):3954–3970

69. Everts J, Krismer F, Van den Keybus J, Driesen J, Kolar JW (2013) Charge-based ZVS soft switching analysis of a single-stage dual active bridge AC-DC converter. In: 2013 IEEE energy conversion congress and exposition, pp 4820–4829
70. Guo Z, Han X (2022) Control strategy of AC-DC converter based on dual active bridge with minimum current stress and soft switching. IEEE Trans Power Electron 37(9):10178–10189
71. Jauch F, Biela J (2016) Combined phase-shift and frequency modulation of a dual-active-bridge AC-DC converter With PFC. IEEE Trans Power Electron 31(12):8387–8397
72. Baranwal R, Castelino GF, Iyer K, Basu K, Mohan N (2018) A dual-active-bridge-based single-phase AC to DC power electronic transformer with advanced features. IEEE Trans Power Electron 33(1):313–331
73. Weise ND, Castelino G, Basu K, Mohan N (2014) A single-stage dual-active-bridge-based soft switched AC-DC converter with open-loop power factor correction and other advanced features. IEEE Trans Power Electron 29(8):4007–4016
74. Sha D, Zhang D, Zhang J (2021) A single stage dual-active-bridge AC-DC converter employing mode transition based on real-time calculation. IEEE Trans Power Electron 36(9):10081–10088
75. Dao ND, Lee D (2020) Modulation of bidirectional AC/DC converters based on half-bridge direct-matrix structure. IEEE Trans Power Electron 35(12):12657–12662
76. Chen T, Yu R, Ma Q, Zhao X, Huang AQ (2018) Optimal control scheme for single-stage dual-active-bridge ac-dc converter. In: 2018 IEEE energy conversion congress and exposition (ECCE), pp 2860–2864
77. Chen T, Yu R, Huang AQ (2022) A bidirectional isolated dual-phase-shift variable-frequency series resonant dual-active-bridge GaN AC-DC converter. IEEE Trans Ind Electron 70(4):3315–3325
78. Wu H, Tang X, Zhao J, Xing Y (2021) An isolated bidirectional microinverter based on voltage-in-phase PWM-controlled resonant converter. IEEE Trans Power Electron 36(1):562–570
79. Guo Z, Zhan W, Chen Z (2024) Single-stage isolated AC-DC converter based on LCL resonant DAB converters with single-loop control. IEEE Trans Power Electron 39(8):9305–9317
80. Lu D, Zhu C, Wang Z, Chen J, Lv X, Xu Q, Hu H (2024) Control of single-stage SR-DAB-based bidirectional AC/DC converter with simplicity and high accuracy. IEEE Trans Power Electron 39(8):9305–9317
81. Zengin S, Boztepe M (2020) A novel current modulation method to eliminate low-frequency harmonics in single-stage dual active bridge AC-DC converter. IEEE Trans Ind Electron 67(2):1048–1058
82. Everts J (2017) Closed-form solution for efficient ZVS modulation of DAB converters. IEEE Trans Power Electron 32(10):7561–7576
83. Singh A, Yadav AK, Khaligh A (2021) Steady-state modeling of a dual-active-bridge AC-DC converter considering circuit nonidealities and intracycle transient effects. IEEE Trans Power Electron 36(10):11276–11287
84. Zhang J, Sha D, Song K (2023) Single-phase single-stage bidirectional DAB AC-DC converter with extended ZVS range and high efficiency. IEEE Trans Power Electron 38(3):3803–3811
85. Jo S-W, Kim N-G, Kim M-H, Kim M-J, Kim M (2022) Highly efficient full-bridge resonant AC/DC converter using T-type voltage doubler. IEEE Trans Ind Electron 69(11):11076–11087
86. Uddin W, Wagaye TA, Kim M (2022) Quasi-single-stage current-fed resonant AC-DC converter having improved heat distribution. IEEE Trans Power Electron 37(11):13499–13512
87. Ma H, Lai J-S, Zheng C, Sun P (2016) A high-efficiency quasi-single-stage bridgeless electrolytic capacitor-free high-power AC-DC driver for supplying multiple LED strings in parallel. IEEE Trans Power Electron 31(8):5825–5836
88. Ma H, Chen G, Yi JH, Meng QW, Zhang L, Xu JP (2017) A single-stage PFM-APWM hybrid modulated soft-switched converter with low bus voltage for high-power led lighting applications. IEEE Trans Ind Electron 64(7):5777–5788
89. Yi J, Ma H, Li X, Lu S, Xu J (2021) A novel hybrid PFM/IAPWM control strategy and optimal design for single-stage interleaved boost-LLC AC-DC converter with quasi-constant bus voltage. IEEE Trans Ind Electron 68(9):8116–8127

90. Forouzesh M, Liu Y-F, Sen PC (2023) A line cycle synchronous rectification strategy based on time-domain analysis for single-stage AC-DC LLC converters. IEEE Trans Power Electron 38(4):5077–5091
91. Luo H, Zang T, Zhao C, Chen S, Zhou Y, Qiu Y, Wang X, Zhou X (2023) Power boundary controlled single-stage LLC power factor correction converter and its optimal parameter design. IEEE Trans Ind Electron 70(12):12219–12232
92. Zhou B, Zhao C, Luo H, Qiu Y, Chen S, Zang T, Zhou Y, Zhou X, Zhao H (2024) Hybrid control strategy for single-stage interleaved totem-pole LLC AC-dc converter with input and output power decoupling. IEEE Trans Power Electron 39(5):5135–5154
93. Kulasekaran S, Ayyanar R (2014) Analysis, design, and experimental results of the semidual-active-bridge converter. IEEE Trans Power Electron 29(10):5136–5147
94. Sha D, Zhang J, Xu Y (2017) Improved boundary operation for voltage-fed semi-DAB With ZVS achievement and nonactive power reduction. IEEE Trans Ind Electron 64(8):6179–6189
95. Lu M, Li X, Guo C (2020) A hybrid control of a semi dual active bridge DC-DC converter with minimum current stress. IEEE Trans Power Electron 35(3):3085–3096
96. Sha D, Zhang J (2019) Multimode control strategy for SiC MOSFETS based semi-dual active bridge DC-DC converter. EEE Trans Power Electron 34(6):5476–5486
97. Li F, Li Y, You X (2018) Optimal dual-phase-shift control strategy of an isolated buck-boost converter with a clamped inductor. EEE Trans Power Electron 33(6):5374–5385
98. Li Y, Li F, Zhao F, You X (2019) Variable-frequency control strategy of isolated buck-boost converter. IEEE J Emerg Sel Top Power Electron 7(3):1824–1836
99. Li Y, Li F, Zhao F, You X, Zhang K, Liang M (2019) Hybrid three-level full-bridge isolated buck-boost converter with clamped inductor for wider voltage range application. IEEE Trans Power Electron 34(3):2923–2937
100. Sha D, You F, Wang X (2016) A high-efficiency current-fed semi-dual-active bridge DC-DC converter for low input voltage applications. IEEE Trans Ind Electron 63(4):2155–2164
101. Sha D, Fu Y (2023) Optimized hybrid control for the current-fed semi dual-active-bridge DC-DC converter. IEEE Trans Power Electron 38(12):15750–15758
102. Wu H, Mu T, Gao X, Xing Y (2015) A secondary-side phase-shift-controlled LLC resonant converter with reduced conduction loss at normal operation for hold-up time compensation application. IEEE Trans Power Electron 30(10):5352–5357
103. Zhou S, Li X, Wang H, Huang P (2022) Modulation of a semi dual-active-bridge resonant converter with soft-switching and zero backflow power. IET Power Electron 15(6):515–530
104. Sha D, Chen D, Khan S, Guo Z (2019) Voltage-fed three-phase semi-dual active bridge DC-DC converter utilizing varying operating modes with high conversion efficiency. IEEE Trans Power Electron 34(10):9447–9458
105. Chen D, Sha D, Sun T (2020) Three-phase current-fed semi dual active bridge DC-DC converter with hybrid operating mode control. IEEE Trans Power Electron 35(2):1649–1658
106. Harrye YA, Ahmed KH, Aboushady AA (2014) Reactive power minimization of dual active bridge DC/DC converter with triple phase shift control using neural network. In: International conference on renewable energy research and application, pp 566–571
107. Wen H, Xiao W, Su B (2014) Nonactive power loss minimization in a bidirectional isolated dc–dc converter for distributed power systems. IEEE Trans Ind Electron 61(12):6822–6831
108. Sha D, Wang S (2021) A single-stage natural power factor corrector based on dual active bridge DC–DC converter without inner current tracking loop. IEEE Trans Power Electron 36(1):342–352
109. Ma P, Sha D (2023) A single-stage AC-DC converter based on semi dual-active-bridge with decoupled inductor current modulation strategy. IEEE Trans Power Electron 38(8):10170–10182
110. Chen T, Yu R, Huang AQ (2022) A bidirectional isolated dual-phase-shift variable-frequency series resonant dual-active-bridge GaN AC–DC converter. IEEE Trans Ind Electron 70(4):3315–3325

Chapter 2
Voltage-Fed Single Phase Semi-DAB DC-DC Converter with Boundary Operation

Abstract In this chapter, a boundary operation is proposed for the voltage-fed semi-dual-active-bridge (VF-SDAB) converter for applications needing only unidirectional power flow capability. The leakage inductor current works in the boundary mode under most operating conditions. The operating principle and comparison with conventional single-phase-shift modulation are presented in detail. The non-active power can be suppressed under most working conditions and the conduction loss for output MOSFETs can be reduced significantly. All the power switches can achieve ZVS while the output side rectifying diodes can obtain ZCS in most operating conditions. The parameter design including the compensated value and the dead-time selection is given. The ZVS achievement can be obtained with reduced non-active power and the closed loop implementation can be simplified. The effectiveness of the control strategy has been validated by the experimental results of 1 kW hardware prototype.

Keywords Semi-dual-active-bridge · Boundary mode · Non-active power · Zero voltage switching · Zero current switching

2.1 Working Principle of the Proposed Modulation

The power circuit configuration of the voltage-fed semi-DAB converter is illustrated in Fig. 2.1. The input side is an active bridge and the output side is a semi active bridge, both are connected by a high frequency transformer T. The input side bridge consists of four MOSFETs S_1–S_4, and the output side bridge is composed of two MOSFETs S_5, S_6 and two diodes D_1, D_2. C_{S1}–C_{S6} and C_{D1}–C_{D2} are the output junction capacitors of switches S_1–S_6 and diodes D_1–D_2 respectively. The transformer turns ratio is defined as n. L_k represents the sum of the external inductance and the primary side is referred as equivalent leakage inductance.

A voltage-fed semi-DAB was proposed with single-phase-shift modulation in [1]. The steady operation map of a semi-DAB under different voltage conversion gains and different load power is given in [2]. Another topology structure of semi-DAB is

D. Sha et al., *High Frequency Isolated Semi Dual Active Bridge DC-DC and AC-DC Converter*, CPSS Power Electronics Series,
https://doi.org/10.1007/978-981-95-1641-4_2

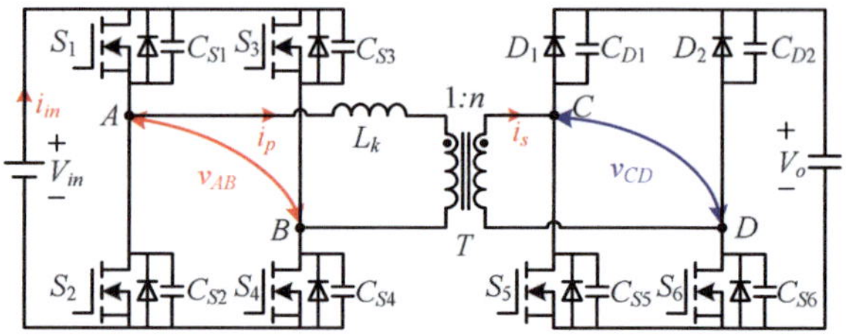

Fig. 2.1 The circuit scheme of voltage-fed S-DAB converter

proposed in [3, 4]. Similarly, with SPS control, the duty cycles for all the switches are fixed as 50%, ZVS can be achieved easily, but the non-active power is large. When the voltage conversion ratio varies, the ZVS range is limited. Besides, high non-active power causes higher peak current and additional conduction loss, thus the conversion efficiency is reduced. To reduce the non-active power, many improved control strategies have been proposed [5, 6]. While the modulation method of the secondary side switches will cause high conduction loss for the body diodes of the output side switches.

Therefore, to suppress the non-active power and reduce the conduction loss for the body diodes of the output side switches, an improved boundary operation is proposed. The leakage inductance current operates in boundary mode in most operating conditions. The current stress in switches and transformer windings can be reduced. With the propose control, ZVS can be obtained for all power switches by proper design of the duty cycles for the both sides. ZCS can be ensured for output diodes in most conditions. Besides, the non-active power can be reduced under most operating conditions. The conduction loss of the output MOSFETs can be reduced dramatically too. The closed loop control is easy to be implemented since there is only one independent variable.

2.1.1 Proposed Improved Boundary Operation

The voltage-fed semi-DAB converter has three control variables: the primary side duty cycle, the secondary side duty cycle and the phase shifted ratio. The primary side of the VF-SDAB converter can generate a high frequency square voltage v_{AB} between the middle point A and B of each leg with phase shifted strategy. The primary side duty cycle refers to the ratio of the high-level voltage (V_{in}) duration of v_{AB} to the switching period T_s. At the same time, the secondary side of the VF-SDAB converter can also generate a high frequency square voltage v_{CD} between the middle point C and D of each leg with PWM strategy. The secondary side duty cycle refers to the ratio of the high-level voltage (V_{o}) duration of v_{CD} to the switching period T_s. The phase shifted ratio refers to the ratio of the time between the rising edge of v_{AB} from zero to high-level and the rising edge of v_{CD} from zero to high-level to the switching period T_s.

The three control variables bring extra control flexibility and complexity. For VF-SDAB topology, the phase shifted ratio is always the variable that the controller uses to control the transferred power of the converter. The primary side duty cycle and the secondary side duty cycle should be constant or constrained with the phase shifted ratio. Otherwise the controller won't know how to regulate them.

For SPS modulation strategy, both the primary side duty cycle and the secondary side duty cycle are fixed at 0.5. This strategy is simple for the controller, but it will affect the efficiency of the converter.

For conventional boundary modulation strategy, the operating principle is shown in Fig. 2.2a and the restriction for the three control variables is given in (2.1). Under this restriction, the leakage inductor current operates in the boundary mode, the peak current is minimized. By employing this strategy, magnetizing current must be used to achieve soft switching for all switches, causing more core loss.

$$\begin{cases} \Phi + nD_1 V_{\text{in}}/V_{\text{o}} = 0.5 \\ D_2 = nD_1 V_{\text{in}}/V_{\text{o}} - D_c \end{cases} \tag{2.1}$$

This paper proposes an improved boundary modulation strategy. Figure 2.2b shows the operating principle of the proposed strategy with duty cycle compensation. The primary side duty cycle, the secondary side duty cycle and the phase shifted ratio are defined as d_1, d_2 and φ. As can be seen in the figure, instead of reducing the secondary side duty cycle in the boundary modulation strategy, the proposed strategy extends the primary side duty cycle and secondary side duty cycle so that every switch can achieve ZVS without the effect of the magnetizing current. Δd_p and Δd_s refer to the duty cycle compensation ratios for the primary side and secondary side respectively.

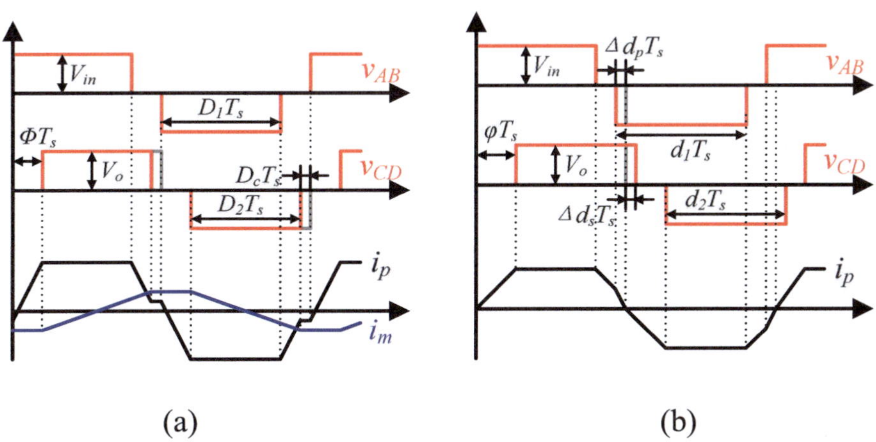

Fig. 2.2 Operating principle **a** conventional boundary strategy, **b** proposed strategy

For the proposed strategy, the restriction for the three control variables is

$$\begin{cases} \varphi + d_2 = 0.5 + \Delta d_p + \Delta d_s \\ d_1 = m(0.5 - \varphi) + (m+2)\Delta d_p \end{cases} \tag{2.2}$$

where m is defined as the voltage conversion ratio

$$m = \frac{V_{\text{o}}}{nV_{\text{in}}} \tag{2.3}$$

2.1.2 Typical Operating Periods

Figure 2.3 shows the steady-state operating waveforms of voltage-fed semi-DAB with the proposed modulation strategy. i_p is the leakage inductance current and i_{in} is the current of the input side voltage source.

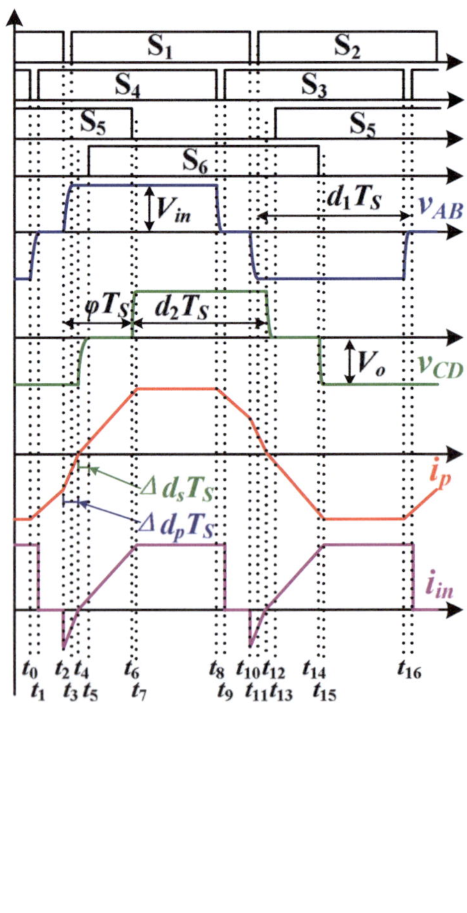

Fig. 2.3 Steady-state operating waveforms of semi-DAB with Proposed control

As seen in Fig. 2.3, an entire switching cycle can be divided into sixteen operating periods from t_0 to t_{16}. Figure 2.4 gives the current flow path of every period in half a switching cycle, which can be discussed as follow.

Period 1 [t_0–t_1]: Prior to t_0, v_{AB} equals $-V_{in}$ and v_{CD} equals $-V_o$. The power is transferred from the input side to the output side.

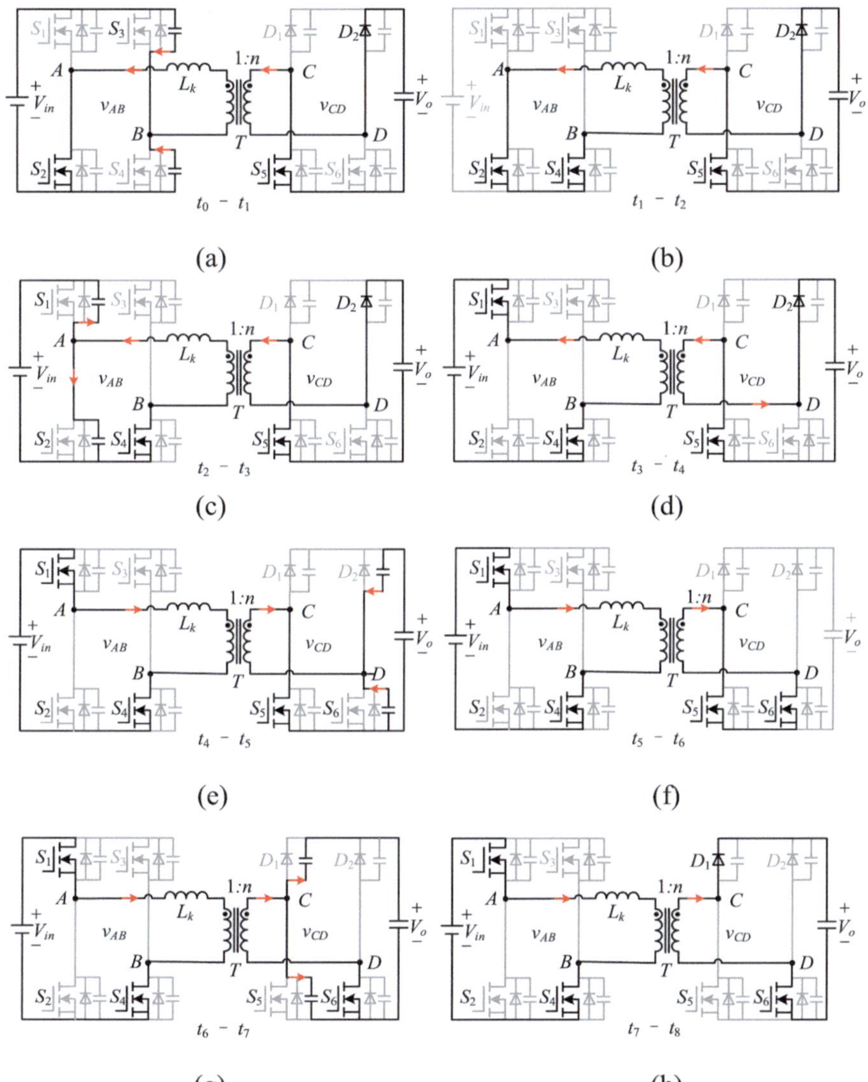

(a) (b) (c) (d) (e) (f) (g) (h)

Fig. 2.4 Typical working modes of the proposed voltage-fed S-DAB

At t_0, S_3 turns off and period 1 begins. During this period, L_k and the parallel connection of C_{S3} and C_{S4} resonate from the initial condition. The current i_p charges/discharges C_{S3}/C_{S4} until u_{ds4} decreases to zero and D_{S4} starts to conduct, so the ZVS condition of S_4 is satisfied.

Period 2 [t_1–t_2]: At t_1, S_4 is turned on with ZVS and period 2 begins. During this period, v_{AB} is equal to zero and v_{CD} remains $-V_o$, causing i_p to increase linearly.

Period 3 [t_2–t_3]: At t_2, S_2 turns off. Then, L_k begins to resonate with C_{S1} and C_{S2}, and the negative current i_p will charge/discharge C_{S2}/C_{S1}.

Period 4 [t_3–t_4]: At t_3, C_{S2} is already completely charged and v_{ds2} equals V_{in}, causing D_{S1} to conduct, so S_1 is turned on with ZVS. Then the value of v_{AB} becomes V_{in}, resulting in i_p increasing linearly with a larger slope.

Period 5 [t_4–t_5]: At t_4, i_p and i_s turn positive. Due to the unidirectional conductivity feature of the diodes, i_s will charge the junction capacitor of D_2 and discharge that of S_6 until v_{ds6} falls to zero, preparing ZVS condition for S_6.

Period 6 [t_5–t_6]: At t_5, S_6 turns on with ZVS and period 6 begins. During this period, v_{AB} equals V_{in} and v_{CD} equals zero, so i_p increases linearly.

Period 7 [t_6–t_7]: At t_6, S_5 is turned off. Then L_k begins to resonate with C_{S5} and C_{D1}. i_s charges C_{S5} and discharges C_{D1}.

Period 8 [t_7–t_8]: At t_7, C_{S5} is completely charged and the voltage across D_1 becomes zero, so D_1 begins to conduct. At the end of the period, S_4 turns off and the other half switching cycle begins, which is similar to this half cycle.

2.2 Analysis and Parameter Design

2.2.1 Output Power Character

The expression of output power P_o is given by

$$P_o = \frac{\int_{t_0}^{t_0+T_s} v_{AB} i_p \mathrm{d}t}{T_s} = \frac{\int_{t2}^{t_4} V_{in} i_p \mathrm{d}t + \int_{t4}^{t_7} V_{in} i_p \mathrm{d}t + \int_{t7}^{t_8} V_{in} i_p \mathrm{d}t}{0.5T_s} \tag{2.4}$$

Based on (2.2), (2.4) can be simplified as

$$P_o = -\frac{T_s V_o^2}{mn^2 L_k}(a\varphi^2 + b\varphi + c) \tag{2.5}$$

where the coefficients a, b and c are given by

Table 2.1 System specifications

Parameter/Device	Value	Parameter/Device	Value
Nominal input voltage	60 V	Rated power	1000 W
Input voltage range	55–65 V	Input side MOSFETs	IPT015N10N5
Nominal output voltage	300 V	Input side switches junction capacitor	1800 pF
Switching frequency	50 kHz	Output side MOSFETs	C3M0065090J
Switching period	20 μs	Output side switches junction capacitor	60 pF
Transformer turns ratio	1:5	Output side diodes	DSEI 8-06AS
Leakage inductance	4 μH	Output side diodes junction capacitor	20 pF

$$\begin{cases} a = m^2 + m + 1 \\ b = -m(2\Delta d_p + m) \\ c = 2\Delta d_p^2 - (1 - m)\Delta d_p + 0.25m(m - 1) \end{cases} \tag{2.6}$$

Furthermore, the maximum output power $P_{o\_\max}$ under each input voltage condition can be yielded as,

$$P_{\text{o\_max}} = -\frac{T_s V_o^2[(m^2 + 2m + 2)\Delta d_p^2 - \Delta d_p - 0.25m]}{mn^2 L_k (m^2 + m + 1)} \tag{2.7}$$

To obtain the rated output power capacity, maximum power for every input voltage must exceed the rated power, which can determine the upper limit for the leakage inductance design. The system specifications of this chapter are listed in Table 2.1.

To achieve rated 1000 W output with 55–65 V input range, the leakage inductance value should satisfy,

$$L_k \leq 5.438\ \mu\text{H} \tag{2.8}$$

Hence, the leakage inductance can be selected as 4 μH. Based on (2.5) and (2.6), the output power curve is plotted in Fig. 2.5.

2.2.2 *Soft Switch Conditions*

To achieve ZVS for all the MOSFETs, the energy stored in the leakage inductance should be high enough to be able to charge/discharge the output junction capacitors completely. At the same time, the dead time should ensure that the resonance process accomplished before the switch is turned on.

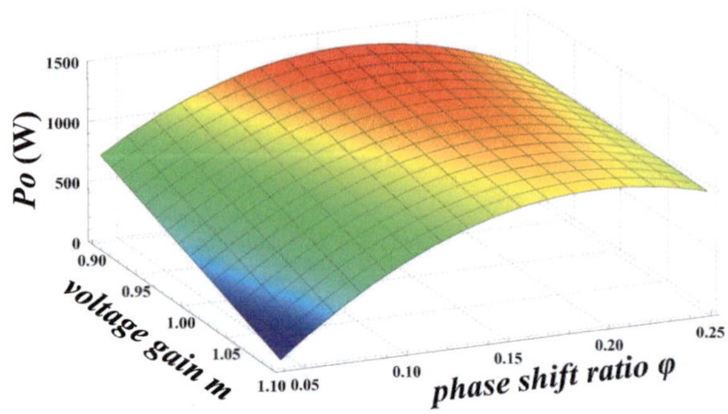

Fig. 2.5 The output power curve versus the phase shift angle and voltage gain

For the input side lagging leg switches, ZVS is easy to accomplish because the leakage inductance current almost approaches its maximum value before the two switches are turned on. However, for input side leading leg switches and output side switches, it is not easy to achieve ZVS. Hence, the time interval lengths of $[t_2, t_4]$ and $[t_4, t_5]$ in the proposed control should be properly designed.

(1) ***ZVS Condition for Input Side Switches***

S_1 is taken as an example to analyze the ZVS condition for input side leading leg switches. Figure 2.6 plots the ZVS condition for S_1. At t_2, S_2 is turned off. Then the leakage inductor L_k begins to resonate with junction capacitors C_{S1} and C_{S2}.

At t_3, C_{S1} has been discharged and C_{S2} has been charged completely. The resonance process is completed and the body diode of S_1 begins to conduct. The current i_p begins to increase linearly. At t_4, i_p reaches zero. L_k will resonate with C_{S1} and C_{S2} again and S_1 will lose ZVS if S_1 is still not turned on. Hence, in order to achieve ZVS for S_1, two basic conditions should be satisfied. Firstly, the leakage inductance at t_2 should be negative and it is large enough to discharge C_{S1} entirely before S_1 is gated on. Secondly, the dead time design should ensure that the resonance process can accomplish while the leakage inductance is non-positive when S_1 is gated on.

Define the time interval length of $[t_2, t_4]$ as

$$\Delta t_p = \Delta d_p \cdot T_s \tag{2.9}$$

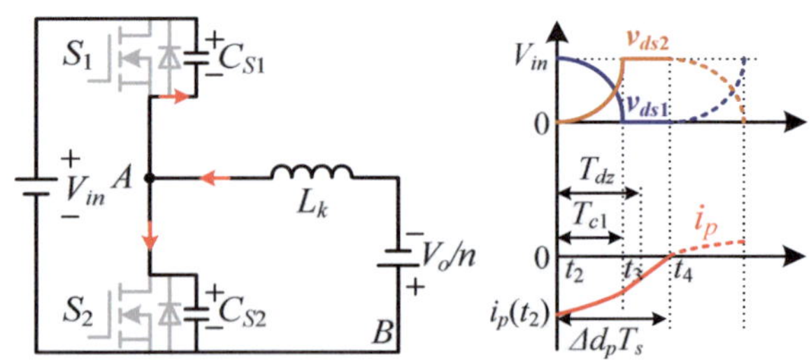

Fig. 2.6 ZVS process of S_1

According to equivalent resonant circuit in Fig. 2.6, the relationship of i_p and v_{ds2} can be expressed as,

$$\begin{cases} L_k \frac{\mathrm{d}i_p}{\mathrm{d}t} = v_{ds2} + \frac{V_o}{n} \\ i_p = C_{S1}\frac{\mathrm{d}(V_{in}-v_{ds2})}{\mathrm{d}t} + C_{S2}\frac{\mathrm{d}v_{ds2}}{\mathrm{d}t} = -2C_{sp}\frac{\mathrm{d}v_{ds2}}{\mathrm{d}t} \end{cases} \tag{2.10}$$

where v_{ds2} is the voltage across S_2 and C_{sp} is the parasitic junction capacitance of input side MOSFETs.

Assuming that the value of $i_p(t_2)$ is a known value i_{p2}, the boundary conditions are

$$v_{ds2}(t_2) = 0, \quad i_p(t_2) = i_{p2} \tag{2.11}$$

Then the solution of the differential equations (2.10) can be written as,

$$\begin{cases} v_{ds2}(t) = \frac{V_o}{n}\cos\left(\frac{t-t_2}{\sqrt{2L_kC_{sp}}}\right) - \sqrt{\frac{L_k}{2C_{sp}}}i_{p2}\sin\left(\frac{t-t_2}{\sqrt{2L_kC_{sp}}}\right) - \frac{V_o}{n} \\ i_p(t) = \sqrt{\frac{2C_{sp}}{L_k}}\frac{V_o}{n}\sin\left(\frac{t-t_2}{\sqrt{2L_kC_{sp}}}\right) + i_{p2}\cos\left(\frac{t-t_2}{\sqrt{2L_kC_{sp}}}\right) \end{cases} \tag{2.12}$$

The charging process is not completed until v_{ds2} reaches V_{in}. From (2.11) and (2.12), the charging time T_{c1} can be solved as

$$T_{c1} = \sqrt{2L_kC_{sp}}\left(\arcsin\frac{V_{in} + \frac{V_o}{n}}{\sqrt{\left(\frac{V_o}{n}\right)^2 + \frac{L_k}{2C_{sp}}i_{p2}^2}} - \arctan\left(\frac{-V_o}{ni_{p2}}\sqrt{\frac{2C_{sp}}{L_k}}\right)\right) \tag{2.13}$$

The value of Δt_p must be large enough to complete the resonance process, indicating that $i_p(t_2 + T_{c1})$ should be negative or zero. Based on (2.12) and (2.13), the critical value of i_{p2} can be obtained,

$$i_{p2} \le -\sqrt{\frac{2C_{sp}}{L_k}\left(V_{in}^2 + 2V_{in}\frac{V_o}{n}\right)} \triangleq i_{p\_cri}(t_2) \tag{2.14}$$

Substituting (2.14) into (2.13) yields the critical resonance time length T_{c1_cri},

$$T_{c1\_cri} = \sqrt{2L_kC_{sp}}\left(\frac{\pi}{2} - \arctan\left(\frac{V_o}{n}\bigg/\sqrt{V_{in}^2 + 2V_{in}\frac{V_o}{n}}\right)\right) \tag{2.15}$$

Therefore, Δt_p should be larger than the maximum value of T_{c1_cri} for the whole input voltage range, meaning that

$$\Delta t_p \ge \max\left(T_{c1\_cri}\right) \tag{2.16}$$

After the resonance process, i_p increases to zero linearly. Hence, the relationship between Δt_p and i_{p2} can be written by

$$\Delta t_p = \sqrt{2L_k C_{sp}}\left(\arcsin\frac{V_{\text{in}} + \frac{V_o}{n}}{\sqrt{\left(\frac{V_o}{n}\right)^2 + \frac{L_k}{2C_{sp}} i_{p2}^2}} - \arctan\left(\frac{-V_o}{n i_{p2}}\sqrt{\frac{2C_{sp}}{L_k}}\right)\right)$$
$$+ \left(\sqrt{\frac{2C_{sp}}{L_k}}\sqrt{\frac{L_k}{2C_{sp}} i_{p2}^2 - V_{\text{in}}^2 - 2V_{\text{in}}\frac{V_o}{n}}\right)\bigg/\left(\frac{V_{\text{in}} + \frac{V_o}{n}}{L_k}\right) \quad (2.17)$$

Once the value of Δt_p is fixed, i_{p2} is only determined by the input voltage V_{in} and it is independent of the load. Thus ZVS for S_1 can be obtained as long as the dead time T_{dz} is selected between T_{c1} and Δt_p. So the dead time value for input side switches should be selected as follows.

$$T_{c1} < T_{dz} < \Delta t_p \quad (2.18)$$

On the other hand, the average non-active power for the proposed control strategy, denoted by $P_{\text{na\_DPPS}}$, can be decided by (2.19), if the resonance process $[t_2, t_3]$ is neglected. Obviously, non-active power increases with the increase of Δt_p.

$$P_{\text{na\_DPPS}} = \frac{\left|\int_{t_2}^{t_4} V_{\text{in}} i_p(t)\text{d}t\right|}{0.5T_s} = \frac{V_o^2(m+1)\Delta t_p^2}{n^2 m^2 L_k T_s} \quad (2.19)$$

Thus, ZVS condition for S_1 can be determined by (2.16) and (2.18). Combining the analysis above, on one hand, Δt_p should satisfy (2.16) to obtain ZVS of S_1. On the other hand, Δt_p should be as small as possible to reduce the non-active power.

The curve of critical resonance time T_{c1_cri} versus the voltage gain m is plotted in Fig. 2.7. The range of T_{c1_cri} is from 137 to 144 ns when m varies between 0.92 and 1.09. Besides, in view of practical situations, the rising time and fall time of the MOSFETs need to be taken into account, so Δt_p can be designed as 200 ns, indicating that Δd_p is designed as 1%.

According to (2.18), the design for T_{dz} is illustrated by Fig. 2.8. T_{dz} is chosen as 150 ns in this chapter.

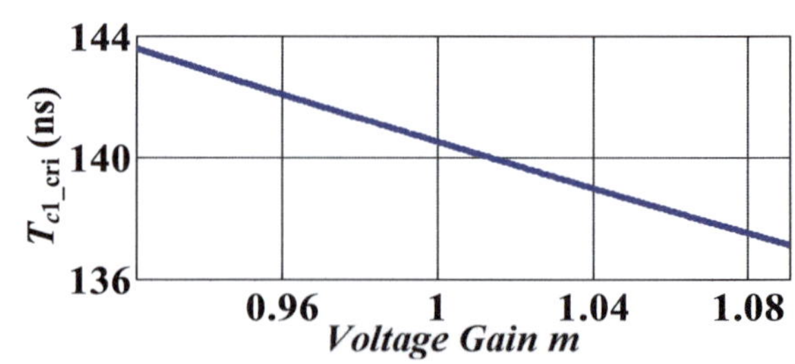

Fig. 2.7 The curve of critical resonance time T_{c1_cri} versus m

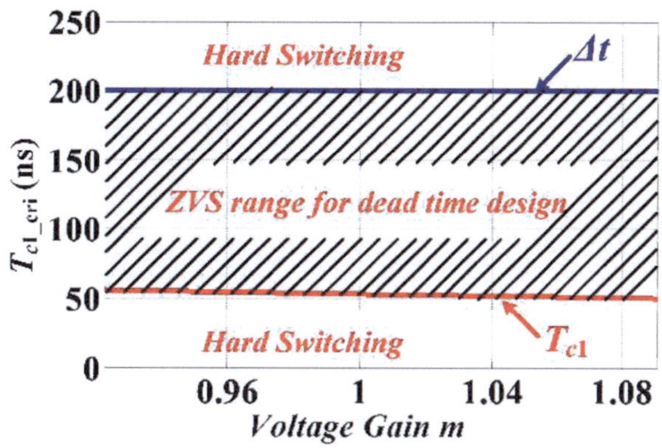

Fig. 2.8 ZVS range for input side dead time design

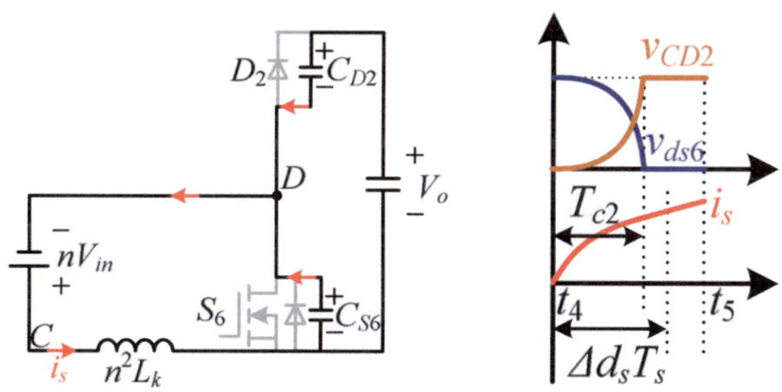

Fig. 2.9 ZVS process of S_6

(2) ***ZVS Condition for Output Side Switches***

S_6 is taken as the example to analyze the ZVS condition. To achieve ZVS for S_6, the secondary side transformer current i_s needs to discharge C_{S6} entirely. Figure 2.9 plots ZVS process for S_6.

The resonance process can be described by the following differential equations,

$$\begin{cases} \left(n^2L_k\right)\frac{\mathrm{d}i_s}{\mathrm{d}t} = nV_{\mathrm{in}} + v_{ds6} \\ i_s = C_{D2}\frac{\mathrm{d}(V_{\mathrm{o}}-v_{ds6})}{\mathrm{d}t} + C_{S6}\frac{\mathrm{d}v_{ds6}}{\mathrm{d}t} = -(C_{ss}+C_d)\frac{\mathrm{d}v_{ds6}}{\mathrm{d}t} \end{cases} \tag{2.20}$$

The initial conditions are

$$v_{ds6}(t_3) = V_{\mathrm{o}}, \quad i_s(t_3) = 0 \tag{2.21}$$

The instantaneous value for v_{ds6} and i_s can be written respectively by

$$\begin{cases} v_{ds6}(t) = (nV_{\mathrm{in}} + V_{\mathrm{o}})\cos\left(\frac{t-t_4}{\sqrt{L_k(C_{ss}+C_d)}}\right) - nV_{\mathrm{in}} \\ i_s(t) = \sqrt{\frac{C_{ss}+C_d}{L_k}}\left(V_{\mathrm{in}} + \frac{V_{\mathrm{o}}}{n}\right)\sin\left(\frac{t-t_4}{\sqrt{L_k(C_{ss}+C_d)}}\right) \end{cases} \tag{2.22}$$

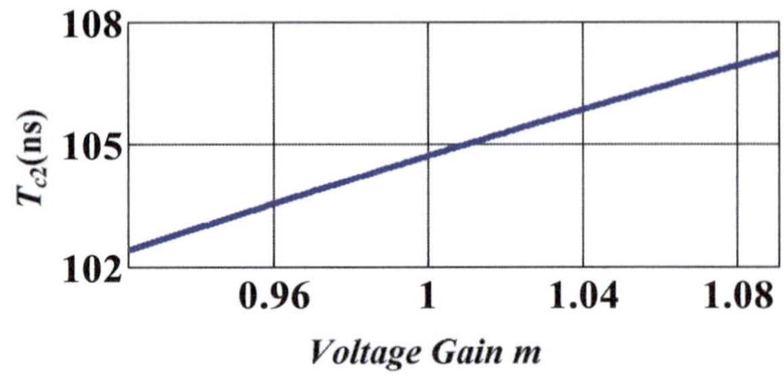

Fig. 2.10 The curve of T_{c2} versus m

The resonance process is completed as v_{ds6} drops to zero. From (2.22), the discharging time T_{c2} can be written as

$$T_{c2} = n\sqrt{L_k(C_{ss} + C_d)} \cdot \arccos\frac{nV_{\text{in}}}{nV_{\text{in}} + V_{\text{o}}} \tag{2.23}$$

The design of Δt_s needs to ensure that the resonance process of L_k, C_{S6} and C_{D2} can be completed before S_6 is turned on regardless of the input voltage V_{in}. So the ZVS condition for S_6 can be described as,

$$\Delta d_s \cdot T_S \geq \max\left(T_{c2\_\text{cri}}\right) \tag{2.24}$$

Besides, the output side current will flow through the MOSFET S_6 instead of its reverse diode once S_6 is gated on, resulting in less conduction loss. So Δd_s should not be too large.

Figure 2.10 illustrates the relationship between T_{c2} and m. The value of T_{c2} falls into the range from 102 to 108 ns when m varies from 0.92 to 1.09. According to (2.24), Δd_s can set as 1%.

2.2.3 ZVS Range Analysis

As discussed in Sect. 2.2, duty cycle compensations for both input side and output side are added in order to achieve ZVS for all the switches and ZCS for output rectifying diodes. To reach the target, some boundary conditions must be satisfied.

For both the input side and output side, the effective duty cycle must be higher than 0 and lower than 0.5. Meanwhile, the lower limit for the phase shift angle is 0 and the upper limit is related to the maximum output power, which is given as (2.7). Therefore, there are three boundary conditions to be satisfied.

$$\begin{cases} 0 \le d_1 \le 0.5 \\ 0 \le \varphi \le \dfrac{\Delta d_p (m+1)^2 + 0.5m^2}{m^2+m+1} \\ 0 \le d_2 \le 0.5 \end{cases} \tag{2.25}$$

If (2.25) is not satisfied, the ZVS condition may not be guaranteed for power switches. Substituting (2.2) into (2.25) yields,

$$\begin{cases} \dfrac{m+2}{m}\Delta d_p + \dfrac{m-1}{2m} \le \varphi \le \dfrac{m+2}{m}\Delta d_p + 0.5 \\ \Delta d_p + \Delta d_s \le \varphi \le \Delta d_p + \Delta d_s + 0.5 \\ 0 \le \varphi \le \dfrac{\Delta d_p (m+1)^2 + 0.5m^2}{m^2+m+1} \end{cases} \tag{2.26}$$

(2.26) can be further simplified as,

$$\begin{cases} \varphi \ge \dfrac{m+2}{m}\Delta d_p + \dfrac{m-1}{2m} \\ \varphi \ge \Delta d_p + \Delta d_s \\ \varphi \le \dfrac{\Delta d_p (m+1)^2 + 0.5m^2}{m^2+m+1} \end{cases} \tag{2.27}$$

Based on (2.27) and (2.5), the ZVS boundary curve can be drawn in Fig. 2.11. It can be found out that the input voltage range with ZVS/ZCS achievement is wide under heavy load and is narrow under light load.

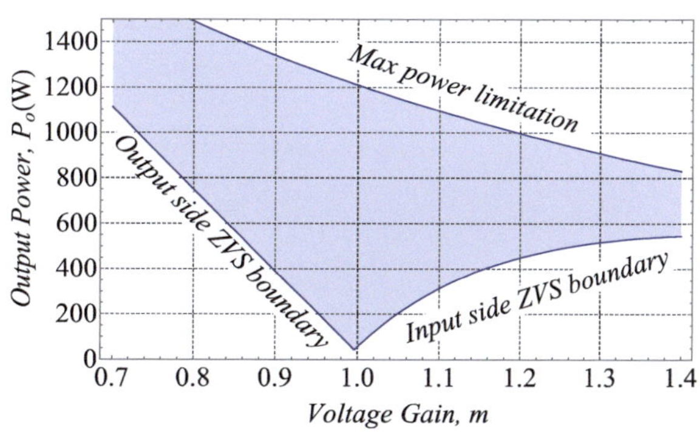

Fig. 2.11 ZVS boundary curve

2.2.4 Simulation Verification for Soft Switch Condition

Simulation is carried out by PSIM to verify the effectiveness of the different values of Δd_p and Δd_s. The circuit parameters of the simulation are the same as those in Table 2.1. The dead time T_{dz} is set as 150 ns. Simulation results are shown in Fig. 2.11.

Figure 2.12a shows the resonance process for S_1 and S_6, in which both Δd_p and Δd_s are set as 1%. It can be noted that the voltage applied across S_1 has already dropped to zero before the gate signal is applied, so ZVS for S_1 can be achieved. This is also effective for S_6.

Figure 2.12b gives the simulation results when Δd_p and Δd_s are set as 0.5% and 1% respectively. As seen, the resonance process for S_6 can be completed before the corresponding gate signal is applied, hence S_6 can achieve ZVS. However, the value

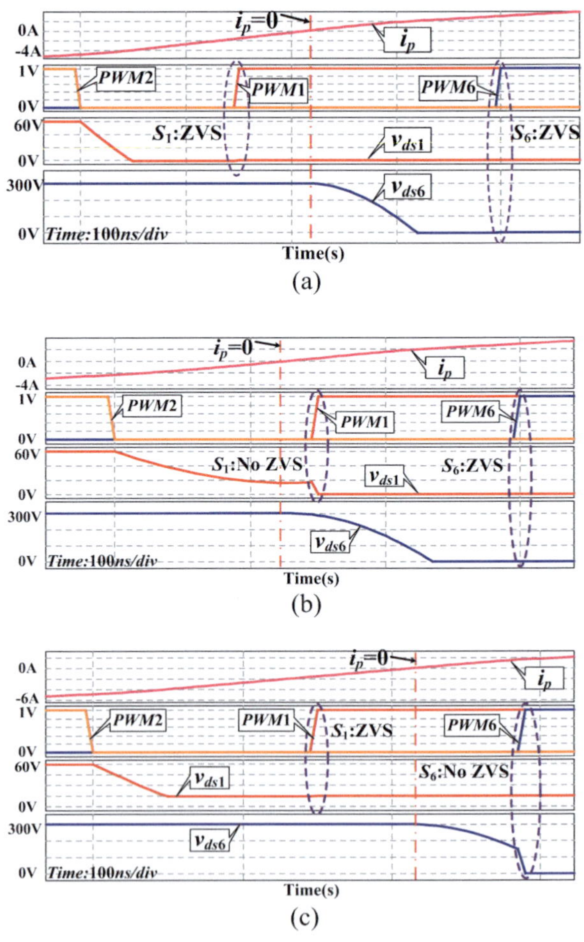

Fig. 2.12 Simulation waveforms for different values of Δd_p and Δd_s. **a** $\Delta d_p = 200$ ns, $\Delta d_s = 200$ ns. **b** $\Delta d_p = 100$ ns, $\Delta d_s = 200$ ns. **c** $\Delta d_p = 200$ ns, $\Delta d_s = 90$ ns

of Δd_p is shorter than the critical time length for S_1 to finish the resonance process, thus ZVS is lost for S_1.

The simulation waveforms when Δd_p equals 1% and Δd_s equals 0.45% are shown in Fig. 2.12c. Since Δd_p is larger than the critical time length for S_1 to finish the resonance process, ZVS for S_1 can be ensured. And after the resonance, the leakage inductance current i_p will increase with a larger slope until i_p crosses zero. But ZVS is lost for S_6 because the gate signal has already been applied before the resonance process is completed.

The effectiveness of the optimal design for Δd_p and Δd_s can be verified by the analysis and simulation above.

2.3 Comparison Between the Proposed Control and SPS Control

2.3.1 *Non-active Power Comparison*

For the same active output power, the larger the non-active power is, the larger the rms value of i_{Lk} will be. So it is necessary to reduce the non-active power. The average non-active power with the proposed control strategy has already been expressed in (2.19). The average non-active power is a constant once Δd_p and m are fixed, so the ratio of non-active power to P_o will decrease as the output power P_o increases.

In comparison, with the SPS control strategy, according to the mode analysis, the non-active power P_{na_SPS} can be calculated as

$$P_{na\_SPS} = \frac{V_o^2 T_s(m+1)}{n^2 m^2 L_k} \frac{\left(m\varphi_{SPS} + \frac{1-m}{2}\right)^2}{(m+2)^2} \tag{2.28}$$

wherein, φ_{SPS} is the phase shift angle in the dimensionless form. Usually, the range selection for φ_{SPS} should obey the basic rule that the P_o increases monotonously as φ_{SPS} increasing. Thus the non-active power increases dramatically as P_o increases.

To illustrate the effectiveness of the non-active power reduction, a comparison by using the proposed control and SPS control is made. The ratio of the non-active power to the active output power is denoted by P_{na_pu}. The curves of P_{na_pu} versus P_o under different input voltages are drawn in Fig. 2.12.

As shown in Fig. 2.13, P_{na_pu} decreases as P_o increasing by using the proposed control. However, it increases as P_o increasing by using the SPS control. P_{na_pu} with the proposed control is much smaller than that with the SPS control when the input voltage is over the nominal input voltage. Only when the input voltage is rather low, P_{na_pu} with the proposed control is a little bit larger than the SPS control under light load conditions.

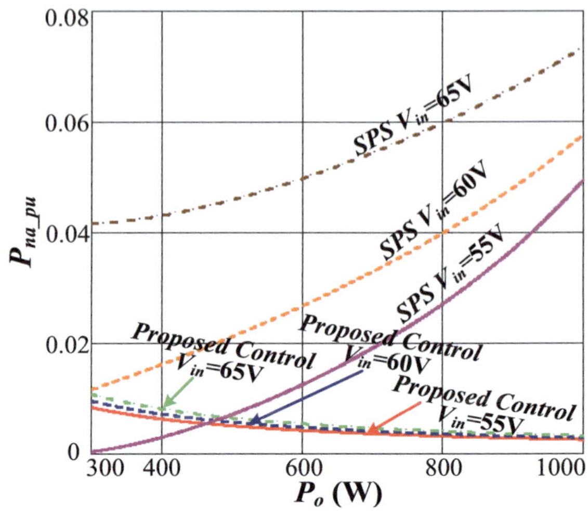

Fig. 2.13 Non-active power comparison curves

2.3.2 *Conduction Loss Comparison for the Output Side Switches*

With the proposed control, the total conduction loss for the output side MOSFETs and rectifying diodes can be reduced obviously within wide load range.

The calculation formulas for each of the output side power semiconductor devices by using the proposed control are listed in Table 2.2. The parameters for output side conduction loss calculation are as follows. The forward voltage across the rectifying diodes V_F is 0.98 V. For output side MOSFETs, the reverse diode forward voltage V_{sd_s} is 4.4 V and the drain-source on-state resistance R_{ds_s} is 65 mΩ.

The conduction loss comparison curves between the two control strategies are plotted in Fig. 2.14. As shown, the total output side conduction loss can be reduced by using the proposed control compared with SPS control at the same operating condition. Especially when the input voltage is rather low, the conduction loss reduction is significant. Thus, even though the proposed control is not competitive as compared with SPS control in view of non-active power when the input voltage is very low

Table 2.2 Conduction loss for output side power semiconductor devices

Component	Average conduction loss
D_1	$\int_{t_7}^{t_{12}} V_F \lvert i_s \rvert \mathrm{d}t \Big/ T_s$
D_2	$\left(\int_{t_0}^{t_4} V_F \lvert i_s \rvert \mathrm{d}t + \int_{t_{15}}^{t_{16}} V_F \lvert i_s \rvert \mathrm{d}t\right) \Big/ T_s$
S_5	$\left(\int_{t_0}^{t_7} i_s^2 R_{ds\_s} \mathrm{d}t + \int_{t_{12}}^{t_{13}} V_{sd\_s} \lvert i_s \rvert \mathrm{d}t + \int_{t_{13}}^{t_{16}} i_s^2 R_{ds\_s} \mathrm{d}t\right) \Big/ T_s$
S_6	$\left(\int_{t_4}^{t_5} V_{sd\_s} \lvert i_s \rvert \mathrm{d}t + \int_{t_5}^{t_{15}} i_s^2 R_{ds\_s} \mathrm{d}t\right) \Big/ T_s$

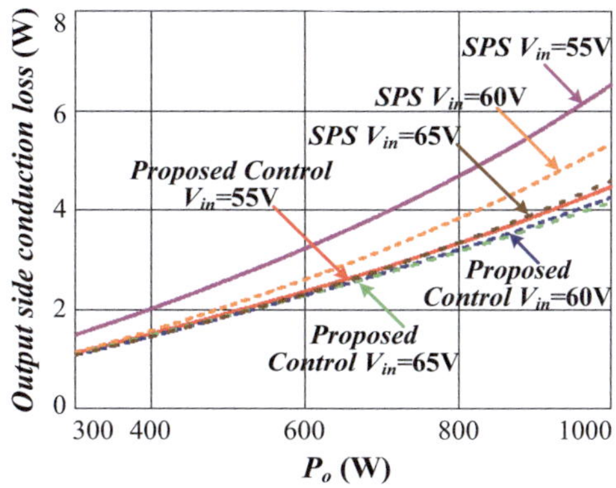

Fig. 2.14 Comparison of power loss for output side switches

under light loads, the conduction loss reduction under this operation condition with the proposed control may improve the total conversion efficiency.

2.4 Experimental Results

2.4.1 Prototype

A 1 kW prototype of the proposed converter has been fabricated. Figure 2.15 shows the photograph of the prototype. The system specifications are listed in Table 2.1. All the control algorithm is implemented by digital signal processor TM320F28335. It is very convenient to change the code to adapt to the variation of the circuit parasitic parameters and delays.

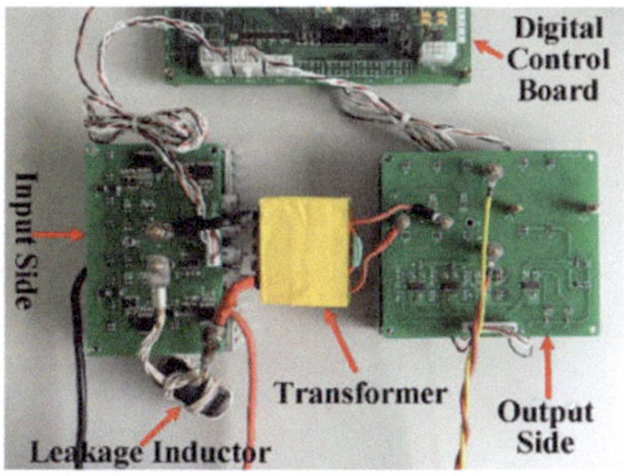

Fig. 2.15 The photograph of the prototype

2.4.2 Experimental Waveforms

Figure 2.16 shows the steady-state operating waveforms for the semi-DAB converter using the proposed control strategy. The converter can work within wide load range for the specified input voltage range 55–65 V.

As can be observed, the stage between the two dashed lines is the non-active power stage. As seen, the non-active power time length is very short and the non-active power is small.

Figure 2.17 shows the switching process of input side leading leg switch S_1 when V_{in} is 60 V. It can be observed that S_1 can achieve ZVS within the wide load range. The leakage inductance current i_p is negative when S_2 is turned off and it is able to

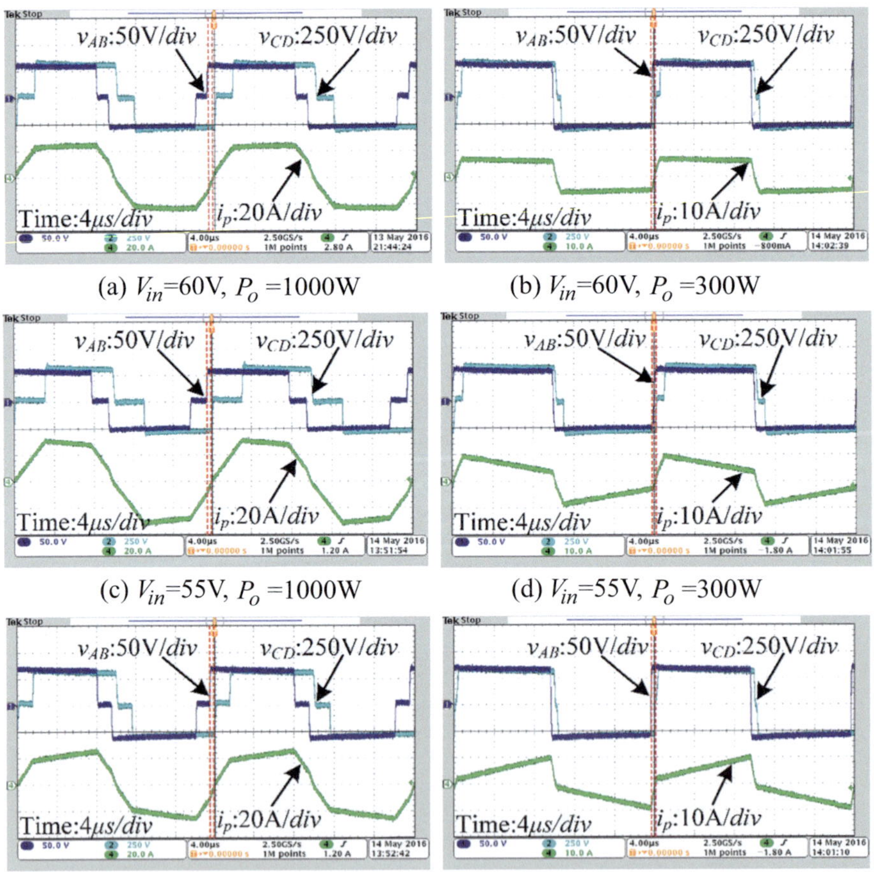

(a) V_{in}=60V, P_o=1000W (b) V_{in}=60V, P_o=300W

(c) V_{in}=55V, P_o=1000W (d) V_{in}=55V, P_o=300W

(e) V_{in}=65V, P_o=1000W (f) V_{in}=65V, P_o=300W

Fig. 2.16 The experimental steady-state operating waveforms: **a** $V_{\rm in}$ = 60 V, $P_{\rm o}$ = 1000 W, **b** $V_{\rm in}$ = 60 V, $P_{\rm o}$ = 300 W, **c** $V_{\rm in}$ = 55 V, P_t = 1000 W, **d** $V_{\rm in}$ = 55 V, $P_{\rm o}$ = 300 W, **e** $V_{\rm in}$ = 65 V, $P_{\rm o}$ = 1000 W, **f** $V_{\rm in}$ = 65 V, $P_{\rm o}$ = 300 W

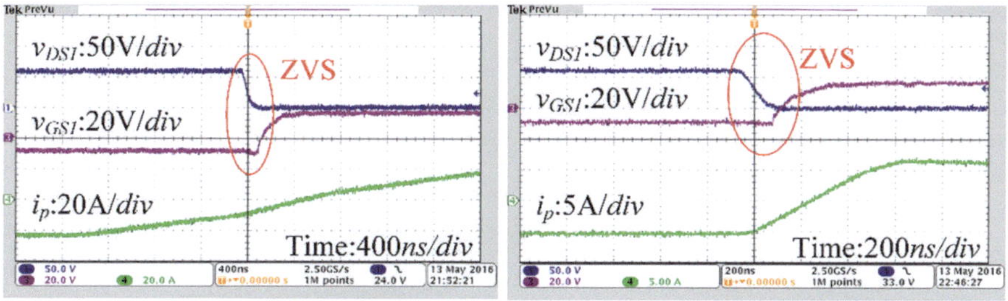

(a) P_o =1000W (b) P_o =300W

Fig. 2.17 Switching waveforms of S_1 at $V_{\text{in}} = 60$ V **a** $P_\text{o} = 1000$ W, **b** $P_\text{o} = 300$ W

discharge C_{S1} completely before S_1 is turned on. Besides, the driving signal of S_1 is given before i_p crosses zero. Thus S_1 can achieve ZVS. The ZVS achievement of S_1 can verify the design of input side duty cycle compensation Δd_p and the dead time.

The switching process of input side lagging leg switch S_3 is shown in Fig. 2.18. As seen, ZVS for S_3 can be obtained under the wide load range. Actually, ZVS for input side lagging leg switches is relatively easy to achieve in most situations, since the resonant current is relatively large and is able to discharge the junction capacitor completely before it is turned on.

Figure 2.19 shows the switching process of output side switch S_5. The output side MOSFETs can achieve ZVS within the wide load range. As seen, the leakage inductance current is negative before S_5 is turned on and it is able to discharge C_{S5} completely, hence S_5 can achieve ZVS, which can verify the design of input side duty cycle compensation Δd_s.

As illustrated in Fig. 2.20, D_1 can obtain ZCS off within wide load range. The reverse recovery issue can be avoided, thus no voltage turn off spike is generated across rectifying diodes and no RC snubber circuit is needed.

In comparison, Fig. 2.21 shows the steady-state operating waveform and switching process waveforms using control strategy without duty cycle compensation at $V_{\text{in}} =$

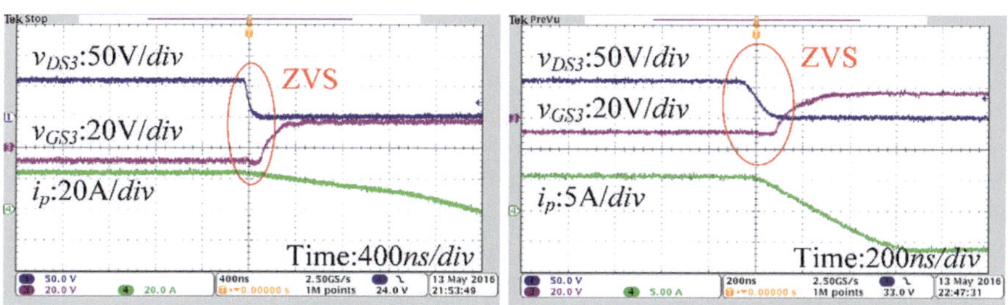

(a) P_o=1000W (b) P_o=300W

Fig. 2.18 Switching waveforms of S_3 at $V_{\text{in}} = 60$ V **a** $P_\text{o} = 1000$ W, **b** $P_\text{o} = 300$ W

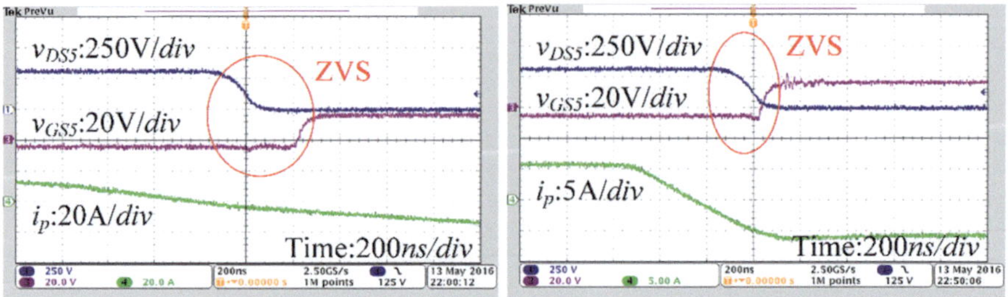

(a) P_o=1000W (b) P_o=300W

Fig. 2.19 Switching waveforms of S_5 at $V_{\text{in}} = 60$ V **a** $P_o = 1000$ W, **b** $P_o = 300$ W

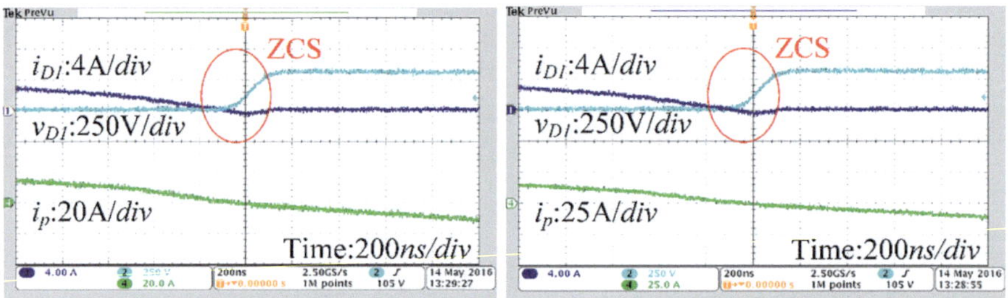

(a) P_o=1000W (b) P_o=300W

Fig. 2.20 Switching waveforms of D_1 at $V_{\text{in}} = 60$ V **a** $P_o = 1000$ W, **b** $P_o = 300$ W

60 V and $P_o = 1000$ W. As can be seen, the semi-DAB converter can work stably. However, ZVS for S_1 and S_5 cannot be achieved. Meanwhile, ZCS for D_1 cannot be secured either. Compared with waveforms of the semi-DAB converter using control strategy with duty cycle compensation shown in Figs. 2.16, 2.17, 2.18, 2.19 and 2.20, the effectiveness of the duty cycle compensation is verified.

2.4.3 *Efficiency Curve*

The efficiency curves for the minimum, nominal, and maximum input voltage are shown in Fig. 2.22. The loss measurement includes all the losses such as power circuits, the driving circuits and the control board. As can be seen, the peak efficiency of the proposed control can reach 97% and the proposed control achieves higher efficiency than SPS control within wide load range for the specified input voltage range.

Figure 2.23 illustrates the system loss breakdown with full load under different input voltage conditions for the proposed control and SPS control. The components specifications that is used for calculation is given in Table 2.3. For each input voltage,

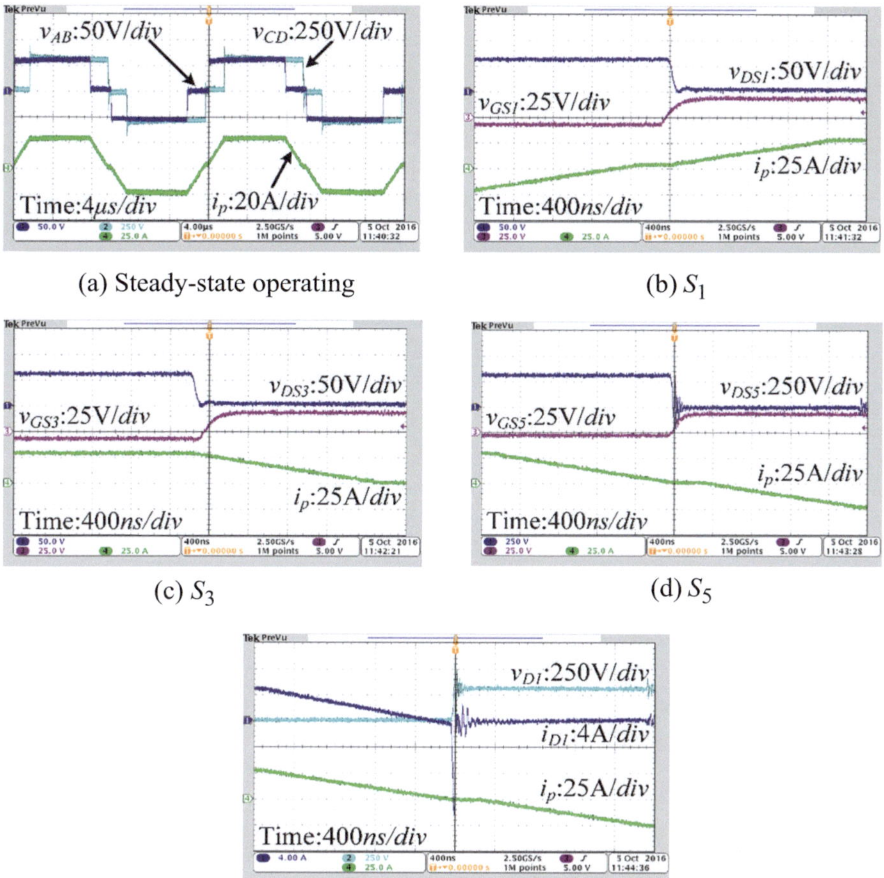

(a) Steady-state operating (b) S_1

(c) S_3 (d) S_5

(e) D_1

Fig. 2.21 Waveforms using control strategy without duty cycle compensation at $V_{\text{in}} = 60$ V and $P_{\text{o}} = 1000$ W **a** steady-state operating, **b** S_1, **c** S_3, **d** S_5, **e** D_1

the left column responds to the proposed control and the right column responds to SPS control.

As can be seen in Fig. 2.23, compared with SPS control, MOSFETs conduction loss and switching loss are reduced significantly with the proposed control. Efficiency improvement results mainly from two reasons. Firstly, when the input voltage is relatively high, the non-active power can be reduced significantly and the output side conduction loss is slightly reduced. Due to the non-active power reduction, the primary side conduction loss can be reduced largely. While when the input voltage is relatively low, the non-active power is slightly increased under light load condition, which can make primary side conduction loss slightly increased. However, it can be compensated by the significant reduction of output side conduction loss. Hence the sum of MOSFETs conduction loss can be reduced. In other words, the proposed

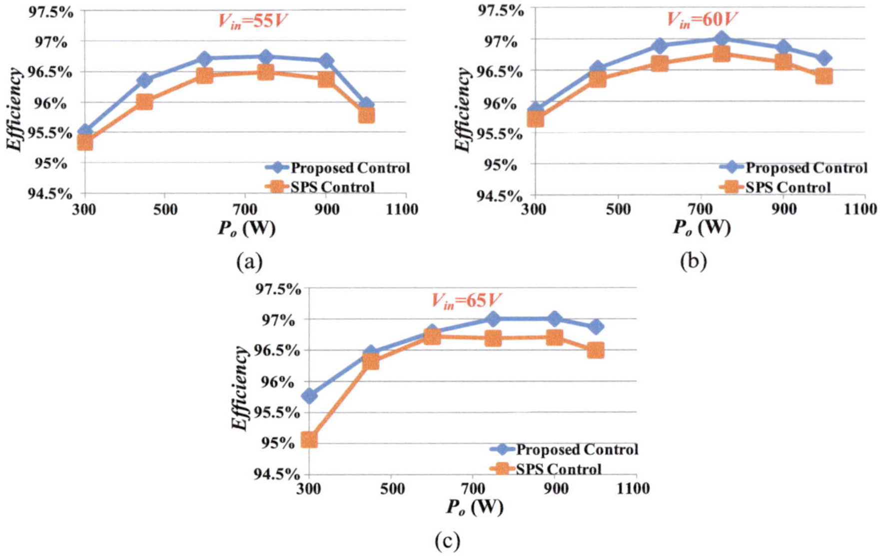

Fig. 2.22 Comparison curves of efficiency between proposed control and SPS control **a** V_{in} = 55 V **b** V_{in} = 60 V **c** V_{in} = 65 V

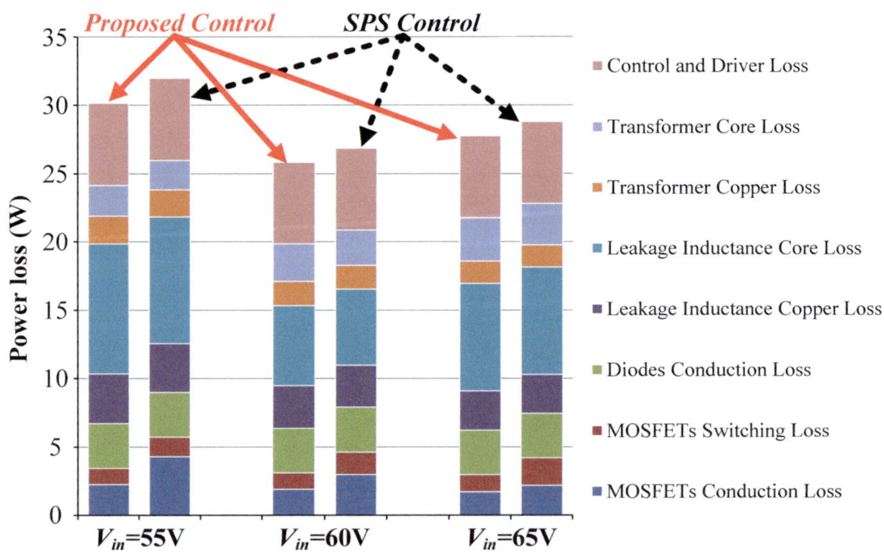

Fig. 2.23 System loss breakdown with full load under different input voltage conditions for the proposed control and SPS control

Table 2.3 Components specifications

Parameter/Device	Value
Leakage inductor core	77439-A7
Leakage inductor turns	6
Leakage inductance	4 μH
Leakage inductor resistance	8.8 mΩ
Transformer core	C_46410EC
Transformer turns ratio	4:20
Transformer primary side resistance	5 mΩ
Transformer secondary side resistance	197 mΩ

control makes a good tradeoff between the non-active power and the output side conduction loss.

2.5 Conclusion

For the voltage-fed semi-DAB DC-DC converter, an improved boundary operation control is proposed. The closed loop implementation can be simplified since there is only one independent control variable. The compensated duty cycle value and the dead time value are discussed in detail in order to achieve ZVS for power switches. Compared with using the conventional single phase shifted control, the proposed control can reduce the non-active power significantly under most operation working conditions. For the output side, the conduction loss can be reduced as well. All the power switches can achieve ZVS under most operating conditions and ZCS can be obtained for rectifying diodes. The conversion efficiency is higher than the SPS control with the same operating condition.

References

1. Kulasekaran S, Ayyanar R (2014) Analysis, design, and experimental results of the semi dual-active-bridge converter. IEEE Trans Power Electron 29(10):5136–5147
2. Lu M, Hu S, Zhou S-Z (2017) The steady-state operation map of a semi-dual-active-bridge converter. In: 2017 12th IEEE conference on industrial electronics and applications (ICIEA), pp 1586–1590
3. Zhang J, Zhang F, Xie X (2004) A novel ZVS DC/DC converter for high power applications. IEEE Trans Power Electron 19(2):420–429
4. Mishima T, Nakaoka M (2011) Practical evaluations of a ZVS-PWM DC-DC converter with secondary-side phase-shifting active rectifier. IEEE Trans Power Electron 26(12):3896–3907
5. Lu M, Li X, Chen G (2020) A hybrid control of a semidual-active-bridge DC-DC converter with minimum current stress. IEEE Trans Power Electron 35(3):3085–3096
6. Li F, Li Y, You X (2018) Optimal dual-phase-shift control strategy of an isolated buck-boost converter with a clamped inductor. IEEE Trans Power Electron 33(6):5374–5385

Chapter 3
Voltage-Fed Single Phase Semi-DAB DC-DC Converter with Multi-mode Control

Abstract In Chap. 2, an improved boundary modulation strategy is proposed for the semi dual active bridge (SDAB) DC-DC converter, but the modulation is suitable only for the middle and high power applications under certain voltage conversion gains. It is not suitable to be used for low power in facing wider voltage gains employing fixed switching frequency control strategy. So in order to meet the full load range operation requirement for semi-DAB DC-DC converter working as a battery charger, a multi-mode control strategy is presented in this chapter. At heavy load, it works with single phase shifted control. At light load, it works with PWM modulation and variable switching frequency. At very light load conditions, it works with PWM modulation and constant switching frequency mode. Light load conversion efficiency can be improved significantly by using the proposed multi-mode control strategy. Detailed loss breakdown is made and compared with different switching frequencies under light load conditions. The implementation of the control strategy was given and seamless transition can be obtained among three operating modes according to the load power. The effectiveness of the control strategy was validated by the experimental results of 1.6 kW hardware prototype.

Keywords DC-DC converter · Semi dual active bridge · Battery charger · Multi-mode control · Variable switching frequency

3.1 Limitations of Conventional Control Strategies

The topology of semi-DAB DC-DC converters is shown in Fig. 3.1. The input side is an active bridge while the output side is a semi active bridge. They are connected by a high frequency transformer T. The input side bridge consists of four MOSFETs S_1–S_4, and the output side bridge is composed of two MOSFETs S_5, S_6 and two diodes D_1, D_2. C_{S1}–C_{S6} and C_{D1}–C_{D2} are the output junction capacitors of switches S_1–S_6 and diodes D_1–D_2 respectively. The transformer turns ratio is defined as n. L_k represents the sum of the external inductance and the secondary side referred equivalent leakage inductance.

D. Sha et al., *High Frequency Isolated Semi Dual Active Bridge DC-DC and AC-DC Converter*, CPSS Power Electronics Series,
https://doi.org/10.1007/978-981-95-1641-4_3

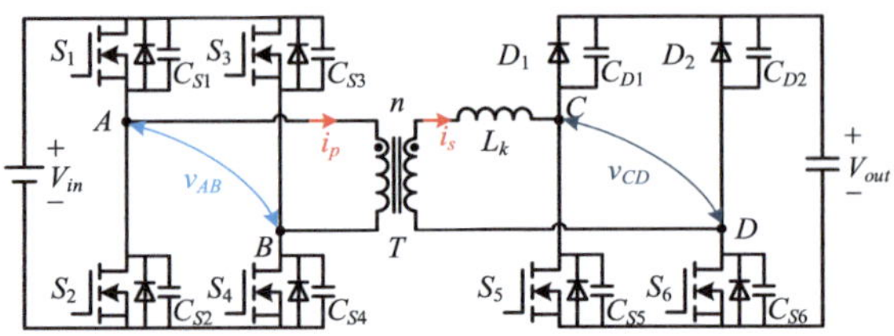

Fig. 3.1 Semi-DAB DC-DC converter topology

With SPS control [1], all the switches can achieve soft switching when the input voltage matches with the output voltage. But when the output voltage varies, soft switching for some switches may be lost under light load conditions, which causes high current spike over switches, large non-active power, external inductors and transformer windings losses. This lowers the conversion efficiency and may bring reliability issues.

To reduce the non-active power and the RMS value of the leakage inductor current, a BCM control strategy was proposed for Semi-DAB converters [2]. However, the effectiveness of this strategy under light load condition is not verified. Theoretically, under light load condition, the output voltage cannot be regulated to track the reference value when the load drops under a certain value. It means that the BCM control strategy is not effective under very light load conditions.

For most DAB and semi-DAB DC-DC converters in literature, their control strategy is a single mode, which means that the control strategy keeps the same when the load varies [3–6]. The switching frequency is always fixed. Thus the conversion efficiency at light load conditions is not high although ZVS have been achieve for all power switches. This paper proposes a novel multi-mode control strategy to face the full load operations required by the battery charging profile. Three working modes work together and seamless transition can be obtained among them once the output power requires. With the proposed control strategy, the semi-DAB DC-DC converter can obtain high conversion efficiency for full load range and wide voltage range operations.

3.1.1 Limitation of SPS Control Strategy

When the semi-DAB converter with SPS control operates in the matching mode (i.e. the voltage gain is 1), it can achieve ZVS turn-on for MOSFETs and ZCS turn-off for diodes within the full load range, thus reducing switching loss and avoiding reverse recovery problems. But when the voltage gain is far away from 1, the ZVS range will be very narrow. Figure 3.2 shows the ZVS range for the Semi-DAB converter with SPS control strategy. The converter specifications for plotting this figure are listed in Table 3.1. The operating waveforms in each region are shown in Fig. 3.3. As seen in the figures, ZVS for all switches can be achieved in region III. But in region I and II, ZVS conditions for some switches are lost.

Fig. 3.2 ZVS range curve for the semi-DAB converter with SPS control strategy

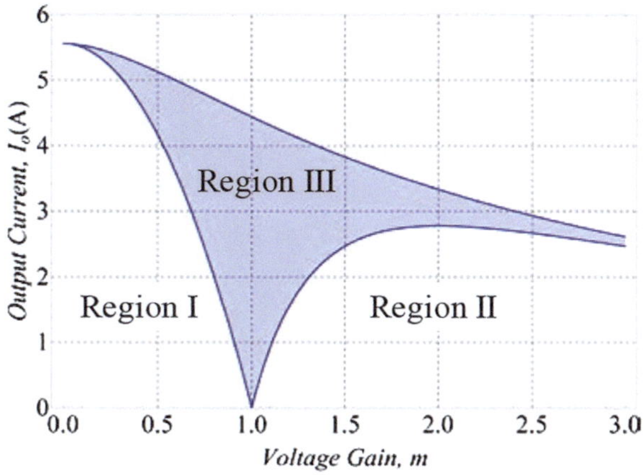

Table 3.1 Converter specifications for Fig. 3.3

Parameter	Value
Input voltage	400 V
Voltage gain	0–3
Switching frequency	100 kHz
Transformer turns ratio	1:1
Leakage inductance	90 μH

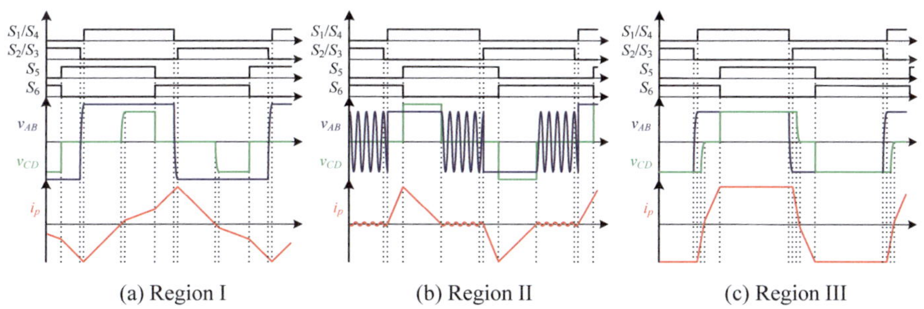

Fig. 3.3 Typical operating waveforms in each regions

This chapter will focus on how to improve the conversion efficiency in region I. When the converter operates in region I, the phase shifted ratio φ is less than 0. That means the operating mode is completely different from that in region III. The operating mode, of which the waveforms are shown in Fig. 3.3a, has several issues. Firstly, ZVS are lost for the secondary side MOSFETs, which causes high turn-on loss. Secondly, the primary side MOSFETs can achieve ZVS on, but they turn off with high current, thus causing high turn-off loss. Finally, the circulating current is very high, which brings significant conduction loss. These issues significantly affect the conversion efficiency by using SPS strategy especially under light load conditions.

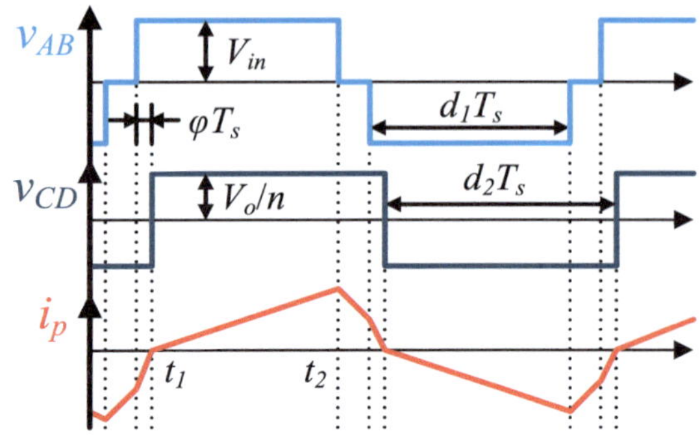

Fig. 3.4 Steady-state waveforms of BCM control at light load condition

3.1.2 Limitation of BCM Control

BCM control is another control strategy available for the semi-DAB converter. Compared with SPS strategy, BCM control regulates the phase shifted ratio and the duty cycles in both sides to control the transferred power. These control variables are coupled strongly following particular equations, which ensure ZVS conditions for all switches. This strategy can also minimize the RMS value of the leakage inductor current, thus reducing the conduction loss. However, when the voltage gain is lower than 1, the boundary condition limits the minimum power capable of transmitting. That implies that the output voltage cannot be regulated to track its reference value.

Figure 3.4 shows the steady-state waveforms for the converter with BCM control when the voltage gain is lower than 1 and the phase shifted ratio equals to its minimum value, with which the converter can transmit its lowest power. As seen, in order to ensure the boundary condition, the phase shifted ratio φ, the primary side duty cycle d_1 and the secondary side duty cycle d_2 are regulated so that $i_p(t_1)$ equals to 0. During the mode $[t_1, t_2]$, since the voltage gain is lower than 1, i_p increase linearly. This leads to a fact that the minimum transferred power $P_{\min}$ is always higher than 0. Therefore, if the output power is lower than $P_{\min}$, the output voltage cannot be regulated to track the reference voltage. Figure 3.5 shows the simulation result for this case. As illustrated, although the phase shifted angle has already reached its minimum value, the output voltage is still higher than its reference voltage under steady state. Thus the BCM control is not suitable to be used in facing full load range and wide voltage range for the semi-DAB DC-DC converter.

3.1.3 Solutions for Full Load Range Operation

One way to extend the output power range of the conventional control strategies is to increase the switching frequency under light load conditions. The increased switching frequency can extend the ZVS range, but switching loss and driving

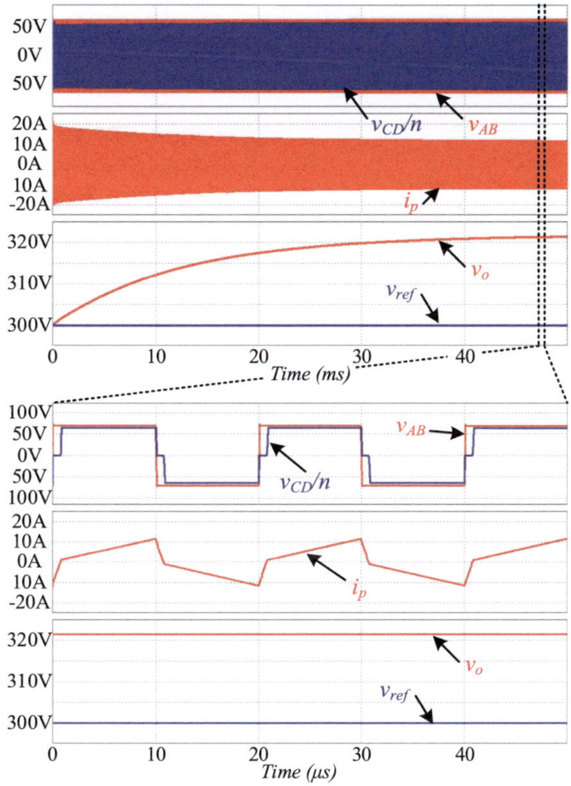

Fig. 3.5 Simulation waveforms with BCM control under very light load condition

loss will increase as well. Compared with fixed frequency control, this strategy can improve the light load operation performance. However, ZVS is still lost under very light load conditions. The increasing driving loss and switching loss lowers the conversion efficiency.

This chapter proposes a multi-mode control strategy in order to achieve full load operating range for semi-DAB. The conversion efficiency with this control strategy can be improved compared with conventional SPS control and BCM control. Three operating modes are used for different load conditions. Under heavy load conditions, the SPS is utilized to ensure ZVS achievement for all switches and extend power transmitted time length. Under medium load conditions, the PWM modulation with variable switching frequency is used to reduce the switching loss and the driving loss. At very light load conditions, the PWM modulation with constant switching frequency is utilized since the switching frequency is reduced to its minimum value. The limited volt-second area over the transformer avoids saturation issue in the transformer core.

3.2 Analysis of the Proposed Control Under Light Load Conditions

For the proposed multi-mode control strategy, the SPS control used for heavy load operation has already been discussed extensively [1]. However, the control strategy used for medium and light load operation needs to be analyzed more. This section focuses the operation modes of PWM modulation with variable and constant switching frequency respectively. The reason why conversion efficiency can be improved with reduced switching frequency is also analyzed.

3.2.1 Typical Operating Modes

The key waveforms of the PWM modulation with constant frequency (CF) mode and variable frequency (VF) mode of the proposed multi-mode control strategy is shown in Fig. 3.6. In this figure, some parameters need to be defined before the analysis. The phase shift duration time length t_p is defined as the period between the rising edge of v_{AB} from 0 to V_{in} and the rising edge of v_{CD} from 0 to V_{o}. The duty cycle duration time length t_d is defined as the period between the rising edge of v_{AB} from 0 to V_{in} and the falling edge of v_{AB} from V_{in} to 0. The switching period is t_s while f_s is the switching frequency.

Figure 3.7 shows the current flow path of each mode in half switching cycle, which can be discussed as follows.

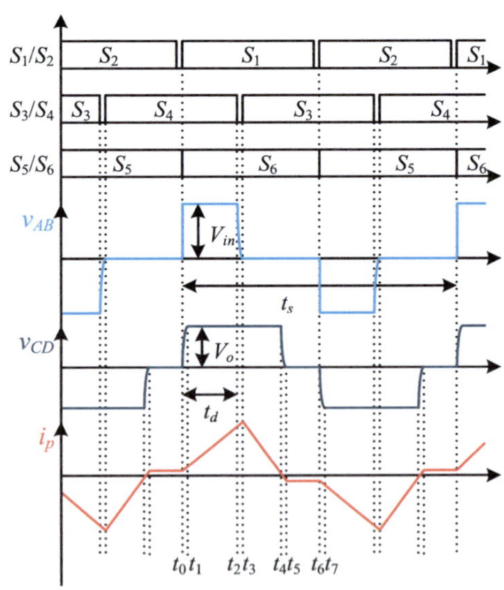

Fig. 3.6 Key waveforms of the PWM modulation

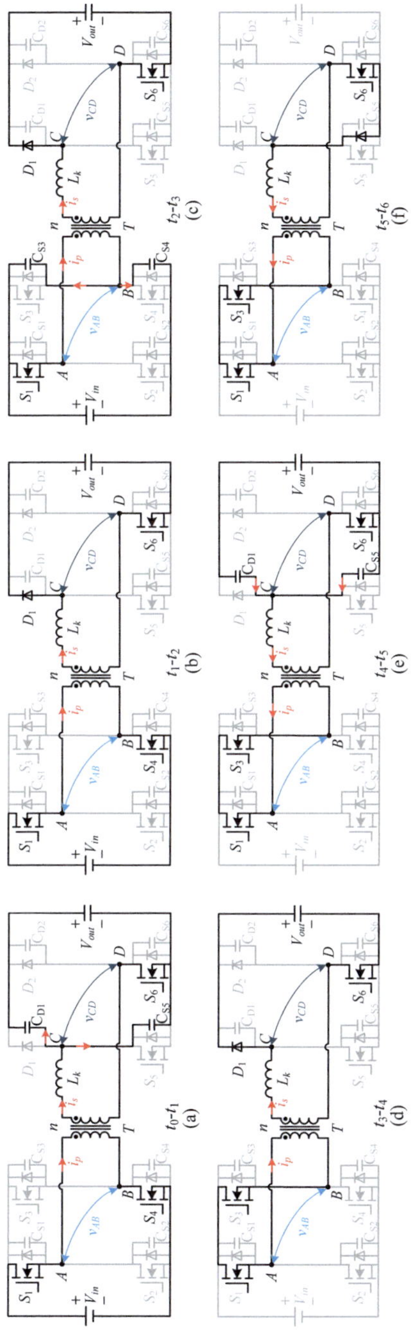

Fig. 3.7 Current flow path of each mode

Mode 1 [t_0–t_1]: Prior to t_0, v_{AB} and v_{CD} equal to zero. i_p stays constant. At t_0, S_1 turns on with ZCS and mode 1 begins. During this mode, i_p and the parallel connection of C_{D1} and C_{S5} resonate from the initial condition. The current i_p charges/discharges C_{S5}/C_{D1} respectively until u_{D1} decreases to zero and D_1 starts to conduct.

Mode 2 [t_1–t_2]: At t_1, D_1 starts to conduct and mode 2 begins. During this mode, v_{AB} equals to V_{in} and v_{CD} equals to V_{o}, causing i_p to increase linearly.

Mode 3 [t_2–t_3]: At t_2, S_4 turns off and mode 3 begins. During this mode, i_p and the parallel connection of C_{S3} and C_{S4} resonate from the initial condition. The current i_p charges/discharges C_{S4}/C_{D3} respectively until u_{DS3} decreases to zero and D_{S1} starts to conduct. This ensures the ZVS condition of S_3.

Mode 4 [t_3–t_4]: At t_3, S_3 turns on with ZVS and mode 4 begins. v_{AB} equals to zero and v_{CD} equals to V_{o}, causing i_p to decrease linearly.

Mode 5 [t_4–t_5]: At t_4, i_p reaches zero and mode 5 begins. i_p and the parallel connection of C_{D1} and C_{S5} resonate from the initial condition. The current i_p charges/discharges C_{D1}/C_{S5} respectively until u_{DS5} decreases to zero and D_{S5} begins to conduct. This ensures the ZVS condition of S_5.

Mode 6 [t_5–t_6]: At t_5, D_{S5} starts to conduct and mode 6 begins. During this mode, v_{AB} and v_{CD} equal to zero, causing i_p to stay constant.

3.2.2 *Expression of the Power Transferred*

For the PWM modulation, t_p is set as 0 while t_d and t_s are regulated to control the transferred power. When t_s is constant, which means the switching frequency is constant, the output power increases as t_d increases. When t_d is constant, the increase of t_s causes the reduction of the output power. Referring to Fig. 3.6, the primary side current i_p can be expressed by (3.1) without taking into account the commutation period,

$$i_p = \begin{cases} \dfrac{V_{\text{in}} - V_{\text{o}}}{L_k} \cdot t, & (t_1 < t < t_2) \\ \dfrac{V_{\text{in}} - V_{\text{o}}}{Lk} \cdot t_s - \dfrac{V_{\text{o}}}{L_k} \cdot t, & (t_3 < t < t_4) \\ nV_{\text{o}}\sqrt{\dfrac{C_{\text{oss\_}S5} + C_{\text{oss\_}D1}}{L_k}}, & (t_5 < t < t_6) \end{cases} \tag{3.1}$$

The output power expression is given by

$$P_{\text{o}} = \int_0^{T_s} v_{AB} \cdot i_p \cdot \mathrm{d}t = \frac{V_{\text{in}} \cdot (n \cdot V_{\text{in}} - V_{\text{o}}) \cdot t_d^2}{L_k \cdot t_s} = k_{po} \cdot \frac{t_d^2}{t_s} \tag{3.2}$$

where

$$k_{po} = \frac{V_{\text{in}} \cdot (n \cdot V_{\text{in}} - V_{\text{o}})}{L_k} \tag{3.3}$$

3.2.3 Switching Frequency Selection

Compared with SPS control strategy, the PWM modulation allows the transformer current to operate in DCM mode, which reduces the conduction loss. However, ZVS for the primary side leading leg MOSFETs is lost. This brings significant switching loss. As the switching loss under the same voltage and current stress is proportional to the switching frequency, reducing the switching frequency is an effective way to reduce the switching loss.

However, when the switching frequency drops, t_d should be increased to enhance the transmitted energy in each switching period. Otherwise, the increased volt-second area may cause magnetic saturation in the transformer core. Therefore, the maximum value of t_d must be limited, thus limiting the maximum value of transferred power. With the proposed multi-mode control strategy, when t_d reaches its upper limit, the operating mode switches to SPS mode in order to transmit higher power. So the transformer must be designed to operate normally under all load conditions. Besides, the decreased switching frequency will also affect the conduction loss and the magnetic loss. So the loss breakdown analysis must be done to select the switching frequency with the highest efficiency.

The loss in the converter includes the switching loss, the conduction loss, the magnetic loss and the driving loss. The methods to calculate these losses are given as below.

(1) *Switching loss*

Referring to Fig. 3.6, the primary side leading leg MOSFETs S_1/S_2 can achieve ZCS on rather than ZVS on. This means high turn-on loss and almost zero turn-off loss. Figure 3.8 shows the equivalent circuit when S_2 turns on. The switching energy of the MOSFETs can be found in the datasheet of the MOSFETs. E_{on} refers to the turn-on energy which varies with the drain-source voltage, the drain current, gate resistance and temperature. Assuming that the gate resistance and temperature are constant, the overall switching loss of the primary side leading leg MOSFETs can be calculated by

$$P_{\text{on}\_p} = 2 \cdot E_{\text{on}}(V_{DS}, I_D) \cdot f_s \tag{3.4}$$

The primary side lagging leg MOSFETs S_3/S_4 can achieve ZVS on. This means only turn-off loss needs to be considered. The turn-off energy E_{off} can be found in the datasheet of the MOSFETs. It varies with the drain-source voltage, the drain current, gate resistance and temperature. Assuming that the gate resistance and temperature

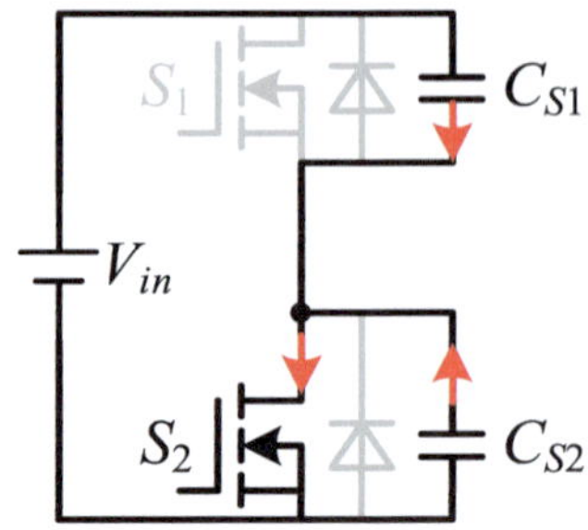

Fig. 3.8 Equivalent circuit when S_2 turns on

are constant, the turn-off loss of the primary side lagging leg MOSFETs can be calculated by

$$P_{\text{off\_}s} = 2 \cdot E_{\text{off}}(V_{DS}, I_D) \cdot f_s \tag{3.5}$$

The secondary side MOSFETs S_5/S_6 can achieve ZVS on and ZCS off, which means no switching loss on S_5/S_6. Similarly, the secondary side diodes can achieve ZCS off, which means no switching loss and no reverse recovery current.

(2) *Conduction loss*

According to Eq. (3.1), the conduction loss in the primary side switches is expressed as

$$P_{\text{cond\_}p} = \int_0^{t_s} i_p^2(t) 2R_{ds\_\text{on}} \mathrm{d}t \tag{3.6}$$

The conduction loss in the secondary side switches is

$$P_{\text{cond\_}s} = \int_0^{t_s} \left[\left(\frac{i_p(t)}{n} \right)^2 R_{ds\_\text{on}} + \frac{i_p(t)}{n} V_f \right] \mathrm{d}t \tag{3.7}$$

where R_{ds_on} refers to the on-state resistance of the MOSFETs and V_f refers to the forward voltage of the secondary side diodes.

(3) *Magnetic loss*

The calculation of the core loss in the transformer uses the improved generalized Steinmetz equation (iGSE) [7–9]. Considering the waveform of v_{AB} shown in Fig. 3.6, the core loss in the transformer can be calculated by

$$P_{\text{core\_}T} = \frac{1}{t_s} \int_0^{t_s} k_i \left| \frac{\mathrm{d}B}{\mathrm{d}t} \right|^{\alpha} (\Delta B)^{\beta-\alpha} \mathrm{d}t V_e \tag{3.8}$$

where ΔB is peak-to-peak flux density, V_e is the volume of the core and

$$k_i = \frac{k}{(2\pi)^{\alpha-1} \int_0^{2\pi} |\cos\theta|^{\alpha} 2^{\beta-\alpha} \mathrm{d}\theta} \tag{3.9}$$

Based on the datasheet of the inductor core, the core loss in the leakage inductor can be calculated by

$$P_{\text{core\_k}} = aB^b f^c \tag{3.10}$$

where B refers to the maximum magnetic flux density in *Tesla* and f refers to the switching frequency in *kHz*. a, b and c is coefficients related with the material.

The copper loss can be calculated by

$$P_{co} = \int_0^{t_s} i_p^2(t) R_m \mathrm{d}t \tag{3.11}$$

where R_m refers to the total resistance of the winding.

(4) *Driving loss*

The driving loss in the switches can be calculated by

$$P_{dr} = 6Q_G(V_{GS\_\text{on}} - V_{GS\_\text{off}}) f_s \tag{3.12}$$

where V_{GS_on} and V_{GS_off} refer to the Gate-to-Source voltage under on-state and off-state respectively. Q_G refers to the gate charge of a MOSFET. The value of Q_G can be found in the datasheet of MOSFET. The number 6 in the equation means that there are 6 active switches in the converter.

(5) *Loss breakdown*

The loss in the converter includes the switching loss, the conduction loss, the magnetic loss and the driving loss. A 1.6 kW prototype was fabricated to evaluate the loss reduction by decreasing the switching frequency with PWM modulation under light load conditions. The system specifications are listed in Table 3.2. Based on these specifications and aforementioned calculating method, the loss breakdown of the Semi-DAB converter is shown in Fig. 3.9.

As seen, the switching loss, including the turn-on loss and the turn-off loss, dominates the total loss when the switching frequency is high, but decreases to a low level when the switching frequency decreases to 20 kHz. The inductor core loss and the driving loss increase as the switching frequency increases, while the conduction loss and the transformer loss do not vary much when the switching frequency varies. Besides, the copper loss is quite few so that it can be neglected.

One 1.6 kW prototype was fabricated to evaluate the conversion efficiency varying with the switching frequency. Figure 3.10 shows the measured conversion efficiency for the prototype with PWM modulation and different switching frequency. The

Table 3.2 System specifications

Parameter	Value
Input voltage	400 V
voltage gain	0–3
Switching frequency	100 kHz
Transformer turns ratio	1:1
Leakage inductance	90 μH
Primary side MOSFETs	C3M0065090J
Secondary side MOSFETs	C3M0065090J
Secondary side diodes	C3D10060G
Transformer material	PC40
Transformer cross section	500 mm^2
Transformer volume	47,000 mm^3
Inductor material	Kool Mμ 26 μ
Inductor cross section	199 mm^2
Inductor volume	21,300 mm^3
Winding resistance	0.1 Ω

figures clearly show that as the switching frequency decreases, the conversion efficiency increases significantly. Therefore, for PWM modulation, the lowest switching frequency can be chosen to achieve high conversion efficiency under light load conditions. However, although lower switching frequency causes the efficiency improvement, the minimum switching frequency is better to be chosen no less than 20 kHz in order to avoid audible noises.

For the aforementioned 1.6 kW prototype, the transformer is designed to operate normally with SPS strategy and 100 kHz switching frequency. As the duty cycle is 50%, the duty cycle duration time length t_d with SPS strategy is 5 μs. Therefore, the duty cycle duration time length t_d under full load range must be no higher than 5 μs to avoid saturation issue in the transformer core. Figure 3.11 shows the value of t_d with PWM modulation and a different switching frequency. As seen, the value of t_d is lower than 5 μs. However, the maximum transferred power is limited with decreased switching frequency. This issue will be discussed in Sect. 3.3.

3.3 Closed Loop Control Strategy Implementation

Section 3.2 have proved that the PWM modulation with decreased switching frequency significantly reduces the loss under light load condition. However, due to the limitation of the volt-second area over the transformer, there should be an upper limit for the duty cycle duration time length t_d, which limits the maximum output power with this modulation strategy. This leaves a gap between the output current

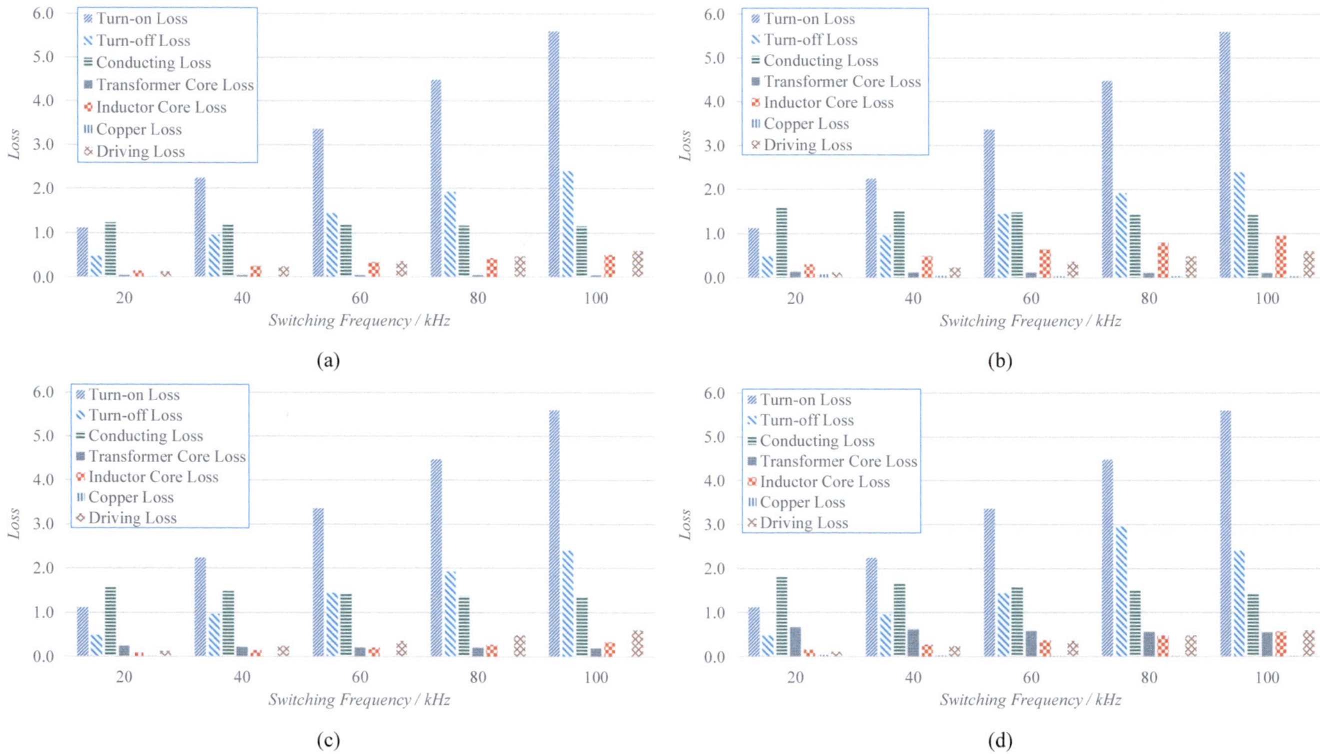

Fig. 3.9 Loss breakdown with PWM modulation: **a** 250 V/0.1 A, **b** 250 V/0.25 A, **c** 350 V/0.1 A, **d** 350 V/0.25 A

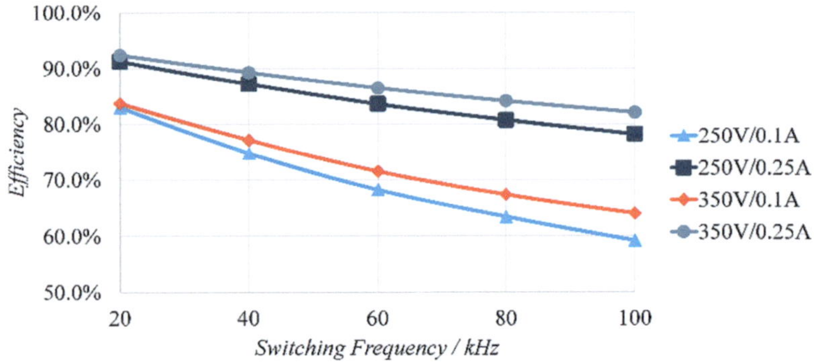

Fig. 3.10 Measured efficiency with different switching frequency

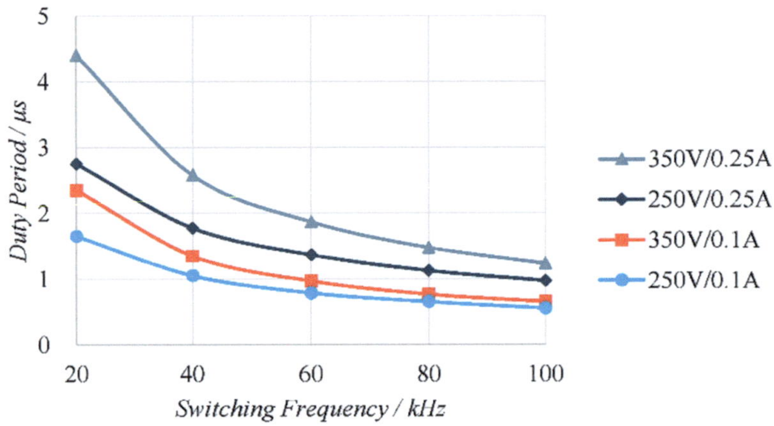

Fig. 3.11 Value of t_d with different switching frequency

range of SPS strategy and PWM modulation with decreased switching frequency. To fill this gap to cover full load range, PWM modulation with variable switching frequency was utilized.

Figure 3.12 shows the output current range of each operating mode. The orange region corresponds to SPS mode, utilizing SPS strategy. The blue region corresponds to PWM plus CF (Constant Frequency) mode, utilizing PWM modulation with decreased switching frequency. The gray region corresponds to PWM plus VF (Variable Frequency) mode, utilizing PWM modulation with variable switching frequency.

When the load current is in the blue region, the converter works in the PWM plus CF mode by employing the PWM modulation with minimum switching frequency. When the load current increases to the case where the duty cycle duration time length t_d has already reached its upper limit, the converter switches to PWM plus VF mode in which t_d is constant as its upper limit and the switching frequency varies. In

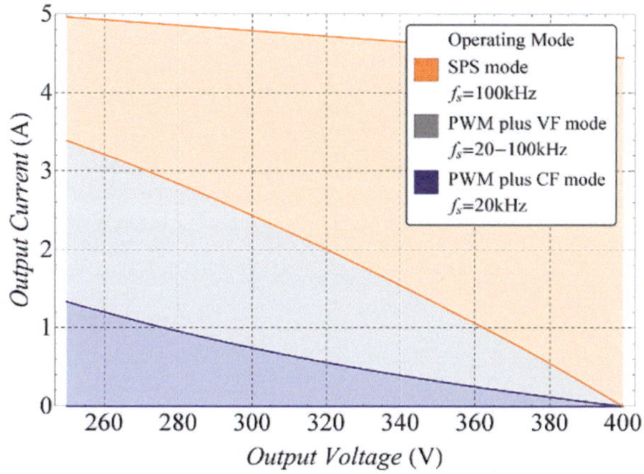

Fig. 3.12 Output current range of each operating mode

this mode, the switching period t_s is regulated to achieve the closed loop control. When the load current increases further and t_s begins to decrease until it is equal to its minimum value (i.e. the switching frequency reaches its maximum value), the converter switches to SPS mode.

To achieve seamless transition among all the three operating modes, the closed loop control strategy is designed as the following. The output of the PI controller is defined as q. If q is lower than q_1, the converter operates in PWM plus CF mode. The switching period t_s and the phase shift duration time length t_p is constant while the duty cycle duration time length t_d varies with q. If q is higher than q_1 and lower than q_2, the converter operates in PWM plus VF mode. The phase shift duration time length t_p and the duty cycle duration time length t_d is constant while the switching period t_s varies with q. If q is higher than q_2, the converter operates in SPS mode. The duty cycle duration time length t_d and the switching period t_p is constant while the phase shift duration time length t_p varies with q. Figure 3.13 shows the flow chart of this closed loop strategy. Figure 3.14 shows the trajectory of three control variables t_d, t_s and t_p versus q.

3.4 Experimental Results

3.4.1 Prototype

To verify the effectiveness of the proposed control strategy, a prototype with the proposed strategy has been fabricated. Figure 3.15 is the photograph of the prototype. The system specifications are listed in Table II. All the control algorithm is implemented by the digital signal processer TMS320F28335.

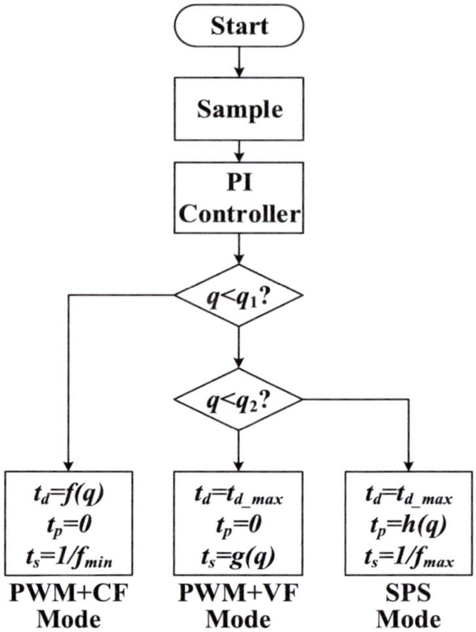

Fig. 3.13 Flow chart of the closed loop strategy

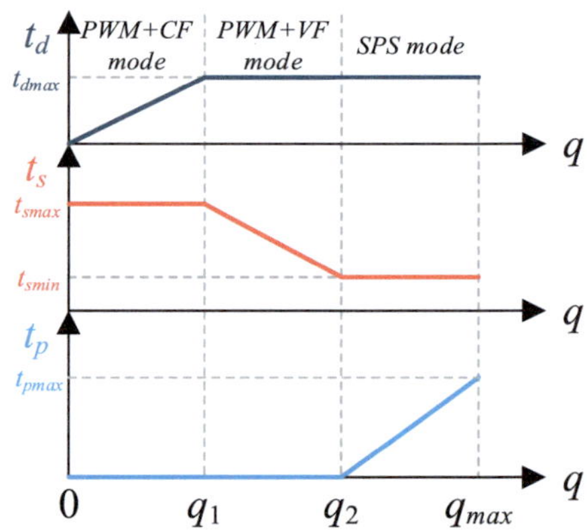

Fig. 3.14 Trajectory of three control variables

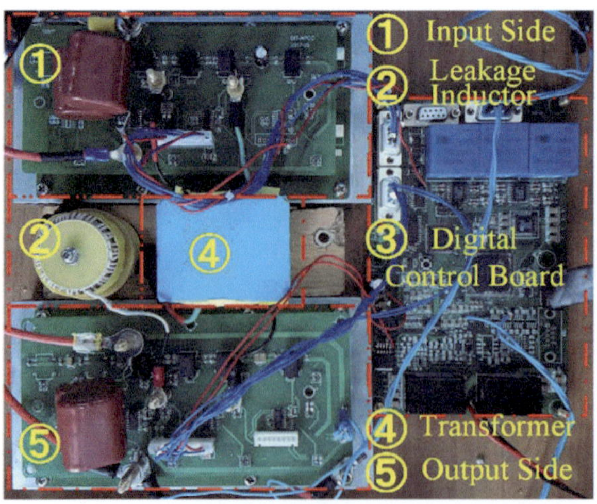

Fig. 3.15 Photograph of the prototype

3.4.2 Experimental Waveforms

Figure 3.16 shows the steady-state operating waveforms of the semi-DAB converter using the proposed multi-mode control strategy under light load condition. The converter works in the PWM plus CF mode. The switching period t_s is always chosen as 50 μs. This means that the switching frequency is 20 kHz. The duty cycle duration time length t_d is regulated to achieve the closed loop control.

Figure 3.17 shows the steady-state operating waveforms of the semi-DAB converter using the proposed multi-mode control strategy under medium load condition. The converter works in the PWM plus VF mode. The duty cycle duration time length t_d is constant with 5 μs. The switching period t_s is regulated.

Figure 3.18 shows the steady-state operating waveforms of the converter using the proposed multi-mode control strategy under heavy load condition. The converter works in SPS mode. The switching period t_s is always kept as 10 μs. This means that the switching frequency is 100 kHz. The duty cycle duration time length t_d, of which the value is 5 μs, is also constant. The phase shift duration time length t_p is used for the regulation.

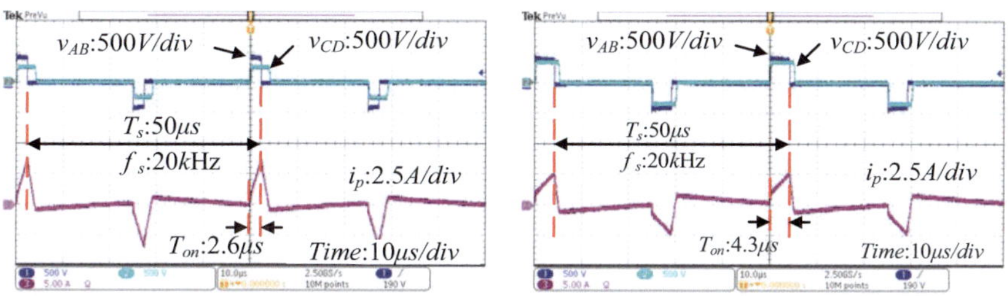

(a) (b)

Fig. 3.16 Steady-state waveforms under light load: **a** $V_o = 250$ V, $I_o = 0.3$ A; **b** $V_o = 350$ V, $I_o = 0.3$ A

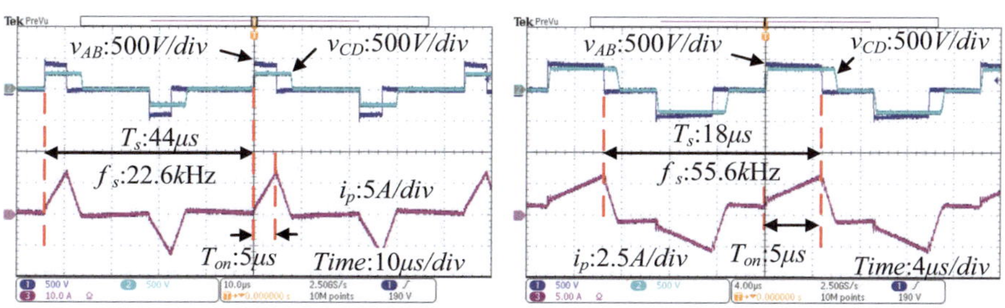

(a) (b)

Fig. 3.17 Steady-state waveforms with medium load: **a** $V_o = 250$ V, $I_o = 1.2$ A; **b** $V_o = 350$ V, $I_o = 1.2$ A

Figures 3.19 and 3.20 show the dynamic response of load transition with the proposed multi-mode control strategy. As illustrated in Fig. 3.19, at light load condition (60 W), the converter works in PWM plus CF mode and the switching frequency is fixed as 20 kHz. As the load power increases to 320 W, the converter switches to PWM plus VF mode. The switching frequency is regulated to about 23.8 kHz while the duty cycle duration time length is fixed as 5 μs. When the load power increases from 320 to 950 W, the converter switches from PWM plus VF mode to SPS mode.

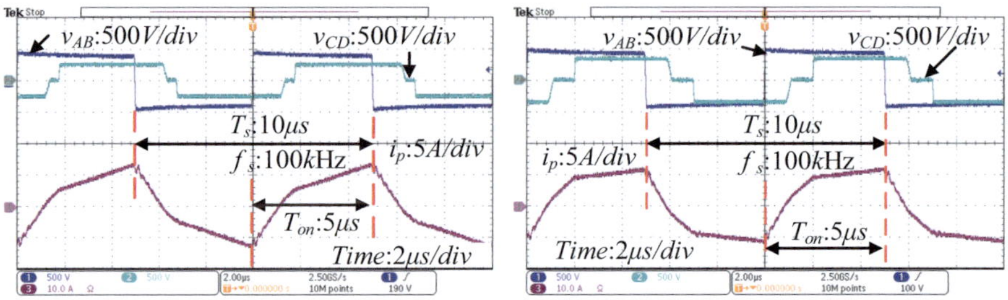

(a) (b)

Fig. 3.18 Steady-state waveforms with heavy load: **a** $V_o = 250$ V, $I_o = 3.6$ A; **b** $V_o = 350$ V, $I_o = 3.6$ A

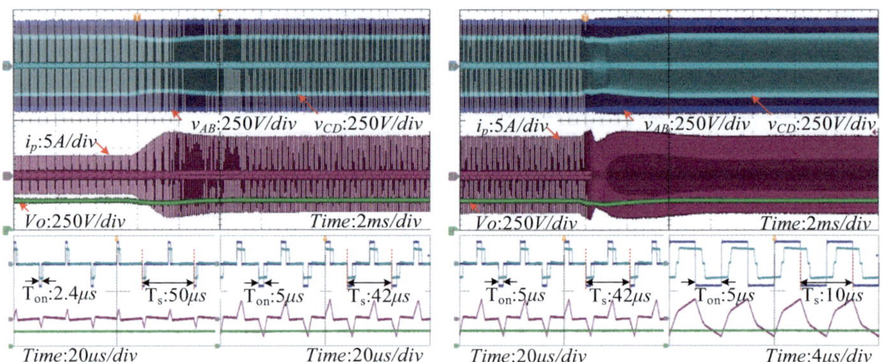

(a) (b)

Fig. 3.19 Dynamic response of load transition: **a** 60–320 W, **b** 320–950 W

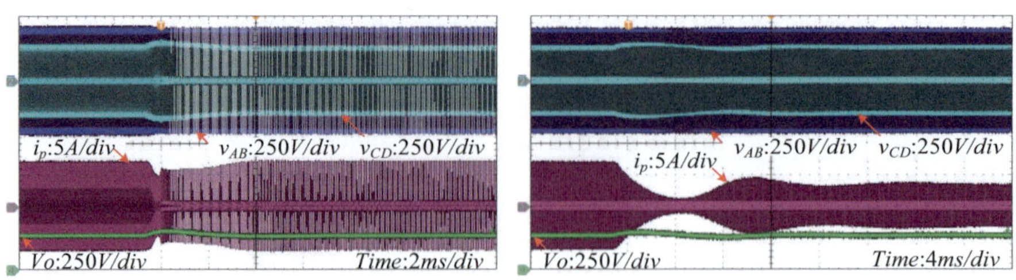

(a) (b)

Fig. 3.20 Dynamic response of load transition: **a** 950–320 W, **b** 320–60 W

The switching frequency is fixed as 100 kHz while the phase shifted ratio is regulated to control the transferred power. As shown in Figs. 3.19 and 3.20, seamless transition among all the three operating modes can be obtained by employing the proposed multi-mode control.

3.4.3 Efficiency Curve

The efficiency curve comparison by using different control strategies for this converter has been given in Fig. 3.21. As illustrated, under heavy load conditions, efficiency of the converter employing the proposed control, the SPS control and the SPS plus PWM with constant frequency control does not vary too much. But under light load conditions, conversion efficiency with the proposed control strategy is the highest.

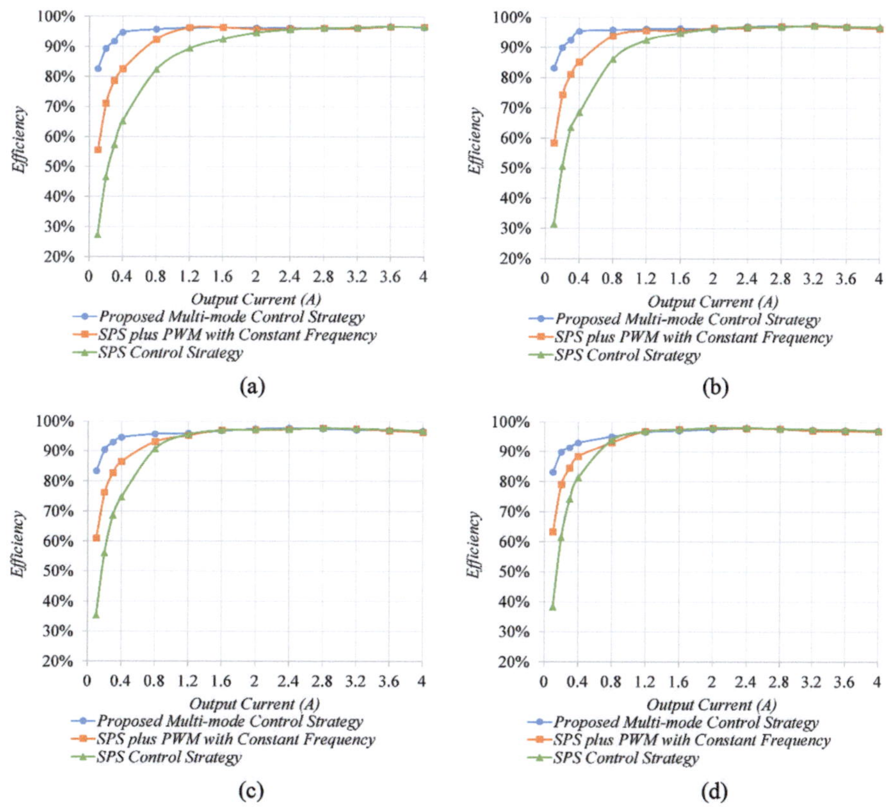

Fig. 3.21 Efficiency curves: **a** $V_o = 250$ V; **b** $V_o = 300$ V; **c** $V_o = 350$ V; **d** $V_o = 400$ V

3.5 Conclusion

To improve the full load operating performance especially conversion efficiency under light load conditions, a multi-mode control strategy was proposed for the semi-DAB DC-DC converter. To improve the conversion efficiency further, all the power switches are SiC MOSFETs. With the proposed control, three working modes can work together and seamless transition can be obtained among them according to the closed loop control requirement. Light load efficiency can be improved dramatically with the proposed control.

References

1. Kulasekaran S, Ayyanar R (2014) Analysis, design, and experimental results of the semi dual-active-bridge converter. IEEE Trans Power Electron 29(10):5136–5147
2. Sha D, Zhang J, Xu Y (2017) Improved boundary operation for voltage-fed semi-DAB with ZVS achievement and nonactive power reduction. IEEE Trans Ind Electron 64(8):6179–6189
3. Zhao B, Song Q, Liu W (2012) Power characterization of isolated bidirectional dual-active-bridge DC-DC converter with dual-phase-shift control. IEEE Trans Power Electron 27(9):4172–4176
4. Krismer F, Kolar JW (2009) Accurate small-signal model for the digital control of an automotive bidirectional dual active bridge. IEEE Trans Power Electron 24(12):2756–2768
5. Mishima T, Nakaoka M (2011) Practical evaluations of a ZVS-PWM DC-DC converter with secondary-side phase-shifting active rectifier. IEEE Trans Power Electron 26(12):3896–3907
6. Xu G, Sha D, Zhang J, Liao X (2017) Unified boundary trapezoidal modulation control utilizing fixed duty cycle compensation and magnetizing current design for dual active bridge DC-DC converter. IEEE Trans Power Electron 32(3):2243–2252
7. Reinert J, Brockmeyer A, De Doncker RWAA (2001) Calculation of losses in ferro- and ferrimagnetic materials based on the modified Steinmetz equation. IEEE Trans Ind Appl 37(4):1055–1061
8. Li J, Abdallah T, Sullivan CR (2001) Improved calculation of core loss with nonsinusoidal waveforms. In: Conference record of the 2001 IEEE industry applications conference. 36th IAS annual meeting (Cat. No.01CH37248), pp 2203–2210
9. Venkatachalam K, Sullivan CR, Abdallah T, Tacca H (2002) Accurate prediction of ferrite core loss with nonsinusoidal waveforms using only Steinmetz parameters. In: 2002 IEEE workshop on computers in power electronics, pp 36–41

Chapter 4
Voltage-Fed Three Phase Semi-DAB DC-DC Converter

Abstract In Chaps. 2 and 3, a single phase voltage-fed semi-dual active bridge (1p-VF-SDAB) DC-DC converter and control strategy have been discussed, but the single phase topology is not preferred to be used for high power application due to the high current stress in switches. So to reduce the current stress for high power applications, a novel high-frequency isolated three phase voltage-fed semi dual active bridge (3p-VF-SDAB) DC-DC converter is proposed for unidirectional power flow with wide input variations and high voltage interface in this chapter. A varying operating mode control independently implemented by phase shift and duty cycle control is subsequently raised to handle the stability and improve conversion efficiency of the converter under the light load conditions. A variety of different operating modes are illustrated and compared in detail. The power expressions and the root mean square (RMS) expressions of the leakage inductor current (LIC) of different modes are derived and the optimal modes are consequently selected. The gating signal falling edge compensation of secondary side switches is implemented to further improve the conversion efficiency. The zero voltage switching (ZVS) conditions based on current polarity is also derived. A 1.6 kW prototype was built to verify the effectiveness of the proposed topology and control strategy. The experimental results show that the proposed converter using the proposed control strategy can achieve the stability throughout the whole input and load variations. Moreover, seamless transition also can be achieved with a good dynamic performance and a high conversion efficiency that can be guaranteed.

Keyword Dual active bridge dc-dc converter · High conversion efficiency · Wide input variations

4.1 Mode Analysis with Proposed Control Strategy

Dual active bridge (DAB) dc-dc converter is very attractive in the field of academia and industry due to its several advantages such as natural soft-switching, galvanic isolation. Compared to 1p-DAB, multi-phase converters have smaller voltage and

D. Sha et al., *High Frequency Isolated Semi Dual Active Bridge DC-DC and AC-DC Converter*, CPSS Power Electronics Series,
https://doi.org/10.1007/978-981-95-1641-4_4

current ripple, smaller filter size and higher power capacity [1–3]. The distinctive characteristics of DAB is very attractive to be used in unidirectional power flow applications. A new concept of 1p-VF-SDAB is proposed in [4–7]. The output side top switches are replaced by diodes. This can simplify the system design and reduce the overall cost. But the current stress is very high when the conversion power is high. Therefore, 1p-SDAB is not suitable for high power application due to the high current stress in switches. Therefore, 1p-SDAB can be extended to 3p-SDAB and the current stress can be minimized. However, 3p-SDAB has never been discussed before in the literature.

In this chapter, 3p-VF-SDAB is discussed for the first time. Most of its possible operating modes are discussed in detail. A varying mode control strategy is proposed to adapt to the wide input voltage range and full load range capability. The uncontrollable issue of SDAB under high input voltage or light load conditions can be dealt with completely. Compared with the SPS control employing negative phase shift angle, the circulating loss can be reduced significantly and thus the conversion efficiency is greatly improved.

The proposed topology is shown in Fig. 4.1. For the primary side, S_1–S_4 are the power switches with body diodes. The AC inductor L_{kA}, L_{kB}, L_{kC} represents the sum of the referred transformer leakage inductance and the external AC inductance for each phase, respectively. In the secondary side, Q_1–Q_3 are the power switches. D_1–D_3 are the power diodes. The transformer is Y–Y connected. The current flow direction is highlighted in Fig. 4.1. The ratio of the on-time of the primary side top switches to the entire period is defined as the duty cycle D. The phase difference between the driving signal of the secondary side switch and the primary bottom switch in each phase is the phase shift angle. The ratio between the phase shift angle and 2π is defined as D_ϕ.

Many articles have already proposed plenty of control methods for the conventional DAB converters composed of full-controllable devices. However, the existing control strategy for SDAB is simple. Moreover, the light load operating conditions of SDAB have not been discussed. Since there exist diodes in the secondary side, the corresponding zero crossing point of the current flowing through diodes rather than be actively controlled. Therefore, there are many complicated modes for 3p-VF-SDAB. The phase shift angle decreases as the input voltage increases or the load decreases.

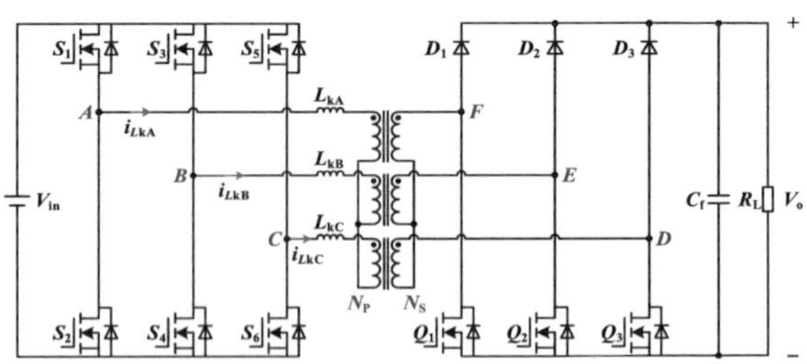

Fig. 4.1 Topology of the 3p-VF-SDAB dc-dc converter

Some typical operating modes in different conditions are shown in Fig. 4.2 and the shaded part indicates the circulating current. G is defined as voltage gain and it is written as $G = nV_o/V_{in}$. Each phase is 120° phase shifted with each other. The driving signals of each leg are complementary. v_{ANp} is the primary side phase to neutral voltage of phase A. v_{FNs} is the secondary side phase to neutral voltage of phase A. The waveform of v_{ANp} only depends on the driving signal, but v_{FNs} is not only determined by driving signals but also the current zero-crossing moment in diodes. i_{LkA} is the primary side leakage inductor current (LIC) of phase A.

The phase shift angle will be reduced to zero as the input voltage increases or output power drops down to a certain value. Thus the input voltage and load conditions will affect the operating performance of the 3p-VF-SDAB dc-dc converter, which may cause an instability issue. This has to be avoided for practical applications. Adjusting the phase shift angle to be negative may help to achieve the operating stability as G is away from unity and the load is light, but at the cost of bringing too much high circuiting current. The schematic diagram when $D_\phi < 0$ is shown in Fig. 4.3.

The shaded part in Fig. 4.3 indicates that the circulating current makes up a very large proportion within one complete cycle. This causes high circulating loss and lowers the conversion efficiency. Therefore, it is not applicable for high input voltage or light load conditions. To solve this issue completely, a varying operating mode control is proposed in this paper. The SPS control with fixed D is used when the input voltage is relatively low or the output power is relatively high. The duty cycle regulation with $D_\phi = 0$ is effective in other cases. All the possible operating modes of duty cycle regulation with $D_\phi = 0$ when $G < 1$ are shown in Fig. 4.4.

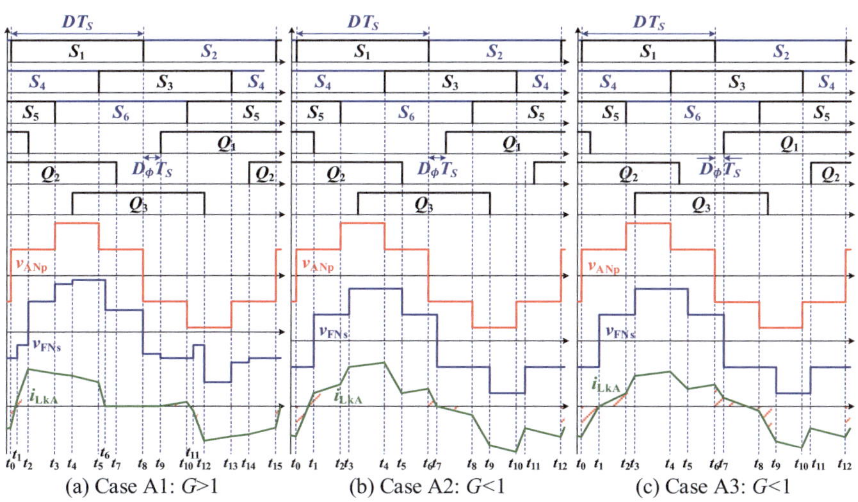

Fig. 4.2 Typical operating waveforms using SPS control when $D_\phi > 0$

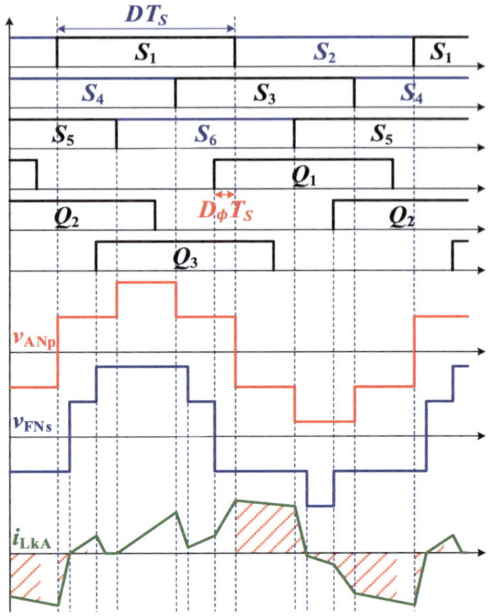

Fig. 4.3 Primary side phase to neutral voltage and phase to neutral current waveforms when $D_\phi < 0$

The shaded part in Fig. 4.4 also represents the circulating current. It is obvious that the circulating loss of Case B1 and Case B2 are significantly higher than the other three cases. It means that these two cases are not good choices. The specific discussions and analysis will be elaborated later.

The modes represented in Figs. 4.2 and 4.4 are defined as group A and group B. Group A is characterized by $D_\phi > 0$ and D is constant. Group B is characterized by $D_\phi = 0$ and D becomes a controllable variable. The transitions between each the adjacent mode are continuous and smooth.

Practically, there are other operating modes in group A. However, those modes are very complicated and difficult to analyze. Thus, this paper only discusses the cases of that $G < 1$. Before the analysis, the following two assumptions are made: All the MOSFETs and diodes are ideal. The effect of dead time is ignored. Case B5 is taken as an example in order to briefly explain its working processes. Other modes are similar. The steady state operating waveforms of Case B5 within one complete period are given in Fig. 4.4e. The detailed working processes will be described as follows.

Stage 1 (before t_0): S_2, S_3, S_5 and Q_1 conduct, and v_{ANp}, v_{FNs}, i_{LrA} are negative. During this stage, the power flows from the primary side to the secondary side.

Stage 2 (t_0–t_1): At t_0, S_1 turns on. v_{ANp} jumps to zero and v_{FNs} maintains at $-2/3 \cdot V_o$. The voltage difference between v_{ANp} and v_{FNs} reflected to the primary side make i_{LkA} increase positively until it reaches zero at time t_1.

Stage 3 (t_1–t_2): At t_1, i_{LkA} turns to be zero. During this stage, i_{LkA} is always equal to zero.

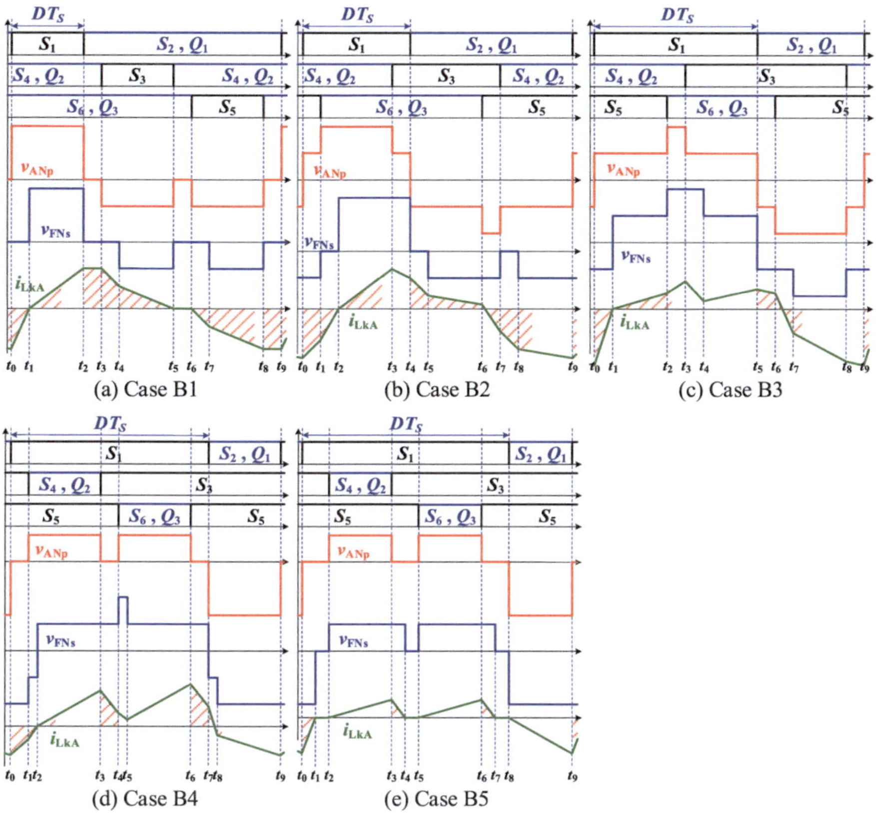

Fig. 4.4 Duty cycle adjustment operation modes when $G < 1$

Stage 4 (t_2–t_3): At t_2, S_4 and Q_2 turn on simultaneously. v_{ANp} and v_{FNs} turn to be $V_{in}/3$ and $V_o/3$, respectively. i_{LkA} increases positively.

Stage 5 (t_3–t_4): At t_3, S_4 and Q_2 turn off simultaneously. v_{ANp} is equal to zero. D still conducts, so v_{FNs} is unchanged. i_{LkA} decreases to be zero at t_4.

Stage 6 (t_4–t_5): At t_4, D_1 is reverse biased. v_{ANp}, v_{FNs} and i_{LkA} are equal to be zero.

Stage 7 (t_5–t_6): At t_5, S_6 and Q_3 turn on simultaneously. v_{ANp} and v_{FNs} turn to be $V_{in}/3$ and $V_o/3$, respectively. i_{LkA} increases.

Stage 8 (t_6–t_7): At t_6, S_6 and Q_3 turn off simultaneously. v_{ANp} jumps to be zero. D_1 still conducts thus v_{FNs} keeps unchanged. i_{LkA} decreases to be zero at t_7.

Stage 9 (t_7–t_8): At t_7, D_1 is reverse biased. v_{ANp}, v_{FNs} and i_{LkA} stay at zero.

Stage 10 (t_8–t_9): At t_8, S_2 and Q_1 turn on simultaneously. v_{ANp} and v_{FNs} turn to be $-2/3 \cdot V_{in}$ and $-2/3 \cdot V_o$, respectively. i_{LkA} decreases negatively. At t_9, S_2 and Q_1 turn off. Since the LIC is negative at this time, the current flows through the body diode of Q_1, so v_{FNs} does not change. The next switching period is repetitive as this cycle, therefore the analysis is the same. The operating principle is the same for the other two phases. It should be pointed out that the sum of the 3p phase to neutral voltages is zero and so is the 3p phase to neutral currents.

4.2 Performance Analysis and Discussion

4.2.1 Derivation of System Output Power

The power transferred by each phase in a 3p converter topology can be determined from the average value of the product of each phase to neutral voltage and its phase to neutral current over a switching period. The total power transferred by the 3p converter is the sum of the power transferred by each phase. Therefore, the power expressions can be summarized in Table 4.1.

As seen from Table 4.1, the transferred power expression of Case A2 is the same as that of Case A3. It is worth noting the power expression is independent of D. It is easy to prove that for Case A2 and Case A3, the transferred power is a monotonically increasing function of D_ϕ. The transferred power in Case A2 and Case A3 versus D_ϕ and G is plotted in Fig. 4.5. The specifications are listed in Table 4.2.

It can be concluded from Fig. 4.5 that high power can be transferred even if D_ϕ is small. It must be pointed out that the output power is not zero even if $D_\phi = 0$ when G deviates away from 1. This indicates that the converter is out of control under low voltage gain or light load working conditions by employing only SPS control. However, the modes of group B can be used to deal with this issue. Case A3 will transit to Case B3 smoothly when D_ϕ is less than 0. For several cases of group A, their demarcation points (lines) are summarized in Table 4.3.

The output power P_O versus D and G of group B is plotted in Fig. 4.6. The specifications are listed in Table 4.2. According to Fig. 4.6 and the output power

Table 4.1 Transferred power expressions

	Output power P_O expressions
Case A2	$P_O = K \cdot \left[D_E^2(G^{-1} - G) + (2D_E - D_\phi)D_\phi G^{-1}\right]$
Case A3	$P_O = K \cdot \left[D_E^2(G^{-1} - G) + (2D_E - D_\phi)D_\phi G^{-1}\right]$
Case B1	$P_O = K \cdot 2D^2(G^{-1} - 1)(2 - G)^{-2}$
Case B2	$P_O = K \cdot G^{-1}\left[2DD_E - (1 - G)^2(2 - G)^{-2}D^2 - D_E^2 - 1\right]$
Case B3	$P_O = K \cdot D_E^2(G^{-1} - G)$
Case B4	$P_O = K \cdot \left[(1 - D)(D - D_E)G^{-1} - D_E^2 G\right]$
Case B5	$P_O = K \cdot 2(1 - D)^2(G^{-1} - 1)G^{-1}$
Notes	$D_E = 1/3 \quad K = V_o^2 n^2 T_S/(2L_k)$

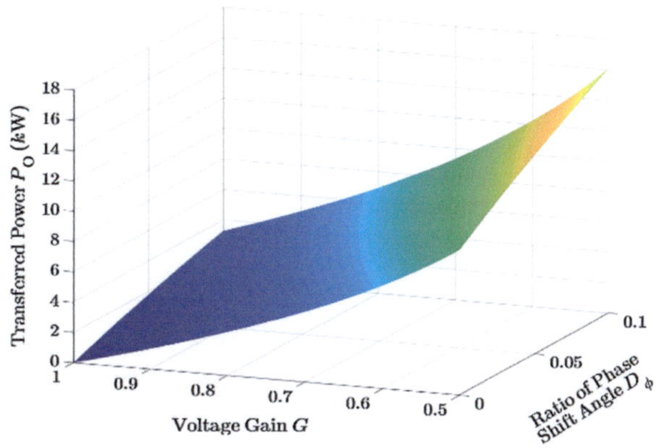

Fig. 4.5 Transferred power versus D_ϕ and G in case A2 and case A3

Table 4.2 System specification

Components	Parameters
Output voltage (V_O)	400 V
Input voltage (V_{in})	200–280 V
Output power (P_{Omax})	1.6 kW
Switching frequency (f_S)	100 kHz
Turn ratio (n)	6: 10
Leakage inductance (L_r)	5 μH
Output capacitance (C_f)	33.6 μF

Table 4.3 Demarcation points or lines within group A

	Case A1 ↔ Case A2	Case A2 ↔ Case A3	Case A3 ↔ Case B3
Demarcation points (lines)	$G = 1$	$D_\phi = D_E(1 - G)$	$D_\phi = 0$

expression of Case B3 in Table 4.2, it can be concluded that the output power is independent of D. This area is marked in red in Fig. 4.6. Furthermore, the maximum transmission power capability of the converter in group B also occurs in this region. Focusing on this region, the transmission power capability decreases on both sides until it reaches zero. This illustrates that the uncontrollable issue when the converter operates in high input voltage or light load conditions caused by SPS control can be solved. The interval length of each section in group B is summarized in Table 4.4.

Fig. 4.6 Output power versus D and G in group B

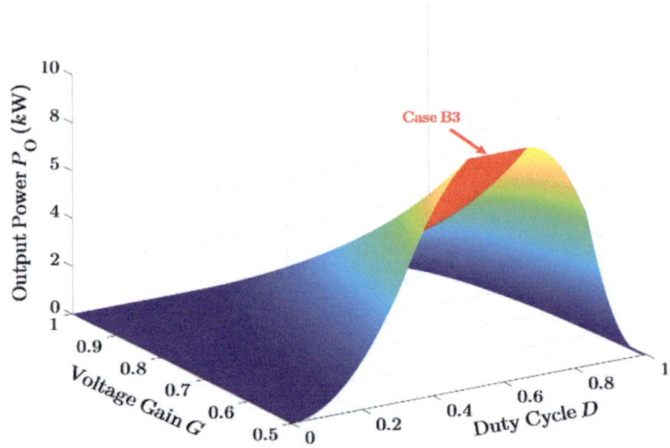

Table 4.4 Demarcation points or lines within group B

	Range of D		Range of D
Case B1	$[0, D_E)$	Case B4	$[2D_E, D_{B2})$
Case B2	$[D_E, D_{B1})$	Case B5	$[D_{B2}, 1]$
Case B3	$[D_{B1}, 2D_E)$		
Notes	$D_{B1} = D_E(2 - G)$ $D_{B2} = 1 - D_E G$		

4.2.2 Leakage Inductor Current Stress Analysis

As aforementioned, the output power is independent of D in Case A2, Case A3 and Case B3. This brings about a question how to choose the optimal duty cycle. Apparently, it cannot be obtained directly from the power expressions. However, the LIC RMS value can be used to reveal the system conduction loss. Hence, it is very useful to analyze the LIC stress. The LIC RMS values of each phase of group A and group B can be easily obtained via MATHCAD, which are summarized in Table 4.5.

The formulas of LIC RMS value of Case A2 and Case A3 are quadratic functions with respect to D. Their symmetry axis can be easily found as shown in (4.1) and (4.2), respectively.

$$Q_{\mathrm{SYM\_Case\ A2}} = 0.5 + G\left[D_E - D_\phi(1 - G)^{-1}\right] \tag{4.1}$$

$$Q_{\mathrm{SYM\_Case\ A3}} = 0.5 - G\left[D_E - D_\phi(1 - G)^{-1}\right] \tag{4.2}$$

$i_{\mathrm{Lk(RMS)}}$ of Case A2 and Case A3 will reach the maximum value when D equals $Q_{\mathrm{SYM\_Case\ A2}}$ and $Q_{\mathrm{SYM\_Case\ A3}}$, respectively. It is well known that if the quadratic function has a local maximum value point, the function value is smaller when the abscissa is further from the symmetry axis. The difference between the distance from $2D_E$ or D_{B1} to $Q_{\mathrm{SYM\_Case\_A}x}$ ($x = 2$ or 3) are examined in this connection, namely

Table 4.5 Leakage inductor current RMS value expressions

	Leakage inductor current RMS value expressions
Case A2	$i_{\mathrm{Lk(RMS)}}=\sqrt{K_2\cdot\left\{\begin{array}{l}DD_E^2\left(3G^{-1}-4-G+2G^2\right)(1-G)^{-2}\\-D_\phi^2\left(D_EG+D_\phi-1\right)(1-G)^{-2}\\+\left[D_E^2(2G+3)-2DD_E\right]D_\phi(1-G)^{-1}\\-\left[D^2D_E+D_E^3(1+G)\right]G^{-1}\end{array}\right\}}$
Case A3	$i_{\mathrm{Lk(RMS)}}=\sqrt{K_2\cdot\left[\begin{array}{l}D_ED_\phi(1-G)^{-1}(2D+2D_EG-1)\\-D_\phi^2(1-G)^{-2}\left(D_\phi+D_EG-1\right)\\-D_EG^{-1}D(D+2D_EG-1)\\-D_E^3G^{-1}D(1-5G)\end{array}\right]}$
Case B1	$i_{\mathrm{Lk(RMS)}}=\sqrt{K_2\cdot 4D^2G^{-1}(2D_E-D)(2-G)^{-2}}$
Case B2	$i_{\mathrm{Lk(RMS)}}=\sqrt{\begin{array}{l}K_2\cdot 3^{-3}G(1-G)^{-1}(2-G)^{-2}\\\left[\begin{array}{l}4G^{-2}(1-G)\left(9D-9D^2-1\right)\\+9D\left(3D^2-3D+1\right)-1\end{array}\right.\end{array}}$
Case B3	$i_{\mathrm{Lk(RMS)}}=\sqrt{K_2\cdot D_E\left[\begin{array}{l}\left(D-D^2-D_E^2\right)G^{-1}\\+D_E(5D_E-2D)\end{array}\right]}$
Case B4	$i_{\mathrm{Lk(RMS)}}=\sqrt{K_2\cdot 3^{-3}(1-G)^{-1}\left[\begin{array}{l}9(3D-1)(D-1)^2G^{-1}\\+3(3D-2)(D-1)\\+G(6D-5)\end{array}\right]}$
Case B5	$i_{\mathrm{Lk(RMS)}}=\sqrt{K_2\cdot 2(1-D)^3G^{-2}}$
Notes	$K_2=n^2V_2^2T_S^2(1-G)^2/\left(9GL_k^2\right)$

$(2D_E - Q_{\mathrm{SYM\_Case\_A}x}) - (Q_{\mathrm{SYM\_Case\_A}x} - D_{\mathrm{B1}})$. Figure 4.7 is then drawn when $P_{\mathrm{O}} = 1$ kW, and the specifications are listed in Table 4.2.

The horizontal intersection axis of the two curves shown in Fig. 4.7 is the mode transition point of Case A2 and Case A3. The left side of the intersection point is Case A3 while the right side is Case A2. Obviously, all the values are greater than zero. It indicates that the LIC RMS value of Case A2 and Case A3 is the minimum value when $D = 2D_E$.

The 3D diagram about $i_{\mathrm{Lk(RMS)}}$ in group B versus D and G is plotted in Fig. 4.8. Its specifications are listed in Table 4.2. As seen, the LIC RMS value decreases as D increases in Case B3 when the transferred power is constant. What's more, the LIC RMS value decreases in both sides of Case B3. It is bound to cause a discussion of the optimal interval selection.

Figure 4.9 shows output power and the LIC RMS value curve of group B when $G = 0.923$. The specifications are listed in Table 4.2. The ordinate on the left side of Fig. 4.9 indicates the transferred power, which corresponds to the red solid curve.

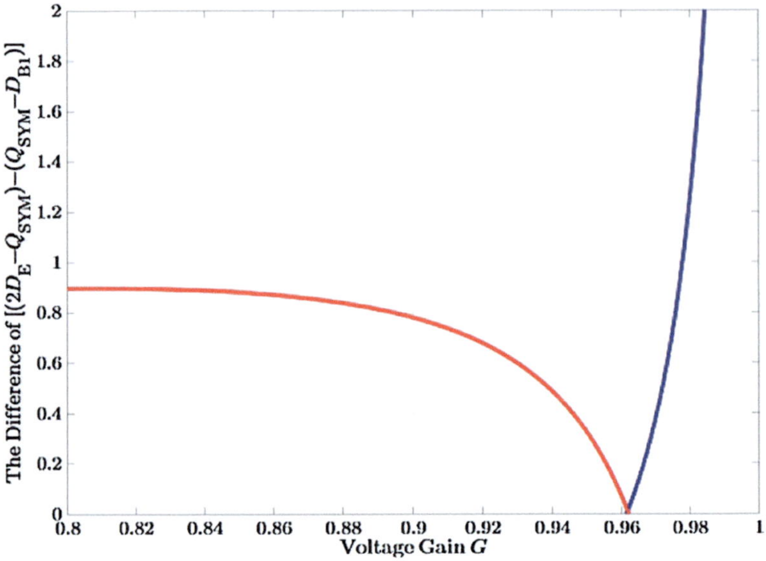

Fig. 4.7 The value of $(2D_E - Q_{\text{SYM\_Case\_A}x}) - (Q_{\text{SYM\_Case\_A}x} - D_{\text{B1}})$ versus G

Fig. 4.8 Leakage inductor current RMS current versus G and D in group B

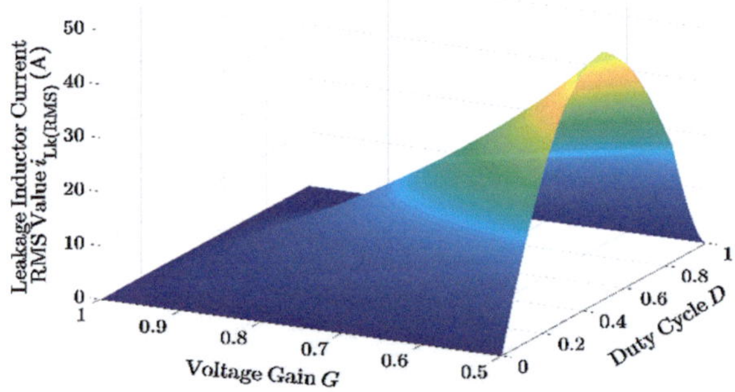

The ordinate on the right side of Fig. 4.9 indicates the LIC RMS value, corresponding to the blue chain line. Apparently, the duty cycle must be different if the converter transfers the same power when G is determined in group B except Case B3. The same level output power is indicated by the green horizontal dashed line in Fig. 4.9, and it intersects with the power curve at two points respectively. LIC RMS values $i_{\text{Lk(RMS)}}$ corresponding to these two points' abscissas are marked by point A and B. Distinctly, the LIC RMS value at point A is significantly higher than that of point B. It indicates that the interval of $D \in [2D_E, 1]$ is better than the interval of $D \in [0, D_{\text{B1}}]$. It can also be proved that the conclusion as mentioned above still works in Case B3 ($D \in [D_{\text{B1}}, 2D_E]$), and it is optimal when $D = 2D_E$.

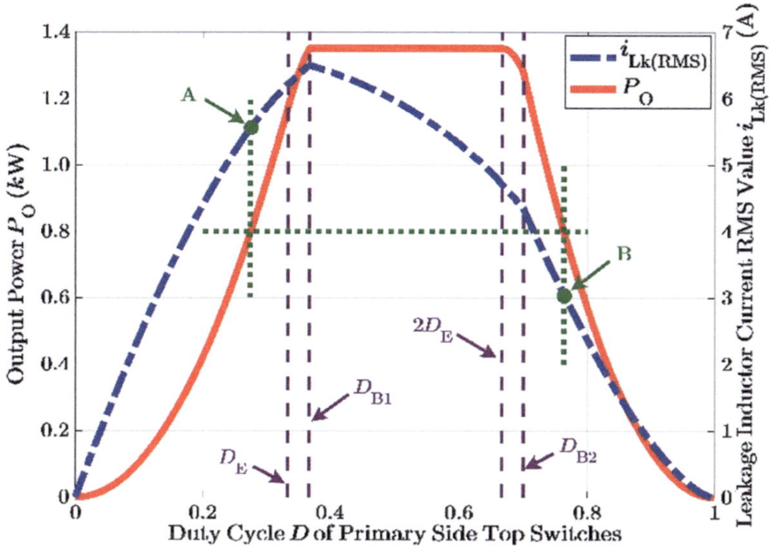

Fig. 4.9 Output power and the leakage inductor current RMS value versus D in group B when $G = 0.923$

Hence, Case B4 and Case B5 are applied in practical implementation.

In summary, D is fixed to be $2D_E$ when $D_\phi > 0$, and D is adjusted from $2D_E$ when D_ϕ reduces to be 0. Therefore, the modes of Case A2 and Case A3 become the state shown in Fig. 4.10.

4.2.3 Soft Switching Conditions

It is well known that ZVS analysis methods can be divided into current based, charging based and energy based. The latter two methods have higher accuracy, but the analysis processes are more complicated. Furthermore, the analysis results are not accurate in view of parasitic parameters of power switches. The current based method is simpler and more effective. It only needs to determine the polarity of the LIC before MOSFETs turn on, so it is appropriate to use the current based method for the qualitative analysis.

The operating state in the ideal conditions of each phase in the 3p converter is the same. Thus, S_1, S_2 and Q_1 can be used as an example for ZVS analysis. The ZVS conditions of primary side switches are summarized in Table 4.6.

By substituting the control variable D_ϕ or D which can be obtained from the power expressions listed in Table 4.3 into the ZVS conditions listed in Table 4.5, it can be inferred that the ZVS of S_1 among the four selected modes can be guaranteed while the S_2 also can achieve the ZVS except that in the Case B4.

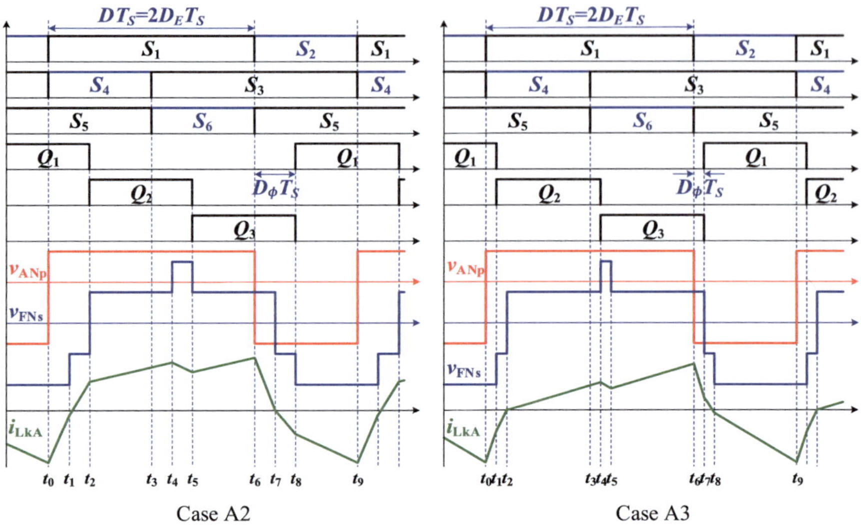

Fig. 4.10 Mode waveforms of group A when $D = 2D_E$

Table 4.6 Soft-switching conditions for primary side

	S_1	S_2
Case A2	$\begin{bmatrix} D_E\left(2 - G^2\right) \\ +G\left(2D_\phi - D_E\right) \end{bmatrix} > 0$	$(1 - G)(2 + G) > 0$
Case A3	$D_E\left(1 - G^2\right) + D_\phi G > 0$	$D_E(1 - G) + D_\phi G > 0$
Case B4	$1 - D + D_E G > 0$	$1 - D - D_E G > 0$
Case B5	$1 - G > 0$	ZCS

For S_2 in Case B4, the ZVS condition can be rewritten as (4.3), and

$$P_O > 2D_E^2 K(1 - G) \tag{4.3}$$

The corresponding ZVS boundary is shown in Fig. 4.11. The specifications are listed in Table 4.2.

It must be pointed out that the ZVS conditions are listed in Table 4.6 is based on the assumption that all the switches are ideal. For practical applications, only the current polarity is not enough to achieve ZVS. Sufficient current is needed to charge and discharge the junction capacitors of MOSFETs completely before they are turned on. For the topology proposed in this paper, the current before S_1 turns on is sufficient. Consequently, ZVS for S_1 is achieved absolutely. However, the current is not enough for S_2. Therefore, although the current polarity of S_2 is satisfied, ZVS is still lost. For Q_1 in the group B, since the LIC polarity is exactly the opposite with that required for S_2 to achieve ZVS. Therefore, the ZVS of Q_1 is lost.

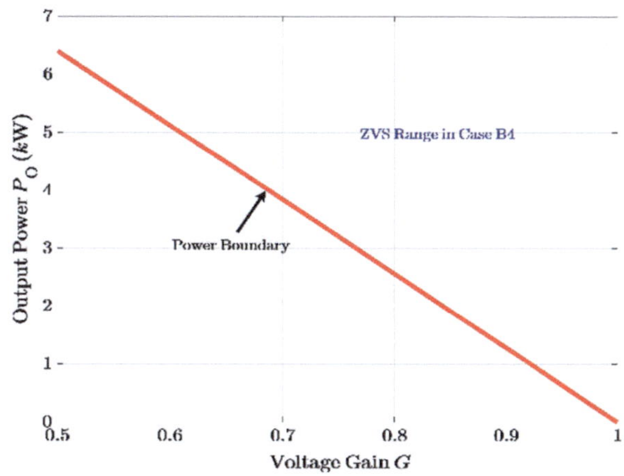

Fig. 4.11 ZVS range in case B4

Although ZVS cannot be maintained for all the switches under all operating conditions, the switching loss is very low due to the use of SiC MOSFETs. The turn-off loss can be reduced due to the relatively small turn-off current.

4.2.4 *Control Strategy Implementation and Small Signal Modeling*

As aforementioned, the output voltage will be out of control as the converter works in high input voltage or light load conditions by utilizing only SPS control. This instability problem can be dealt with by extending D_ϕ to be negative, but it causes higher circulating loss and lower conversion efficiency. In order to solve this issue, a varying mode control strategy is proposed in this paper. The block diagram of this method is shown in Fig. 4.12.

As illustrated in Fig. 4.12, V_o is the feedback value of the output voltage while its reference is V_{ref}. The value that $V_{ref} - V_o$ serves as an input to the PI compensator, and the output of PI regulation is restricted by the limiter. When the limiter output is positive, D is set as $2D_E$ and the output voltage is only regulated by D_ϕ. On the contrary, if the output of the limiter is negative, D_ϕ is set as 0 and the output voltage is only regulated by D. The falling edge compensation is employed to reduce the

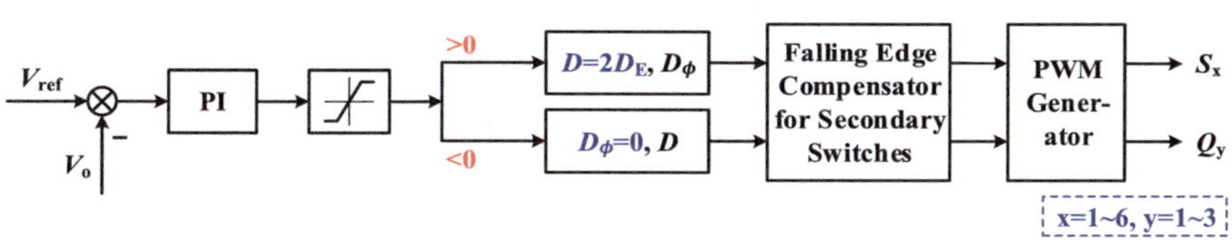

Fig. 4.12 Phase shift and duty cycle segmentation control block diagram

conducting duration length of secondary side MOSFETs body diodes. Finally, the information of D_ϕ and D are used to produce the gating signals for power switches.

It must be pointed out that the gating signals of secondary side switches is only phase shifted with those of the primary bottom switches. Most of the time, the reverse current will flow through the body diode of the MOSFETs after Q_x ($x = 1$–3) turns off in Case B4 and Case B5. This causes higher conduction loss and turn off loss for the secondary side due to the high voltage drop on body diodes and high turn off current. Therefore, the gate signal falling edge compensator is only used to delay the secondary side driving signals. Figure 4.13 illustrates the waveforms of Case B4 after falling edge compensation.

The orange shaded part in Fig. 4.13 is the time interval compensation of the falling edge of secondary side driving signals. Q_1 should have been turned off simultaneously with S_2 without compensation, but now it is purposely delayed until i_{LkA} reduces to zero. The rising edge of Q_1 is the same as that for S_2.

The 3p-VF-SDAB proposed in this paper has a relatively complicated circuit structure, which contains a large number of power devices and magnetic components.

Therefore, device-level modeling methods are not feasible. To simplify the analysis, the 3p-VF-SDAB dc-dc converter circuit can be viewed as a two-port network as shown in Fig. 4.14, where R_o is the load resistance, and C_f is the output capacitance. The nonlinear factors of the converter are ignored. Since Case A2 and Case B5 occupy a large interval time, in concise consideration, only Case A2 and Case B5 are taken as examples for analysis. The low frequency assumptions, small ripple assumptions, and small signal assumptions in conventional modeling methods are still true. The required transfer function can be obtained just by analyzing the relationship between the input and output in Fig. 4.14. And the effect of the ratio of phase shift angle or the duty cycle on the output voltage are only discussed. And the transfer functions in different cases are shown in (4.4) and (4.5), respectively.

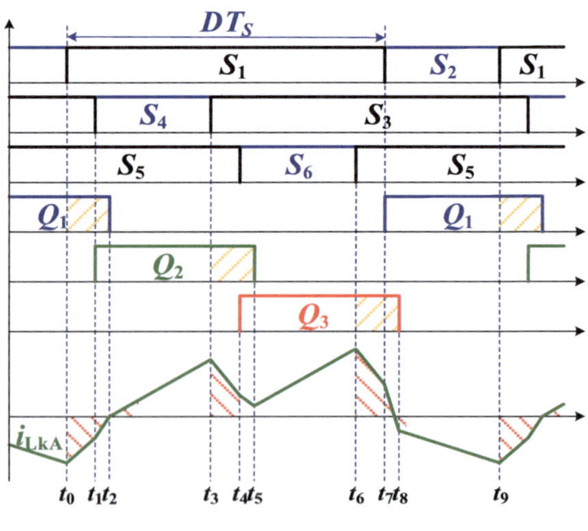

Fig. 4.13 Gating signal falling edge compensation of secondary side in Case B4

$$G_{\text{CaseA2}}(s) = \frac{\widehat{v_o}}{\widehat{d_\phi}} = \frac{2nR_oV_{in}^2T_S(D_E - D_\phi)}{2L_kC_fR_oV_{in}s + 2L_kV_{in} + R_oT_SD_E^2n^3V_o} \tag{4.4}$$

$$G_{\text{CaseB5}}(s) = \frac{\widehat{v_o}}{\widehat{d}} = \frac{-2V_{in}V_oR_oT_S(1-D)(V_{in} - V_on)}{3L_kC_fR_oV_o^2s + 3L_kV_o^2 + R_oT_SV_{in}^2(1-D)^2} \tag{4.5}$$

where $\widehat{v_o}$, $\widehat{d_\phi}$ and $\widehat{d}$ are the AC small signal components of output voltage, ratio of phase shift angle, and duty cycle, respectively. And other capitalized symbols represent the steady state quiescent values. Since in the Case B5, the larger the duty cycle, the lower the output voltage, so the transfer function has a negative sign. The control block diagrams of the converter in different cases are then shown in Fig. 4.15.

Where $G_{\text{PI}}(s)$ is the proportion-integral (PI) compensator, it can be explained as $K_p + K_i/s$. And $H(s)$ is the sampling coefficient. In the experiment, $H(s) = 1$. According to Fig. 4.15, the open loop transfer function of the converter system in different cases can be written as follows, respectively.

$$G_{A2\_OL}(s) = G_{\text{PI}}(s) \cdot G_{\text{CaseA2}}(s) \cdot H(s) \tag{4.6}$$

$$G_{B5\_OL}(s) = -G_{\text{PI}}(s) \cdot G_{\text{CaseB5}}(s) \cdot H(s) \tag{4.7}$$

They are the minimum phase system. Figure 4.16 shows the Bode diagram corresponding to (4.6) and (4.7). Table 4.2 shows the specifications.

According to Fig. 4.16, the proposed varying mode control can achieve stability and good dynamic response in the full load and wide input voltage range theoretically.

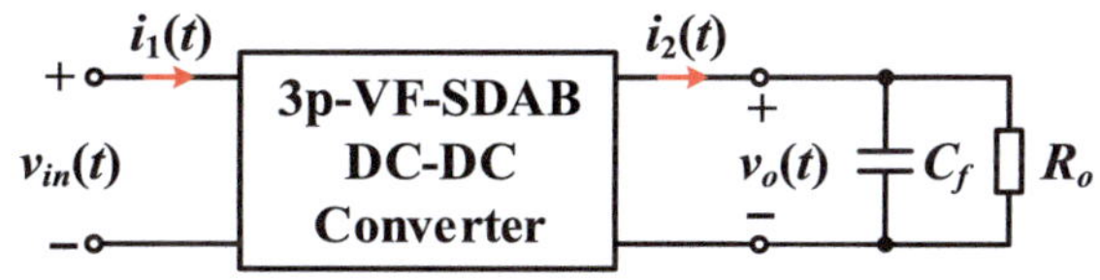

Fig. 4.14 Two-port network simplified model of 3p-VF-SDAB dc-dc converter

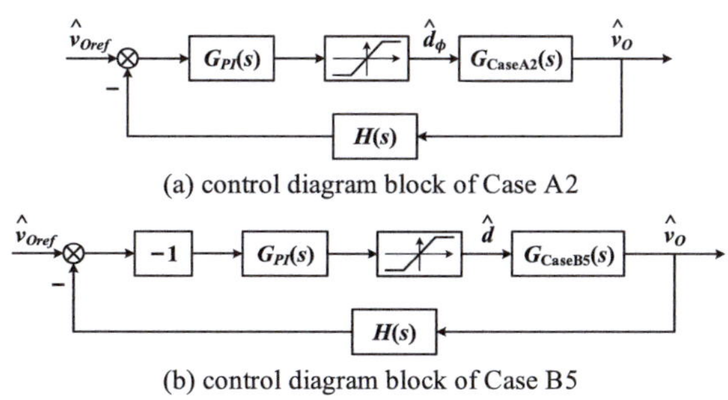

Fig. 4.15 Control block diagram in different cases

Fig. 4.16 Bode diagram in different cases and different operating conditions

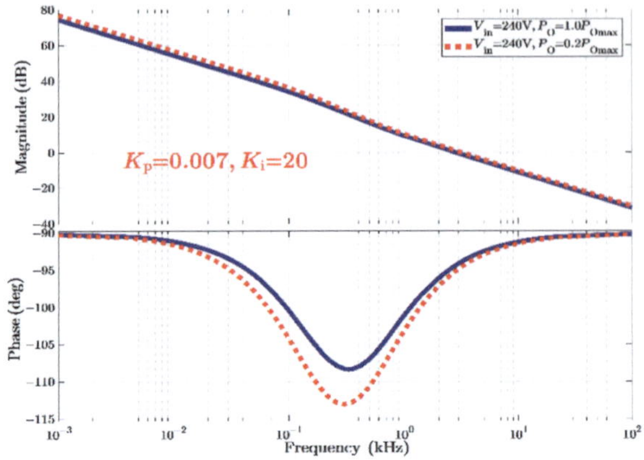

(a) bode diagram of Case A2

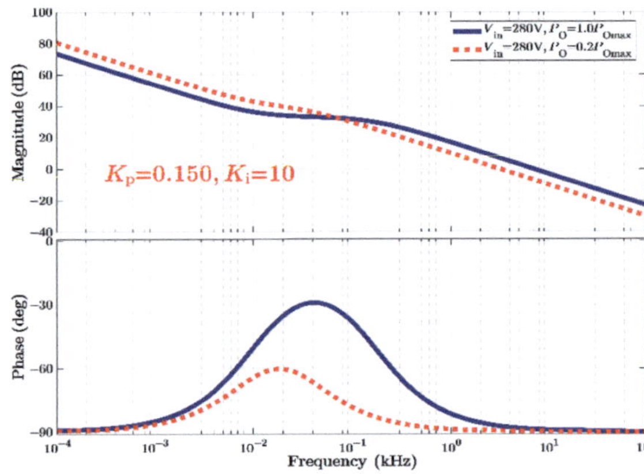

(b) bode diagram of Case B5

4.3 Experimental Results

4.3.1 Prototype

A 1.6 kW experimental prototype has been built to verify the effectiveness of the 3p-VF-SDAB dc-dc converter employing the proposed control strategy. The prototype photograph is shown in Fig. 4.17. The control chip is TMS320F28335, the drive chip is Si8233, all the switches used are SiC MOSFETs typed C3M0065090J, the diodes type is C3D10060G and the high frequency transformer core is EE42A. The power circuit is covered by the driving boards. The system specifications are illustrated in Table 4.2.

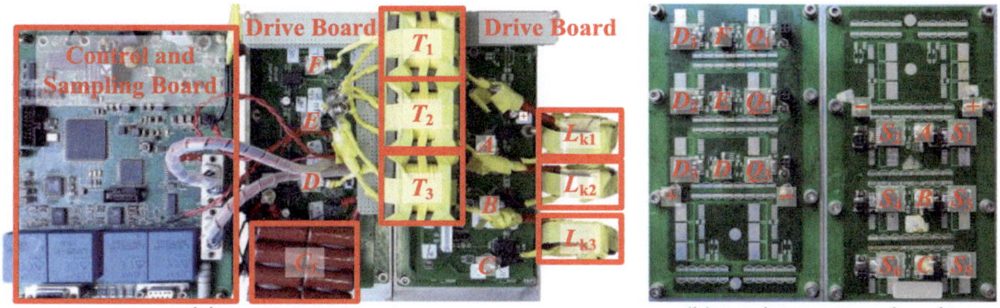

(a) converter system (b) main power circuits

Fig. 4.17 Photograph of the test prototype

4.3.2 Steady State Experimental Results

Figures 4.18 and 4.19 show the experimental waveforms of the proposed converter under different voltage gains and different loads with the proposed control strategy. It can be seen that the phase to neutral voltage waveforms (v_{ANp} and v_{FNs}) in Figs. 4.18 and 4.19 are not so regular, which is a little different from the ideal analysis.

This is due to the fact that the neutral point of the primary and secondary side is floating. The asymmetry of each phase parameters results in a very irregular neutral point voltage waveform. This affects the phase to neutral voltage waveforms. Meanwhile, the effect of the parasitic parameters and dead time in the actual circuits will

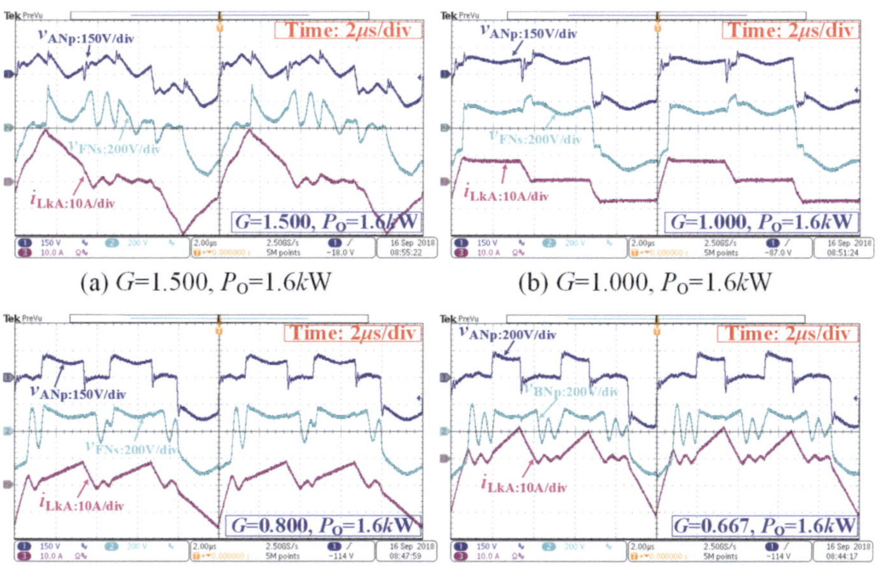

(a) G=1.500, P_O=1.6kW (b) G=1.000, P_O=1.6kW

(c) G=0.800, P_O=1.6kW (d) G=0.667, P_O=1.6kW

Fig. 4.18 Steady state experimental waveforms in 1.6 kW with different voltage gains

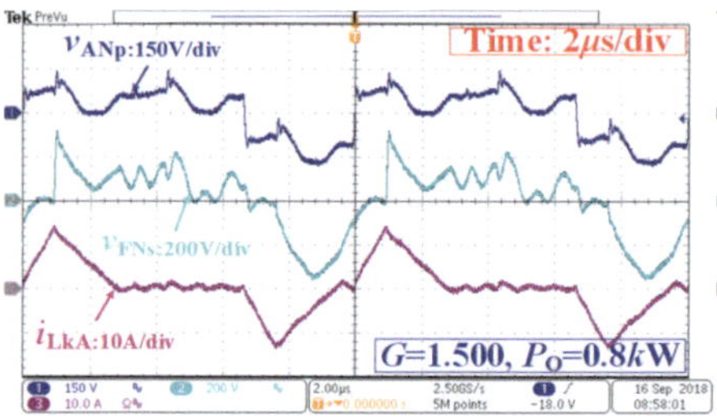

(a) G=1.500, P_O=0.8kW

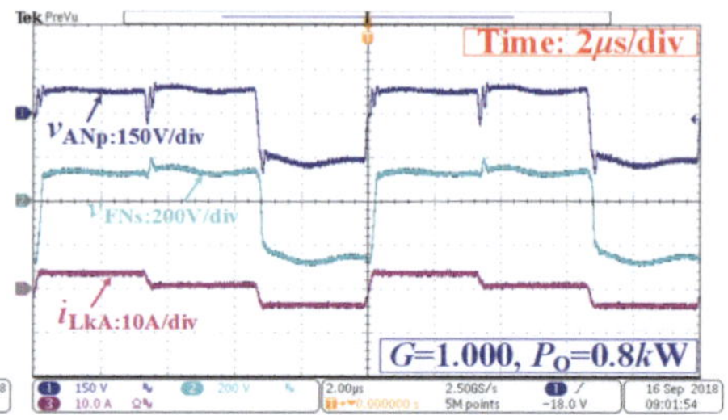

(b) G=1.000, P_O=0.8kW

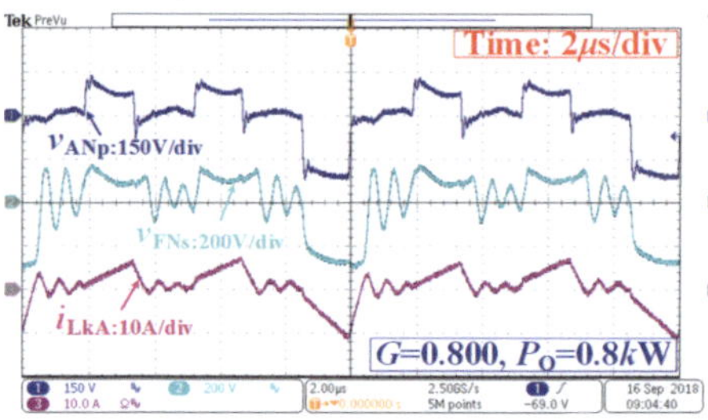

(c) G=0.800, P_O=0.8kW

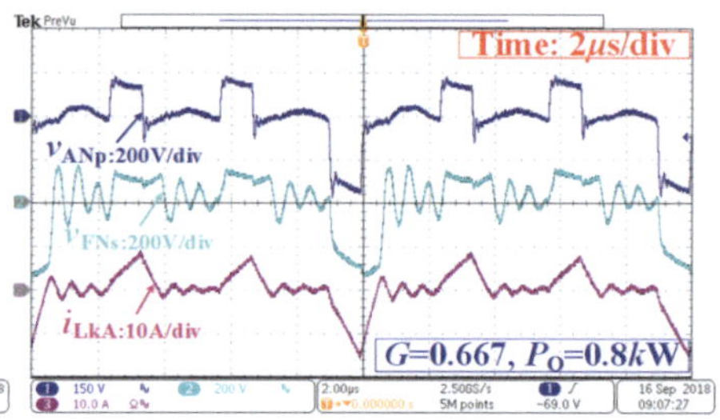

(d) G=0.667, P_O=0.8kW

Fig. 4.19 Steady state experimental waveforms in 0.8 kW with different voltage gains

be the other factors affecting the irregularity of phase to neutral voltages. In comparison, the experimental current waveforms is relatively regular and does not differ that much with their theoretical ones. At the same time, the voltage gain is as large as 1.5 and as low as 0.667 to verify the wide input voltage range operating capability with the proposed converter and the proposed control strategy as shown in Fig. 4.20.

Figure 4.21 shows the experimental results in $G = 0.923$ ($V_{in} = 260$ V) with negative D_ϕ under different loads. It can be seen that the non-active current loss is pretty high, especially at light load conditions. Furthermore, the output voltage is out of control when the output power drops down to $0.2P_{Omax}$. It implies that SPS control with negative D_ϕ only can achieve stability by extend the input voltage range and load range to a certain value. The operating stability within the full load range cannot be solved completely.

In comparison, Figs. 4.22 and 4.23 show the operating waveforms of the converter under no-load conditions. As seen from these pictures, it can be concluded that the proposed converter can work very stably throughout the full load range even at no load condition by employing the proposed control strategy.

4.3.3 Dynamic Response Experimental Results

In order to verify that seamless transition between different working modes by employing the proposed control, load transient experiments have been made.

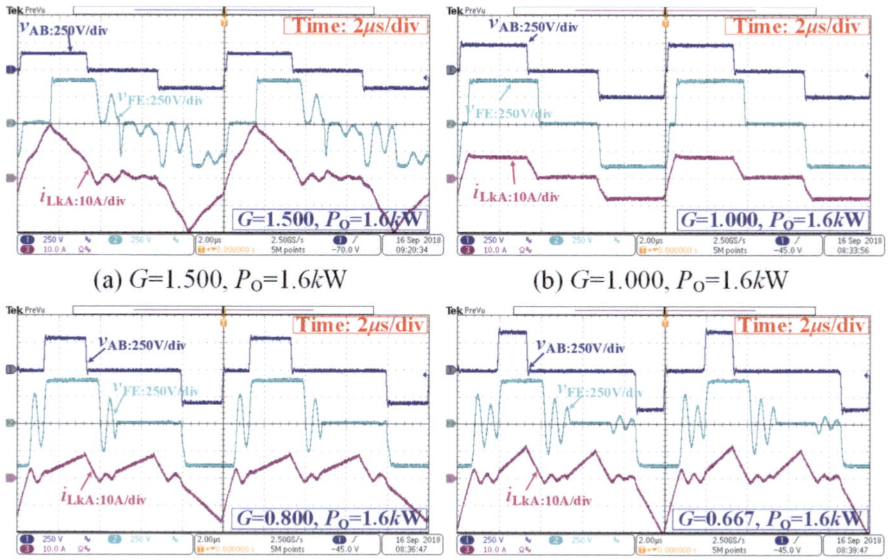

(a) G=1.500, P_O=1.6kW (b) G=1.000, P_O=1.6kW

(c) G=0.800, P_O=1.6kW (d) G=0.667, P_O=1.6kW

Fig. 4.20 Primary and secondary side line to line voltage and phase to neutral current waveforms

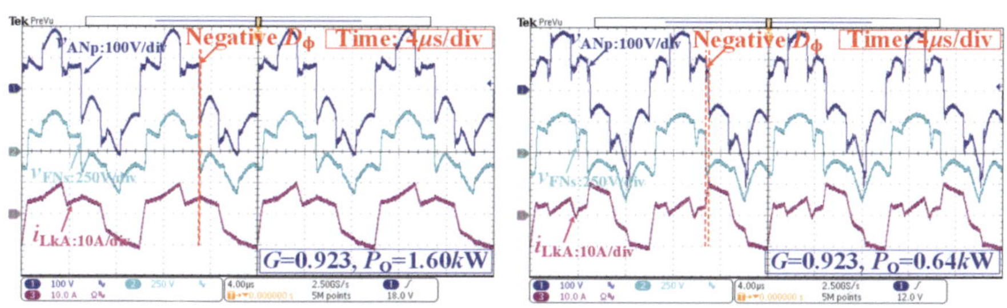

(a) G=0.923, P_O=1.60kW (b) G=0.923, P_O=0.64kW

Fig. 4.21 Steady state experimental waveforms in $G = 0.923$ with negative D_ϕ

Figure 4.24 shows the experimental results about the load switching back and forth between 0.8 and 1.6 kW under different voltage gains. As seen from Fig. 4.24, the transition between the two different operating modes is smooth and the output voltage changes only a little bit in facing the load transient. It indicates that the proposed converter that uses the proposed control strategy has a very good dynamic response.

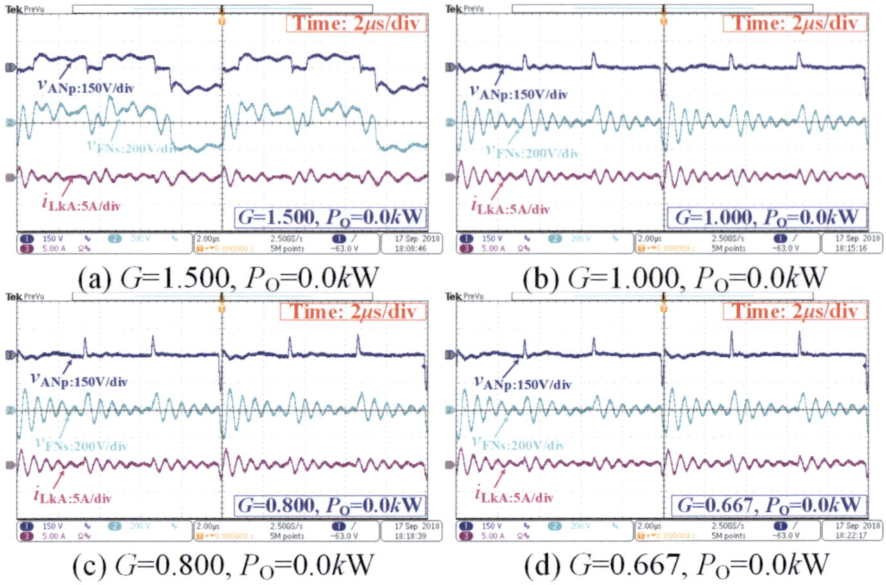

(a) G=1.500, P_O=0.0kW (b) G=1.000, P_O=0.0kW

(c) G=0.800, P_O=0.0kW (d) G=0.667, P_O=0.0kW

Fig. 4.22 Steady state experimental waveforms in 0.0 kW with different voltage gains

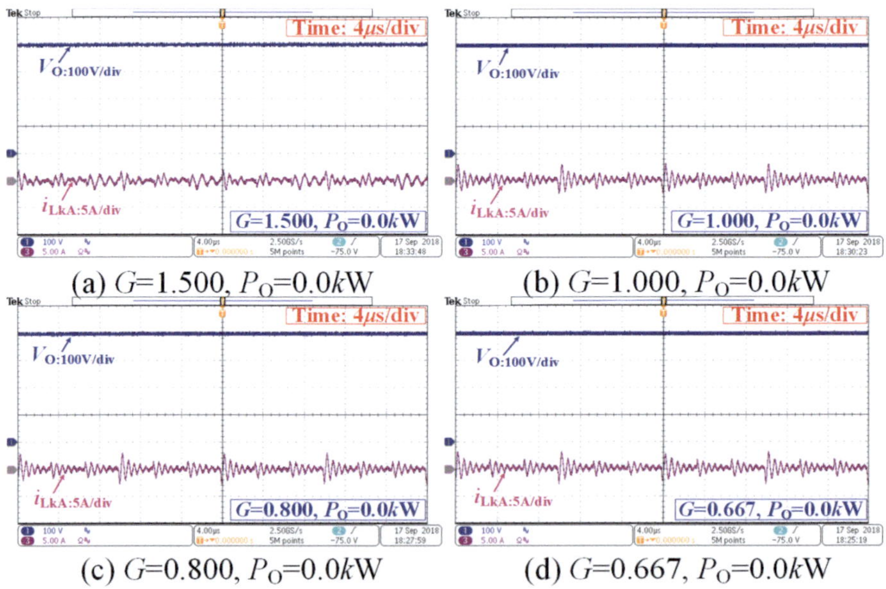

(a) G=1.500, P_O=0.0kW (b) G=1.000, P_O=0.0kW

(c) G=0.800, P_O=0.0kW (d) G=0.667, P_O=0.0kW

Fig. 4.23 Output voltage experimental results under different voltage gains with no load conditions

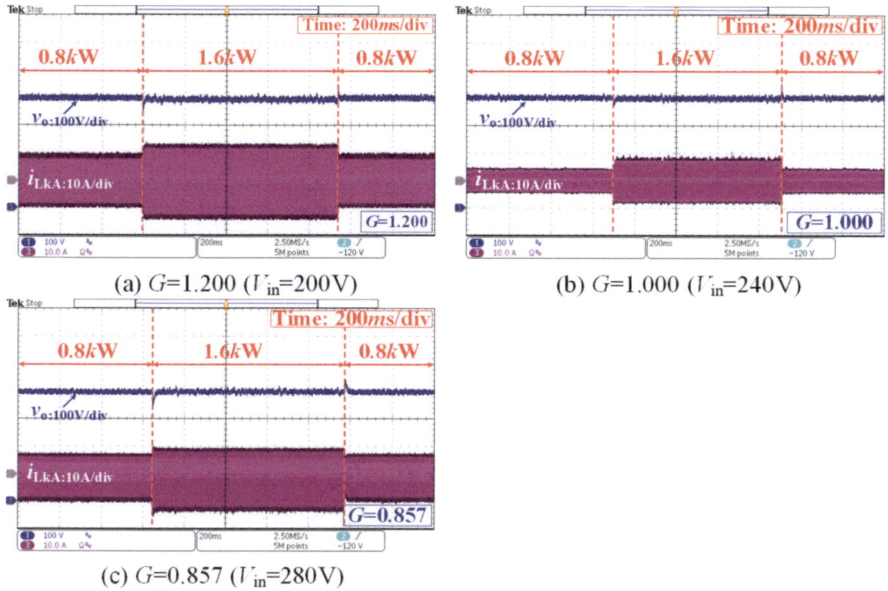

(a) G=1.200 (V_{in}=200V)

(b) G=1.000 (V_{in}=240V)

(c) G=0.857 (V_{in}=280V)

Fig. 4.24 Dynamic responses under different voltage gain

4.3.4 Conversion Efficiency and Loss Breakdown

Figure 4.25 shows the measured efficiency curves under different input voltages (voltage gain) and different output inductor core loss in the case $0.6P_{\mathrm{Omax}}$ when $V_{\mathrm{in}} = 160$ V is dramatically increased. In the case of $0.8P_{\mathrm{Omax}}$, the core loss decreases. Figure 4.26 shows the power loss breakdown analysis under different input voltage. Figure 4.27 shows the measured efficiency curve comparison under different control strategies when $G = 0.857$ ($V_{\mathrm{in}} = 280$ V). The blue solid line with the square dots represents the proposed control strategy with the falling edge compensation while the orange dashed line with the circular dots represents the proposed control strategy without the compensation. The green chain line with the triangle dots is the SPS control. It can be seen from the figure that under different load conditions, the proposed control with the compensation method has the highest conversion efficiency, followed by the proposed control without compensation, and SPS control with negative phase shift angle has the lowest conversion efficiency. At heavy load, conversion efficiency using SPS control strategy does not differ that much. But under light load conditions, the relatively higher circulating loss and the higher core loss by SPS control results in lower conversion efficiency (Fig. 4.26).

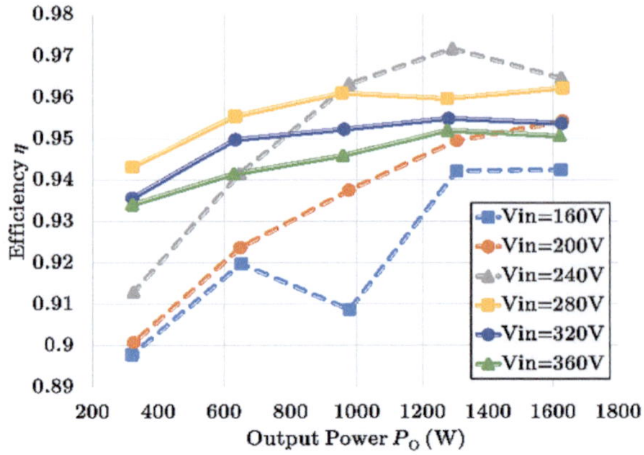

Fig. 4.25 Power conversion efficiency

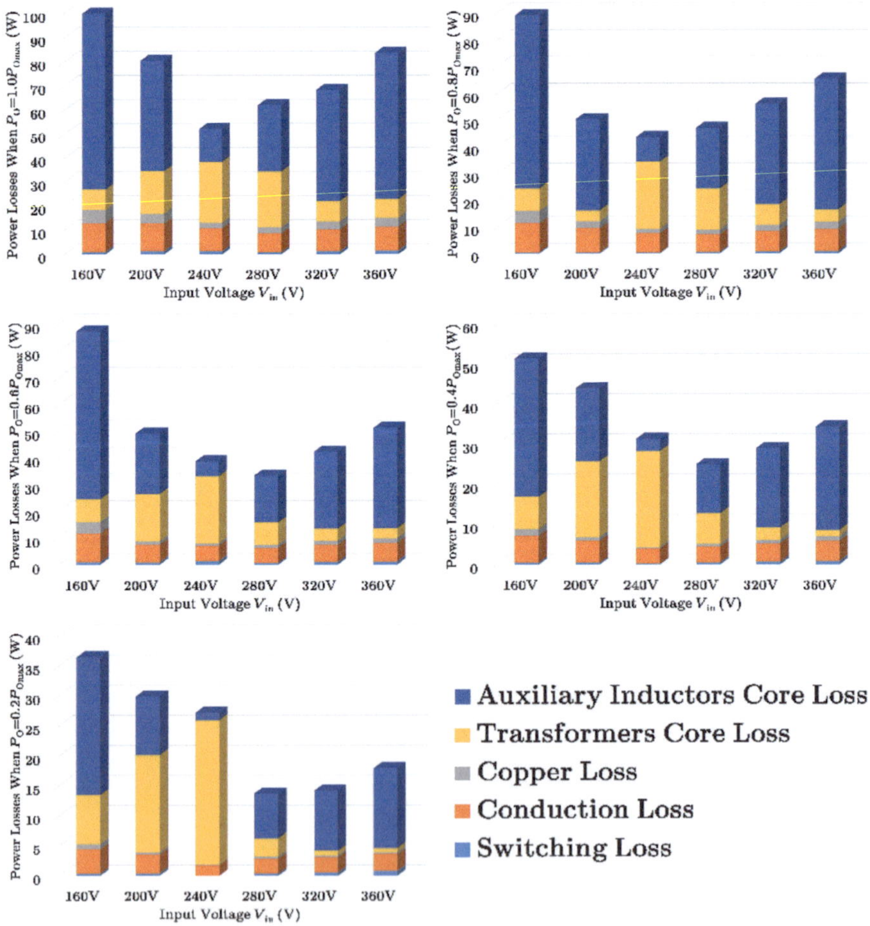

Fig. 4.26 Power loss breakdown analysis

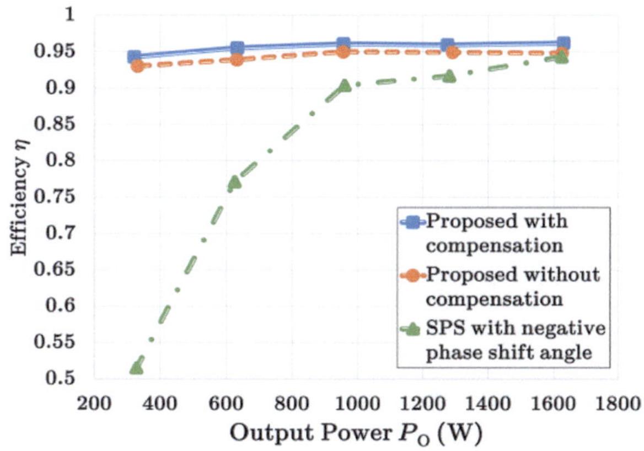

Fig. 4.27 Efficiency curve under different control methods with $G = 0.857$ ($V_{in} = 280$ V)

4.4 Conclusion

A high-frequency isolated unidirectional 3p-VF-SDAB dc-dc converter is proposed for wide input voltage range applications. Most of the possible combinations of switching patterns and the natural current zero crossing points of secondary side diodes when $G < 1$ have been studied extensively. The ideal power transferred expressions are derived for the analysis of power transmission characteristics. The ideal LIC RMS expressions are derived for optimal duty cycle selection as the phase shift angle is zero. The ideal soft-switching conditions based on current polarity has been given. In order to deal with the operating instability issue when the load is light or the input voltage is much higher than the matching point, a varying mode control is proposed. SPS control is implemented as the load is heavy or the voltage gain is close to 1. Duty cycle control is employed as the SPS control is not enough to stabilize the output voltage even when the phase shift angle is equal to zero. This control method can significantly reduce the non-active power loss. Thus the conversion efficiency is increased significantly compared with that obtained by setting the phase shift angle to be negative at light load conditions. At the same time, a fixed compensation of secondary side driving signal falling edge is also implemented to reduce the conduction loss of the body diode of MOSFETs. The validity of the proposed converter and the proposed control strategy has been verified by experiments of a 1.6 kW prototype using SiC MOSFETs and SiC diodes. The converter system has good steady state and dynamic performance even in mode transition. The converter with the proposed control can achieve full load range operating stability in facing the wide input voltage range. Its overall high conversion efficiency is high throughout the full load range with a wide input voltage input.

References

1. Segaran D, Holmes DG, McGrath BP (2008) Comparative analysis of single and three-phase dual active bridge bidirectional DC-DC converters. In: Universities power engineering conference, pp 1–6
2. Wang Z, Li H (2010) Three-phase bidirectional DC-DC converter with enhanced current sharing capability. In: Proceedings of the IEEE energy conversion congress and exposition (ECCE), pp 1116–1122
3. van Hoek H, Neubert M, Kroeber A, De Doncker R (2012) Comparison of a single-phase and a three-phase dual active bridge with low-voltage high-current output. In: International conference on renewable energy research and applications, pp 1–6
4. Kulasekaran S, Ayyanar R (2014) Analysis, design, and experimental results of the semidual-active-bridge converter. IEEE Trans Power Electron 29(10):5136–5147
5. Sha D, You F, Wang X (2016) A high-efficiency current-fed semi-dual active bridge DC–DC converter for low input voltage applications. IEEE Trans Ind Electron 63(4):2155–2164
6. Sha D, Zhang J, Xu Y (2017) Improved boundary operation for voltage-fed semi-DAB with ZVS achievement and non-active power reduction. IEEE Trans Ind Electron 64(8):6179–6189
7. Sha D, Zhang J, Sun T (2019) Multi-mode control strategy for SiC MOSFETs based semi dual active bridge DC-DC converter. IEEE Trans Power Electron 34(6):5476–5486

Chapter 5
Current-Fed Single Phase Semi-DAB DC-DC Converter with Dual PWM and Phase Shifted Control

Abstract For voltage-fed semi dual active bridge (VF-SDAB) DC-DC converter topology, modulation and control strategy in previous chapters, once the voltage is far away from the matched value, the ZVS range is limited and the peak current of leakage inductor will be relatively high. What is more, the current ripple becomes high especially at low input voltages. So in this chapter, a current-fed semi dual active bridge (CF-SDAB) DC-DC converter with symmetrical dual PWM and phase shifted control was proposed. All the power switches can achieve zero voltage switching (ZVS) and the output side diodes can achieve zero current switching (ZCS) naturally. The converter has lower conduction loss, lower peak current and higher efficiency. The typical operating modes were analyzed. The parameter optimal design criterion was given to achieve soft switching and circulation loss reduction. Simulation and experimental results for a 1 kW prototype were made to verify the effectiveness of the proposed topology employing the proposed control.

Keywords DC-DC · Current-fed semi dual active bridge · Zero voltage switching · Zero current switching · Circulation loss

5.1 The Proposed Current-Fed Semi-DAB Circuit Topology

For voltage-fed DAB and semi-DAB, if the voltage amplitudes of the transformer primary and secondary sides are not matched, the current stress may be high, the non-active power and conduction are large. These are disadvantages for the voltage-fed DAB and semi-DAB. Besides, its current ripple is comparatively high [1].

To reduce the current ripple, a novel current-fed DHB was introduced [2]. As the input voltage varies, the peak current of the leakage inductance current may be high. To deal with this issue, a PWM plus shifted control was proposed, in which the low voltage side (LVS) converter is PWM modulated. Thus the active clamp voltage can be regulated according to the high voltage side (HVS) voltage and the turns-ratio. Thereby, during the power transfer stage, the leakage inductance current slop can be maintained to be zero despite the voltage conversion ratio varies [3–5]. However,

D. Sha et al., *High Frequency Isolated Semi Dual Active Bridge DC-DC and AC-DC Converter*, CPSS Power Electronics Series,
https://doi.org/10.1007/978-981-95-1641-4_5

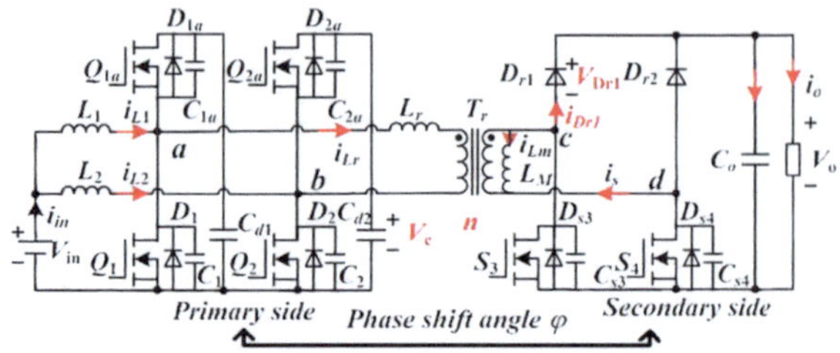

Fig. 5.1 The proposed current-fed semi dual-active-bridge circuit topology

the current spike and circulation loss are large especially when the current-fed side voltage is rather low. To adapt to a wide range of LVS voltage and to suppress the current spike, the equivalent transformer leakage inductance cannot be designed to be rather small. Normally an additional inductor is required to be series connected with the transformer winding. But, as the equivalent leakage inductance value increases, the circulation stage length increases as well and the power transfer stage length shrinks. The maximum power that can be transmitted is limited due to the large value of the leakage inductance. Besides, the very high current spike generated in the non-power transfer stage may damage power switches, causing reliability issues. To reduce the circulation loss and suppress the high current spike significantly, PWM can be employed for both sides of current-fed DAB [6, 7].

Similarly, for the semi-DAB, to reduce the input current ripple and non-active power, a current-fed semi-DAB is proposed in this chapter. The proposed converter schematic is shown in Fig. 5.1. The input side is dual boost-half bridge composed of four MOSFETs. The DC inductor L_1 and L_2 have the same inductance. Q_{1a}, C_{d1} and Q_{2a}, C_{d2} constitute the active clamp circuit. The AC leakage inductor L_r represents the total value of the primary-referred transformer leakage inductance and L_m is the secondary-referred magnetizing inductance. The secondary side is composed of a semi-active bridge using two diodes D_{r1}, D_{r2} and two switches S_3, S_4. V_c is the primary side active clamp voltage. D is duty cycle of switches, φ is defined as the phase shift angle between Q_2 and S_4. The transformer turns ratio is n.

5.2 Working Principle of the Modulation

The operation principle can be divided into two cases according to the current waveform of the leakage inductance current in the discontinuous conduction mode in which $\varphi < 2\pi(D - 0.5)$ or continuous conduction mode where $\varphi > 2\pi(D - 0.5)$.

Figure 5.2 illustrates the operation principle of the proposed converter when $\varphi < 2\pi(D - 0.5)$. As shown, the duty cycles for power switches Q_1, Q_2, S_3 and S_4 are equal to D. The value of the duty cycle D varies with V_{in}, making the clamp voltage V_c to be matched with the output voltage V_o. Thus the slew rate of the leakage inductance current during the matched period is zero in spite of the input voltage and load variation. Since D is determined by V_{in}, the output power varies as the phase shift angle φ. Generally, output power increases with the increase of φ. One complete period can be divided into sixteen stages from θ_0 to θ_{16}. Among all the working stages, some stages can be defined as current circulating stages in which the power is not transmitted to the output. For example, during the range $[\theta_2, \theta_6]$ or $[\theta_{10}, \theta_{14}]$, v_{cd} is zero, so the transmitted power is zero. Other stages can be defined as power transfer stages in which both v_{cd} and i_s are not zero. For example, one of the power transfer stages is $[\theta_6, \theta_{10}]$. To achieve ZVS for the output side switches, phase shifted PWM gating sequence is used.

To minimize the circulation loss, PWM control is employed for the output side just like the input side with the same duty cycle. As seen in Fig. 5.2, v_{ab} leads v_{cd}, during the stage $[\theta_0, \theta_2]$, i_{Lr} drops to zero, then stays as a small bias current within the stage range $[\theta_3, \ \theta_4]$ until v_{ab} becomes positive. Thus, the *rms* value of i_{Lr} is very small during the circulation stage and the circulation loss can be minimized. To simplify the analysis, the effect of the magnetizing inductance is neglected and the detailed mode analysis can be described as follows.

The typical working modes are shown in Fig. 5.3. They can be described as follows:

Stage 1 (θ_0–θ_1): Before θ_0, Q_1, Q_{2a}, S_3 and D_{r2} conduct, power is transmitted to the output side. The primary side voltage v_{ab} equals to nv_{cd}, where v_{cd} is the secondary voltage of the transformer. Thus before θ_0, the slew rate of leakage inductance i_{Lr} is zero. At θ_0, Q_{2a} is turned off. The difference between the current i_{Lr} and DC

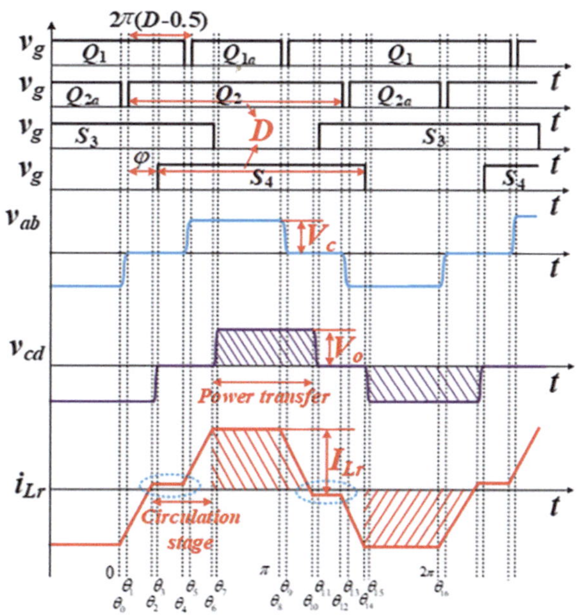

Fig. 5.2 Operation principle with $\varphi < 2\pi(D - 0.5)$

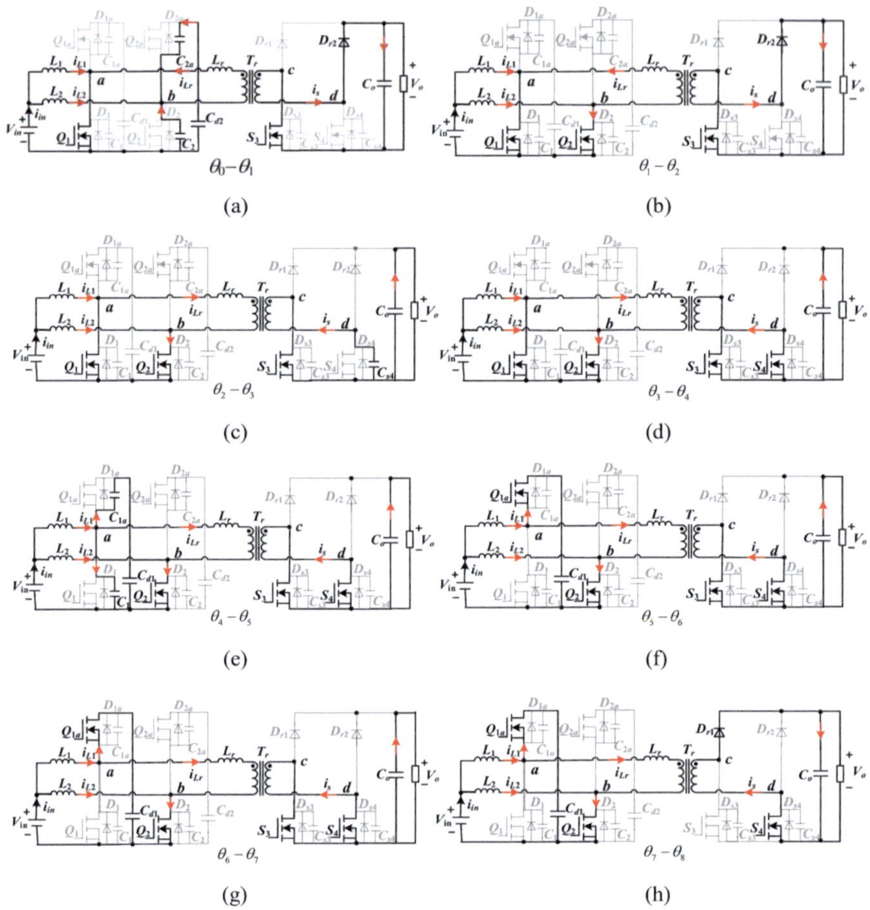

Fig. 5.3 Typical working modes with $\varphi < 2\pi(D - 0.5)$

inductance current i_{L2} charges C_{2a} and discharges C_2 until the body diode D_2 begins to conduct. After this, Q_2 can be gated on with ZVS.

Stage 2 (θ_1–θ_2): At θ_1, Q_2 is turned on with ZVS. The voltage v_{ab} becomes zero, and voltage v_{cd} is $-V_o$, so i_{Lr} begins to decrease.

Stage 3 (θ_2–θ_3): At θ_2, i_{Lr} decreases to zero, so the secondary rectifier diode D_{r2} is turned off at ZCS. Then L_r and the junction capacitor C_{s4} begin to resonate until the body diode D_{S4} begins to conduct. Hence, i_{Lr} flows in the opposite direction, and stays as a current bias. S_4 can be turned on with ZVS during [θ_3, θ_4], but in order to realize minimum conduction loss, S_4 is turned on at θ_3.

Stage 4 (θ_3–θ_4): At θ_3, S_4 is turned on with ZVS. v_{ab} and v_{cd} are zero, so i_{Lr} is the bias current I_{bias}. During the stage, no power is transmitted from the primary side to the secondary side.

Stage 5 (θ_4–θ_5): At θ_4, Q_1 is turned off. The difference of i_{L1} and i_{Lr} charges C_1 and discharges C_{1a} until the body diode D_{1a} begins to conduct. Then, v_{ab} becomes V_c and v_{cd} is zero, so i_{Lr} will increase with the slew rate V_c/L_r.

Stage 6 (θ_5–θ_6): At θ_5, Q_{1a} is turned on at ZVS. v_{ab} and v_{cd} are the same as stage 5. i_{Lr} increases linearly.

Stage 7 (θ_6–θ_7): At θ_6, S_3 is turned off, i_s charges C_{s3} until the rectifier diode D_{r1} begins to conduct.

Stage 8 (θ_7–θ_8): At θ_7, D_{r1} begins to conduct current. Then v_{cd} is V_o, During the stage, v_{ab} equals to $n{*}v_{cd}$, causing the slew rate of i_{Lr} to be zero. The first half cycle is over, and the next half cycle is similar to the first half cycle.

The key waveforms when $\varphi > 2\pi(D - 0.5)$ are shown in Fig. 5.4. There are also sixteen stages in one period. Most of the stages are similar to the stages when $\varphi < 2\pi(D - 0.5)$. Its difference can be observed in stages [θ_2, θ_5] and [θ_{10}, θ_{13}].

The special working modes under $\varphi > 2\pi(D - 0.5)$ situation are shown in Fig. 5.5, which can be described as,

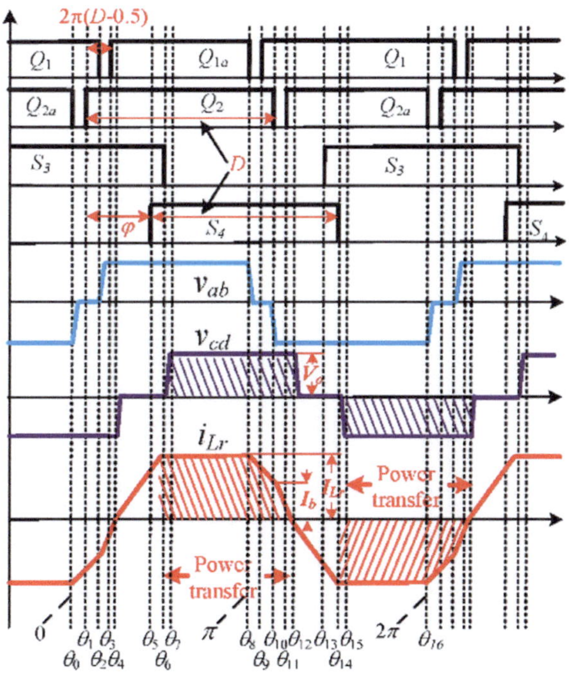

Fig. 5.4 Operation principle with $\varphi > 2\pi(D - 0.5)$

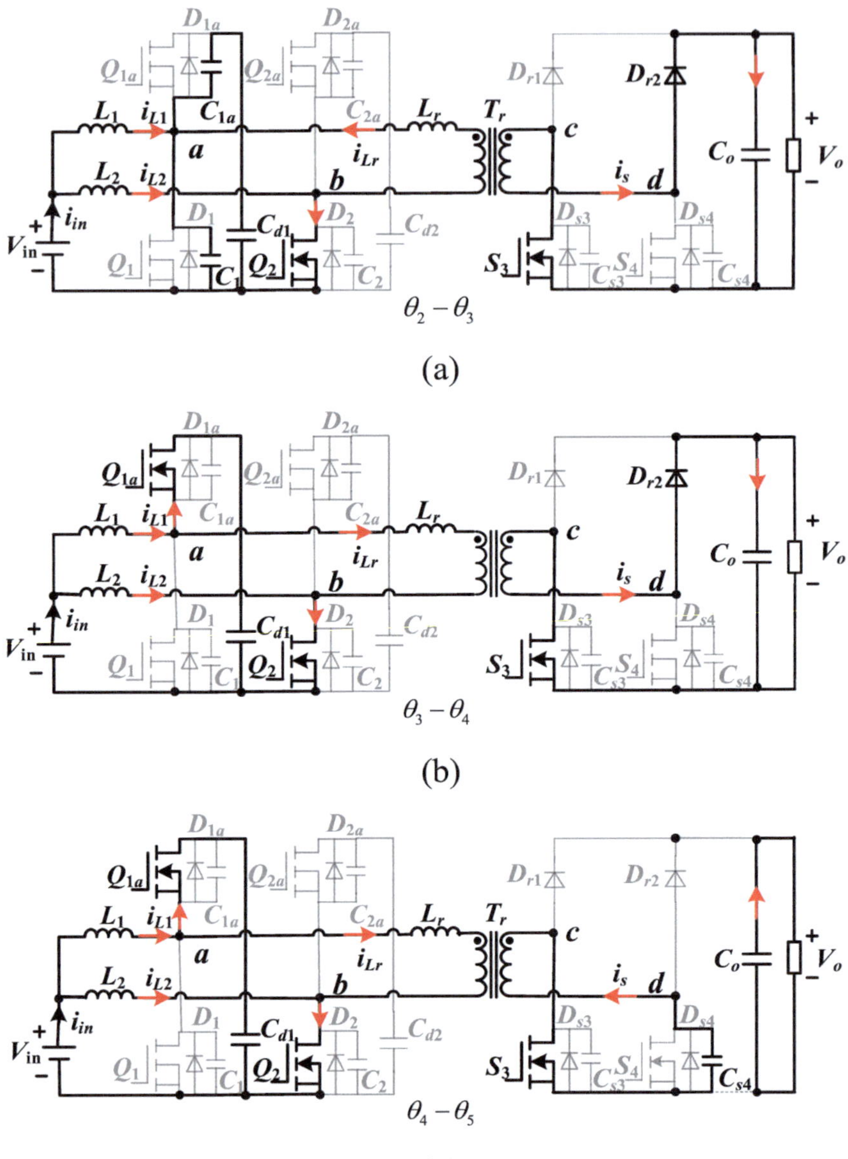

Fig. 5.5 Special working modes with $\varphi > 2\pi(D - 0.5)$

Stage 3 (θ_2–θ_3): Before θ_2, Q_1, Q_2, S_3, and D_{r2} conduct. v_{ab} is zero and v_{cd} is equal to $-V_o$. Hence the leakage current i_{Lr} decreases with the slew rate V_c/L_r. At θ_2, Q_1 is turned off. The difference between DC inductance current i_{L1} and i_{Lr} charges C_1 and discharges C_{1a} until the body diode D_{1a} conducts. Then v_{ab} is equal to the clamp voltage V_c, and v_{cd} is equal to $-V_o$, causing the current i_{Lr} to decrease linearly with the slew rate of $2V_c/L_r$.

Stage 4 (θ_3–θ_4): When $\varphi > 2\pi(D-0.5)$, Q_{1a} leads S_4. So, at θ_3, Q_{1a} is turned on with ZVS. The voltages v_{ab} and v_{cd} remain unchanged. The current i_{Lr} decreases until it drops to zero.

Stage 5 (θ_4–θ_5): At θ_4, i_{Lr} drops to zero. D_{r2} is turned off at ZCS. L_r and output junction C_{s4} begins to be resonant until the body diode D_{s4} begins to conduct, then v_{cd} is zero. Hence, i_{Lr} flows in the opposite direction, and increases linearly with the slew rate of V_c/L_r. At θ_5, S_4 is turned on at ZVS.

5.3 Parameter Optimization Design

5.3.1 Output Power Curve Based on System Specifications

The output power expression when $\varphi < 2\pi(D - 0.5)$ as shown in Fig. 5.2 can be obtained by

$$P_o = \frac{\int_0^{T_s} n i_{Lr} \times v_{cd}\,\mathrm{d}t}{T_s} = \frac{\int_{\theta_7}^{\theta_8} n I_{Lr} \times V_c \mathrm{d}\theta + \int_{\theta_8}^{\theta_{10}} n i_{Lr} \times V_c \mathrm{d}\theta}{\pi} \tag{5.1}$$

P_o can be further written by

$$P_o = \frac{2nI_{Lr}V_o[2\pi(1-D)-\varphi] + nI_{Lr}V_o\varphi}{2\pi} \tag{5.2}$$

The peak current I_{Lr} and the clamp voltage V_c can be expressed by respectively

$$\begin{cases} I_{Lr} = \dfrac{V_c T_s \varphi}{2\pi L_r} \\ V_c = nV_o \end{cases} \tag{5.3}$$

Substituting (5.3) into (5.2) yields,

$$P_o = -\frac{V_c^2 T_s \varphi^2}{4\pi^2 L_r} + \frac{V_c^2 T_s (1-D)\varphi}{\pi L_r} \tag{5.4}$$

Table 5.1 System specifications

Parameter	Value	Parameter	Value
V_o	300 V	V_{in}	18–28 V
n	1:5	V_c	60 V
f_s	50 kHz	L_1	22 μH
C_{d1}, C_{d2}	30 μF	L_2	22 μH

Likewise, when $\varphi > 2\pi(D - 0.5)$ as shown in Fig. 5.4, P_o is calculated by

$$P_o = -\frac{5V_c^2 T_s \varphi^2}{18\pi^2 L_r} + \frac{V_c^2 T_s (5.5 - 5D)\varphi}{9\pi L_r} + \frac{V_c^2 T_s (-14D^2 + 20D - 6.5)}{18L_r} \tag{5.5}$$

Table 5.1 gives the system specifications. The power curve can be obtained based on Eqs. (5.4) and (5.5) and the system specifications shown in Table 5.1.

The relationship among the output power P_o, the duty cycle D and phase shift angle φ is illustrated in Fig. 5.6a in 3-D form. Figure 5.6b plots the output power curve versus the phase shift angle at given input voltage or duty cycle D in 2-D form. It is worth noting that the maximum phase angle value must be set as the input voltage varies.

5.3.2 ZVS Condition for Secondary Side Switches

According to the relationship between φ and D, the soft switching condition for the secondary side switches can be divided into two situations. Figure 5.7 plots the ZVS condition for S_3 respectively for the two situations. As illustrated, once i_{Lr} reflected to the output side is equal to the magnetizing current i_{Lm}, D_{r1} turns off at ZCS naturally.

Then the output junction capacitor C_{S3} begins to be resonant with L_r, discharging C_{S3}. Because the secondary-referred magnetizing inductance L_m is large enough, during the resonant process, the magnetizing current i_{Lm} can be regarded as constant, i.e., $i_{Lm(t1-t4)} = i_{Lm(t1)} = I_{Lm}$. Figure 5.7a shows the situation when $\varphi < 2\pi(D - 0.5)$. The leakage inductance current i_{Lr} stays at the bias current I_{bias} because both v_{ab} and v_{cd} are equal to zero since the resonance is complete. During the time range $[t_1, t_3]$, Q_1 and Q_2 conduct the relationship between i_{Lr} and the drain to source voltage v_{ds3} of S_3 is given by,

$$\begin{cases} -I_{Lm} + n i_{Lr} = C_{\text{ossH}} \dfrac{\mathrm{d}v_{dsS3}}{\mathrm{d}t} \\ -n v_{dsS3} = L_r \dfrac{\mathrm{d}i_{Lr}}{\mathrm{d}t} \end{cases} \tag{5.6}$$

The initial values for v_{dsS3} and i_{Lr} are:

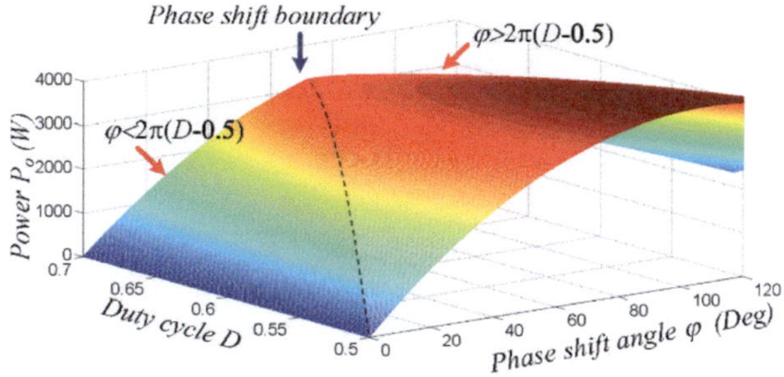

(a) The output power versus the phase shift angle and duty cycle D in 3-d form.

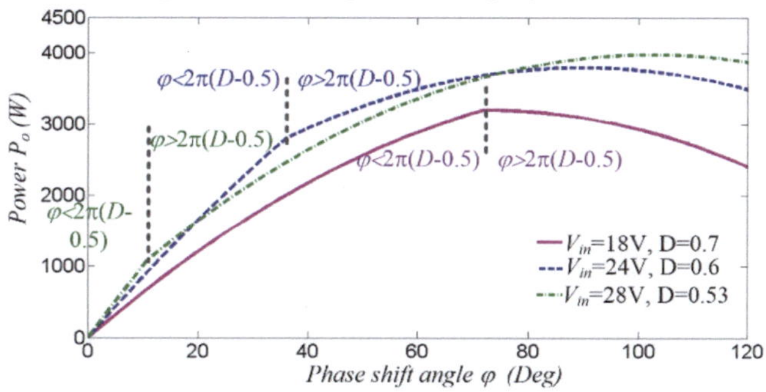

(b) The output power versus the phase shift angle φ and duty cycle D.

Fig. 5.6 The output power versus the phase shift angle φ and duty cycle D

Fig. 5.7 Soft switching curve for secondary side switch S_3

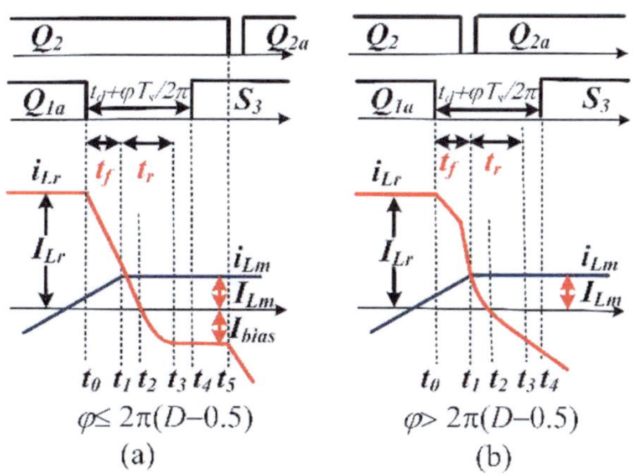

$$v_{dsS3}(0) = V_o,\ i_{Lr}(0) = \frac{1}{n} I_{Lm} \tag{5.7}$$

Based on (5.6) and (5.7), i_{Lr} and v_{dsS3} are obtained as,

$$\begin{cases} i_{Lr}(t) = -V_o\sqrt{\dfrac{C_{\text{ossH}}}{Lr}}\sin(\omega_1 t) + \dfrac{1}{n} I_{Lm} \\ v_{dsS3}(t) = V_o\cos(\omega_1 t) \end{cases} \tag{5.8}$$

where ω_1 is the resonant angular frequency, C_{ossH} is the effective output junction capacitance of S_3. As soon as the resonant process is complete, i_{Lr} is fixed and equal to the bias current I_{bias}. ω_1 and I_{bias} are respectively written by,

$$\begin{cases} \omega_1 = n\sqrt{\dfrac{1}{C_{\text{ossH}} L_r}} \\ I_{\text{bias}} = V_o\sqrt{\dfrac{C_{\text{ossH}}}{L_r}} - \dfrac{1}{n} I_{Lm} \end{cases} \tag{5.9}$$

At t_3, the voltage over C_{s3} is zero, and the body diode of S_3 conducts. To achieve ZVS for S_3, the following must be satisfied,

$$t_f + t_r <= t_d + \frac{\varphi}{2\pi} T_s \tag{5.10}$$

where t_f represents the length from the turn-off of Q_{1a} to the time interval when i_{Lr} drops exactly to I_{Lm} and t_r is the resonant process length. Then, the dead time t_d can be obtained as,

$$t_d > \frac{\pi\sqrt{C_{\text{ossH}} L_r}}{2n} - \frac{L_r T_s}{2 L_m G n} \tag{5.11}$$

Figure 5.7b plots the curve as $\varphi > 2\pi(D - 0.5)$. As soon as the resonance is complete, v_{ab} is equal to V_c while v_{cd} is zero. Hence, i_{Lr} increases. During the time range $[t_1, t_4]$, the switches Q_{2a} and Q_1 conduct, so the relationship between i_{Lr} and v_{dsS3} can be given by,

$$\begin{cases} -I_{Lm} + n i_{Lr} = C_{\text{ossH}} \dfrac{dv_{dsS3}}{dt} \\ -n v_{dsS3} = -V_C + L_r \dfrac{di_{Lr}}{dt} \end{cases} \tag{5.12}$$

The initial values of v_{dsS3} and i_{Lr} have been given in (5.7), the solutions of the differential equations are given by,

$$\begin{cases} i_{Lr}(t) = -2V_o\sqrt{\frac{C_{\text{ossH}}}{L_r}}\sin(\omega_1 t) + \frac{1}{n}I_{Lm} \\ v_{dsS3}(t) = 2V_o\cos(\omega_1 t) - V_o \end{cases} \tag{5.13}$$

At t_3, the voltage over C_{s3} is zero, thus the body diode D_{s3} conducts. As mentioned previously, Eq. (5.10) should be satisfied to achieve ZVS for S_3, thereby the value of the dead time t_d can be calculated as follows,

$$t_d > \frac{(G-2)T_s}{3G} - \frac{\varphi T_s}{3\pi} - \frac{T_s L_r}{4L_m G n} + \frac{\pi\sqrt{C_{\text{ossH}}L_r}}{3n} \tag{5.14}$$

where $G = nV_o/V_{\text{in}}$ is the voltage ratio. To achieve ZVS for the secondary side switches S_3 and S_4, they must be turned on until the resonance between their individual junction capacitors and L_r is completed. To sum up, in order to achieve ZVS condition for secondary side switches in spite of input voltage and load variation, the dead time t_d should be designed as,

$$t_d > \max\left(\frac{(G-2)T_s}{3G} - \frac{\varphi T_s}{3\pi} - \frac{T_s L_r}{4L_m G n} + \frac{\pi\sqrt{C_{\text{ossH}}L_r}}{3n}, \frac{\pi\sqrt{C_{\text{ossH}}L_r}}{2n} - \frac{L_r T_s}{2L_m G n}\right) \tag{5.15}$$

Besides, MOSFET IXFN420N10T is used to work as the input side switch, its junction capacitance C_{ossL} is 4090 pF. The output side switch is IXFB44N50P, its junction capacitor C_{ossH} is 639 pF. The output side rectifying diode is DSEI29-06A. The inductance L_m is 1 mH.

Figure 5.8 plots the minimum value curve of the dead time as the input voltage and output power vary. When $\varphi < 2\pi(D - 0.5)$, ZVS for output side switches is more difficult to be achieved than that under the situation where $\varphi > 2\pi(D - 0.5)$. So the dead time t_d is designed to be fixed as 250 ns to realize ZVS for output side switches within the full range of the input voltage and the load.

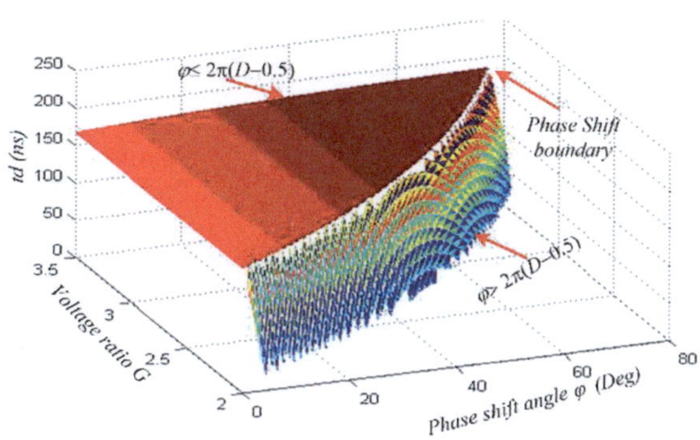

Fig. 5.8 Minimum dead time t_d versus φ and voltage ratio G

5.3.3 ZVS Condition for Primary Side Switches

Since it is easier for the primary side upper switches Q_{1a} and Q_{2a} to achieve ZVS than lower switches Q_1 and Q_2, the key is to obtain ZVS for primary side lower switches.

Figure 5.9 illustrates the typical soft switching condition for Q_1. I_{Lr} represents the peak value of i_{Lr} while $i_{L1\min}$ is the minimum value of the input DC side inductor current i_{L1}. They are expressed as follows,

$$\begin{cases} I_{Lr} = \dfrac{V_c(1-D)T_s}{L_r} - \sqrt{\left(\dfrac{V_c(1-D)T_s}{L_r}\right)^2 - \dfrac{P_oT_s}{L_r}} \\ i_{L1\min} = \dfrac{P_o}{2V_c(1-D)} - \dfrac{V_cD(1-D)T_s}{2L} \end{cases} \tag{5.16}$$

Achieving ZVS for Q_1 implies that the difference between $i_{L1\min}$ and i_{Lr} must be able to discharge its parasitic capacitor voltage to zero within the dead time t_d. This can be classified into two cases. One case is under heavy load situation where $I_{Lr} > V_ct_d/L_r$. The ZVS condition in this case can be described as,

$$I_{Lr} - \frac{V_c}{2L_r}t_d - i_{L1\min} > 2C_{\mathrm{ossL}}\frac{V_c}{t_d} \tag{5.17}$$

Another case is under light load situation where $I_{Lr} < V_ct_d/L_r$. Achieving ZVS in this case means

$$\frac{L_r}{2V_ct_d}I_{Lr}^2 - i_{L1\min} > 2C_{\mathrm{ossL}}\frac{V_c}{t_d} \tag{5.18}$$

where C_{ossL} is the equivalent output junction capacitance of Q_1, the boundary between (5.17) and (5.18) is $I_{Lr} = V_ct_d/L_r$, when the load is so light that I_{Lr} is less than this

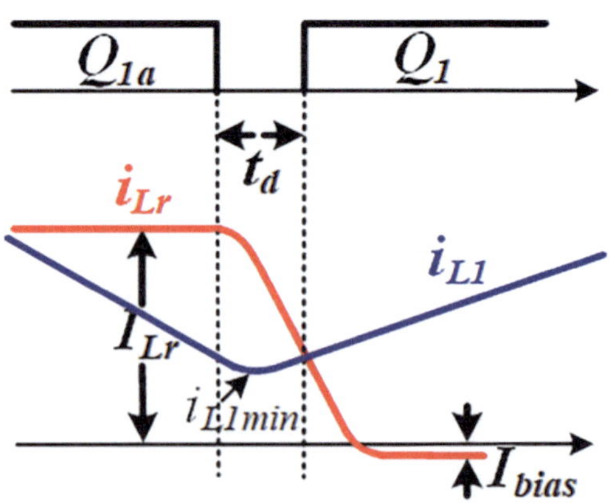

Fig. 5.9 Soft switching condition for input side switch Q_1

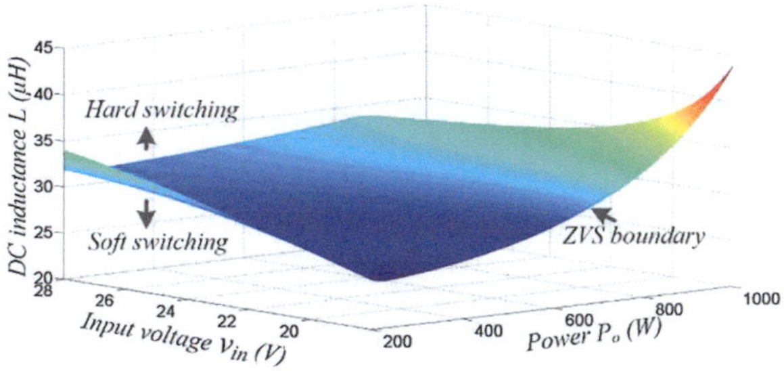

Fig. 5.10 DC inductance value versus V_{in} and P_{o}

value, the leakage current i_{Lr} decays to zero during the dead time and remains at zero. In order to obtain ZVS for Q_1, the value of the DC inductance L should be designed as the following,

$$\begin{cases} L < \dfrac{V_c D(1-D)T_s}{2\times\left[2C_{\text{ossL}}\frac{V_c}{t_d} - I_{Lr} + \frac{V_c}{2L_r}t_d + \frac{P_o}{2V_c(1-D)}\right]} & : I_{Lr} > V_c t_d/L_r \\ L < \dfrac{V_c D(1-D)T_s}{2\times\left[2C_{\text{ossL}}\frac{V_c}{t_d} - \frac{L_r}{2V_c t_d}I_{Lr}^2 + \frac{P_o}{2V_c(1-D)}\right]} & : I_{Lr} \le V_c t_d/L_r \end{cases} \tag{5.19}$$

Figure 5.10 indicates that ZVS can be achieved for all the input side bottom switches as the DC inductance value is less than the boundary value. As seen, it is the most difficult to achieve ZVS when $V_{\text{in}} = 18$ V at no load condition. Thus, in view of the ZVS condition and the input current ripple reduction, the DC inductance value is selected as $L_1 = L_2 = 22$ μH.

5.3.4 *Optimal Design of* L$_r$ *to Minimize Circulation Loss*

The circulation current can be optimized by designing a proper value of L_r according to the power circuit specification. The optimal design is given as follows.

Figure 5.11 illustrates the typical waveforms of circulation stage. When $\varphi < 2\pi(D - 0.5)$, the leakage inductance current can be divided into two parts. One is the bias current I_{bias} and the other is the current transition stage where i_{Lr} increases or decreases linearly. The target is to keep both low values of the bias current I_{bias} and the length of the current transition stage t_{up}. The peak value of the magnetizing current I_{Lm} is

$$I_{Lm} = \frac{V_c(1-D)T_s}{2L_m} \tag{5.20}$$

As seen in (5.18), as L_r increases, I_{bias} decreases. At the lowest input voltage, shown in Table 5.1, $D = D_{\max} = 0.7$, $V_c = nV_o = 60$ V, $C_{\text{ossH}} = 639$ pF, $V_o = 300$ V.

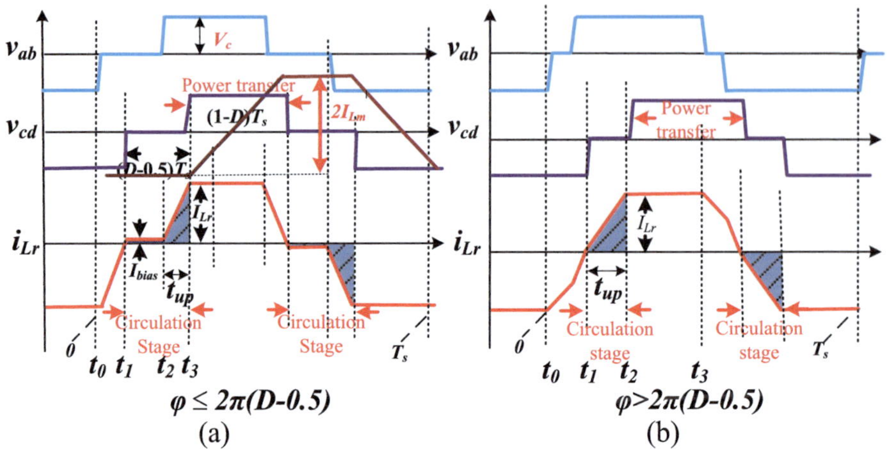

Fig. 5.11 Circulation current minimization curve

The magnetizing inductance L_m is supposed to be 1 mH. If $I_{\text{bias}} < 1.5$ A, according to (5.9) and (5.20), the leakage inductance must be designed as follows

$$L_r > 1.597\,\mu\text{H} \tag{5.21}$$

But the current transition length t_{up} is given by

$$t_{\text{up}} = \frac{L_r I_{Lr}}{V_c} \tag{5.22}$$

At maximum load, when the input voltage is the minimum, the duty cycle D is maximum. The leakage inductance current peak value I_{Lr} is maximum as well. Compared with the power transfer stage length, the current transition length can be neglected. Thus P_{o} can be expressed as,

$$P_{\text{o}} = 2I_{Lr} \times V_c \times (1 - D_{\max}) \times \eta \tag{5.23}$$

where η is the conversion efficiency. Substituting the expression of I_{Lr} obtained by (5.23) into (5.22) leads to

$$t_{\text{up}} = \frac{L_r P_{\text{o}}}{2V_c^2(1 - D_{\max})\eta} \tag{5.24}$$

If t_{up} is designed as follows, $t_{\text{up}} < 5\%T_s = 1\ \mu\sigma$, and substituting the circuit specifications where $V_{\text{inmin}} = 18$ V, $D_{\max} = 0.7$, $\eta = 0.97$, $V_c = 60$ V, $P_{\text{o}} = 1000$ W into (5.24), the leakage inductance must be designed as,

$$L_r < 2.095\,\mu\text{H} \tag{5.25}$$

According to (5.21) and (5.25), the leakage inductance can be designed within the range shown as follows

$$1.597\,\mu\text{H} < L_r < 2.09\,\mu\text{H} \tag{5.26}$$

When $\varphi > 2\pi(D - 0.5)$ shown in Fig. 5.11b, as L_r decreases, the length t_{up} decreases as well. This reduces the circulation loss. As seen, by selecting L_r properly, the circulation loss can be optimized to make a trade-off between the bias current loss and current transition loss. In this paper, the value of L_r is designed to be 1.8 μH.

5.4 Implementation of the Control Strategy

The control block diagram is shown in Fig. 5.12. All the control part is implemented by TI-28335 DSP. The PWM control loop is designed to regulate the clamp voltage V_c according to V_o and the turns-ratio n. For the phase shift control loop, the outer loop is used to stabilize the output voltage V_o, the output of the outer loop works as the reference for the inner current loop while the inner current loop output is used to generate the phase shift φ. Since the duty cycle D is calculated by the PWM control loop, its value is determined by V_{in} and V_c.

For improving the reliability of the converter, the protection scheme is necessary. As shown in Fig. 5.12, the protection scheme includes the over-current, over and under-voltage protection for the input side while over-voltage and short-circuit protection for the output side. Once either protection occurs, the DSP disables all the PWM output by the logic signal PS.

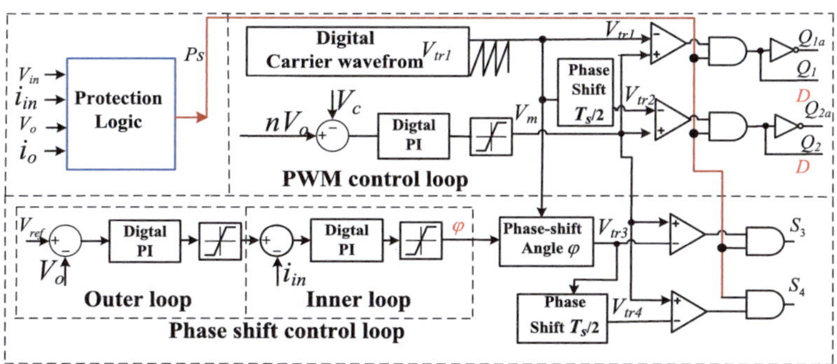

Fig. 5.12 Block diagram of the control strategy using digital control

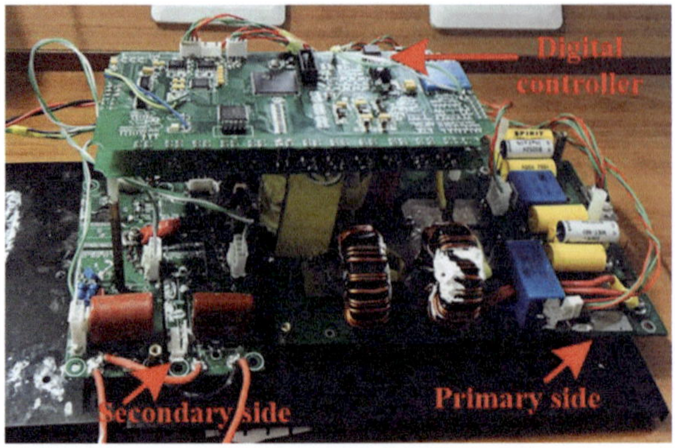

Fig. 5.13 Prototype picture

5.5 Experimental Results

5.5.1 Prototype

A 1 kW prototype controlled by a TM320F28335 digital signal processor is built to verify the theoretical analysis of the proposed converter. The system specifications are listed in Table 5.1 and the power stage elements are the same as those in Sect. 5.3.

Figure 5.13 illustrates the prototype picture, it should be noted that since the current stress can be suppressed in spite of the input voltage and load variation with the proposed control, the equivalent leakage inductance L_r can be designed rather small and no additional series inductance is needed any more.

5.5.2 Experimental Waveforms

Figure 5.14 shows the steady state waveforms of the S-DAB converter. With the proposed control strategy, the leakage current has no current spike. During the circulation stage, the bias current is rather small and the current transition stage length is also short. The circulation loss is reduced within the circulating stage.

As seen in Fig. 5.15, the input current ripple can be suppressed for the current-fed S-DAB by using the interleaving PWM technology.

Figure 5.16 shows the switching waveforms for the Q_1 with different output powers when $V_{\text{in}} = 24$ V. As seen, Q_1 can be turned on under ZVS.

Figure 5.17 gives the soft switching waveforms for the output side switch S_3. As seen, ZVS can be obtained for the output switches within the wide load range.

As shown in Fig. 5.18, ZCS of the output diodes can be obtained naturally with wide load range, thereby the reverse recovery issue can be avoided and there is no reverse recovery loss. No RC snubber circuit is needed here. Besides, for the rectifying diodes, their voltage stress is the just the output voltage in spite of the

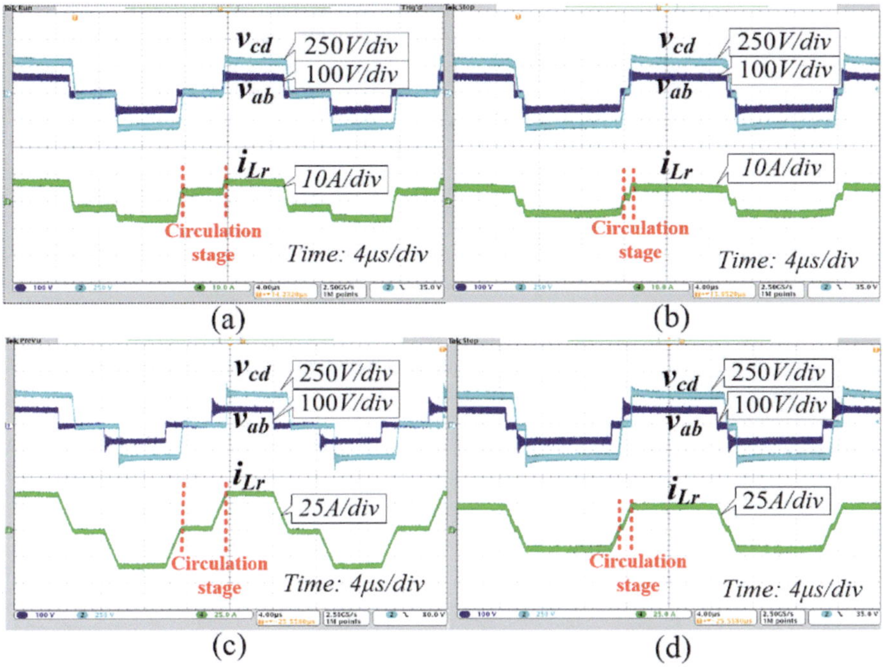

Fig. 5.14 Experimental waveforms of v_{ab}, v_{cd} and i_{Lr} **a** $V_{\text{in}} = 18$ V, $P_{\text{o}} = 200$ W, **b** $V_{\text{in}} = 28$ V, $P_{\text{o}} = 200$ W, **c** $V_{\text{in}} = 18$ V, $P_{\text{o}} = 1000$ W, **d** $V_{\text{in}} = 28$ V, $P_{\text{o}} = 1000$ W

Fig. 5.15 Input current ripple

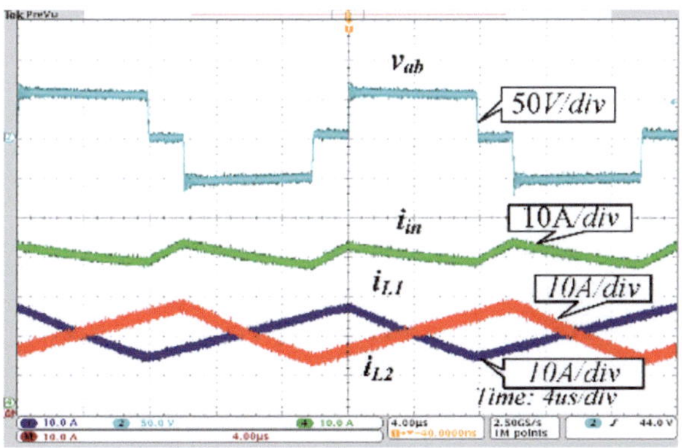

variation of the input voltage. Since the output voltage is 300 V. Not very high voltage rating fast recovery diode can be used here. For diodes, their forward voltage drop decreases as their voltage rating decreases. This causes the reduction of the rectifying conduction loss.

Efficiency curves of the proposed converter are given in Fig. 5.19 for input voltages of 18, 24 and 28 V. At rated load for nominal 24 V input, the efficiency is over 96%.

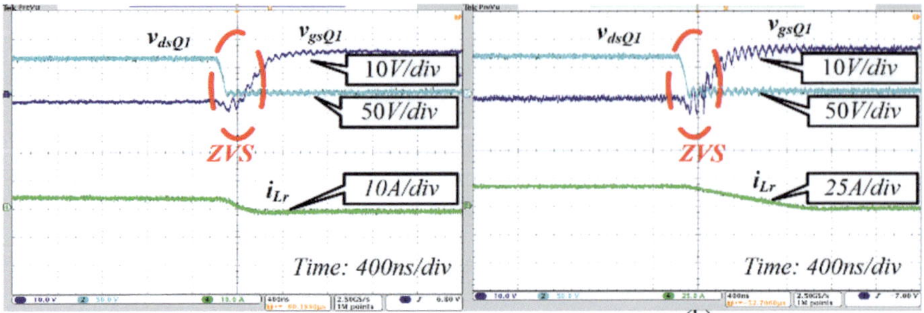

(a) (b)

Fig. 5.16 Switching waveforms of Q_1 at 24 V input with different output power **a** $P_o = 200$ W, **b** $P_o = 1000$ W

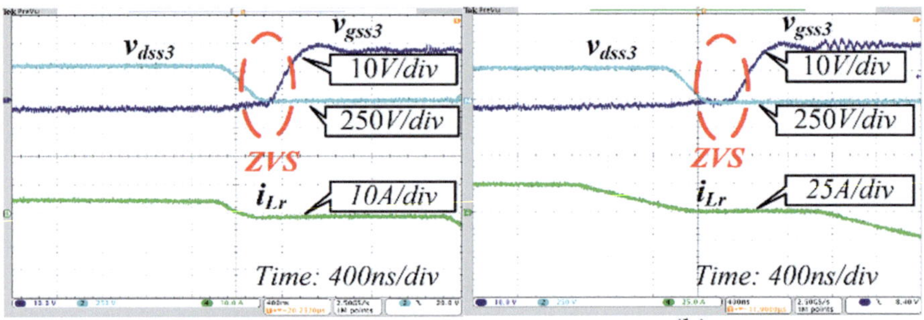

(a) (b)

Fig. 5.17 Switching waveforms of S_3 at 24 V input with different output power **a** $P_o = 200$ W, **b** $P_o = 1000$ W

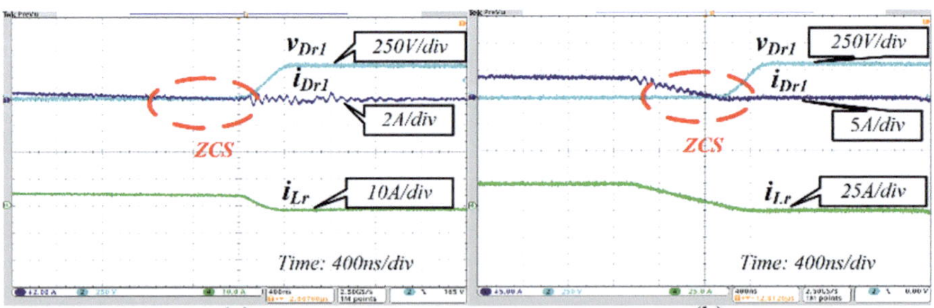

(a) (b)

Fig. 5.18 Switching waveforms of D_{r1} at 24 V input with different output power **a** $P_o = 200$ W, **b** $P_o = 1000$ W

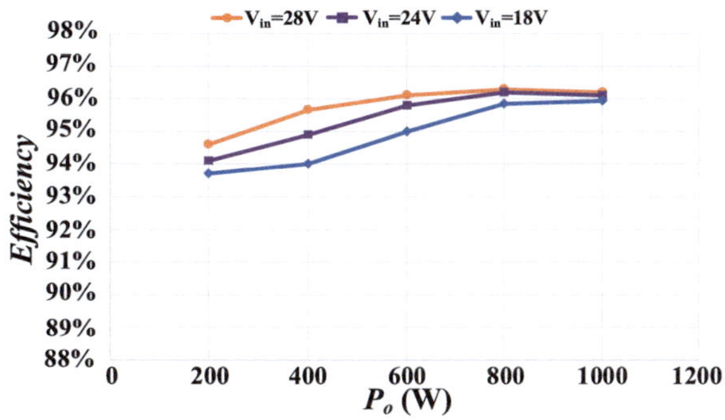

Fig. 5.19 Measured efficiency curves for different input voltage

In comparison, by using a boost type full bridge DC-DC converter [8], current fed ZVS isolated boost converter [9] and PPS control current-fed DAB [4], the efficiency at 100 W output is less than 88% despite the input voltage. At rated load (1000 W), the efficiency of the aforementioned three converters at 24 V input is less than 94%. With the proposed converter employing the proposed modulation strategy, the overall efficiency under light load and heavy load is high.

5.6 Conclusion

As a solution for unidirectional power flow applications where the input voltage is wide and low input current ripples are required, a current-fed S-DAB and associated symmetrical dual PWM and phase shifted control strategy has been proposed. The leakage inductance of the high frequency transformer can be used to work as the power transmission agent. The peak value of the leakage inductance current can be suppressed in spite of the input voltage and the load variation. To achieve ZVS for all the power switches, the key parameter optimization including the DC inductance design and the dead time are analyzed in detail. With the proposed system parameter optimization, all the power switches can achieve ZVS while the secondary side rectifying diodes obtain ZCS naturally within wide load range. Besides, the circulating loss can be reduced dramatically by optimal design of transformer leakage inductance. The overall efficiency is high even at light load. The effectiveness of the proposed converter, the associated control strategy and the parameter optimization have been validated by simulation and experimental results.

References

1. Kulasekaran S, Ayyanar R (2014) Analysis, design, and experimental results of the semi dual-active-bridge converter. IEEE Trans Power Electron 29(10):5136–5147
2. Peng F, Li H, Su G (2004) A new ZVS bi-directional dc–dc converter for fuel cell and battery applications. IEEE Trans Power Electron 19(1):54–65

3. Xu D, Zhao C, Fan H (2004) A PWM plus phase-shift control bidirectional dc–dc converter. IEEE Trans Power Electron 19(3):666–675
4. Xiao H, Guo L, Xie S (2007) A ZVS bidirectional dc-dc converter with phase-shift plus PWM control scheme. In: Proceedings of the IEEE applied power electron conference, pp 943–948
5. Li W, Wu H, Yu H (2011) Isolated winding-coupled bidirectional ZVS converter with PWM plus phase-shift (PPS) control strategy. IEEE Trans Power Electron 26(12):3560–3570
6. Shi Y, Li R, Xue Y (2015) Optimized operation of current-fed dual active bridge DC-DC converter for PV applications. IEEE Trans Ind Electron 62(5):3316–3324
7. Wang Z, Li H (2012) A soft switching three-phase current-fed bidirectional DC-DC converter with high efficiency over a wide input voltage range. IEEE Trans Power Electron 27(2):669–684
8. Chen R-Y, Liang T-J, Chen J-F (2008) Study and implementation of a current-fed full-bridge boost DC–DC converter with zero-current switching for high-voltage applications. IEEE Trans Ind Appl 44(4):1218–1226
9. Kim H, Yoon C, Choi S (2010) An improved current-fed ZVS isolated boost converter for fuel cell applications. IEEE Trans Power Electron 25(9):2357–2364

Chapter 6
Current-Fed Single Phase Semi-DAB DC-DC Converter with Optimized Hybrid Control

Abstract In Chap. 5, a current-fed semi dual active bridge (CF-SDAB) DC-DC converter with symmetrical dual PWM and phase shifted control is discussed, which can extend ZVS range of switches with voltage match control strategy, but the switches turn off at very high current, so high turn-off loss is increased especially as the input voltage is low. In addition, the clamp voltage is controlled to be constant and the LVS duty cycle range is limited. This brings unstable operation issue especially under very light load conditions. To deal with the aforementioned issues, this chapter proposes a variable clamp voltage control strategy for the current-fed semi-dual active bridge dc-dc converter. For the low-voltage-side (LVS), the duty cycle of the bottom switches will be changed with the load variation. Full load operation stability and full range of ZVS can be achieved. An instantaneous turn-off current tracking loop is designed to achieve duty cycle control and ZVS. The dc inductor current ripple can be designed smaller and the turn-off loss can be reduced. For the high-voltage-side (HVS), the duty cycle will be calculated to avoid additional loss caused by body diodes. Besides, coupled inductor is adopted to reduce core loss. The ZVS resonant process is analyzed in detail. The proposed control is verified to be effective by a 1 kW experimental prototype.

Keywords Variable clamp voltage · Current-fed · Semi dual active bridge · Couple inductor

6.1 Proposed Hybrid Control Strategy

For DAB or semi-DAB topology, when the voltage is far away from the matched value, the ZVS range is limited and the peak current of leakage inductor will be relatively high [1, 2]. What is more, the current ripple becomes high especially at low input voltages. CF-DAB has been used due to its unique advantages [3, 4]. For unidirectional power flow conditions, a current-fed semi-dual active bridge (CF-SDAB) topology with symmetrical dual-pulse width modulation and phase-shift control is proposed [5, 6]. Compared with bidirectional CF-DAB topology, the overall

D. Sha et al., *High Frequency Isolated Semi Dual Active Bridge DC-DC and AC-DC Converter*, CPSS Power Electronics Series,
https://doi.org/10.1007/978-981-95-1641-4_6

cost can be reduced and the shoot through issue on HVS is avoided. All the switches can achieve ZVS and the rectifying diodes can achieve ZCS naturally. A clamp voltage closed loop is implemented by using PWM control in order to match the output voltage. The dc inductor current ripple has to be designed large enough to achieve ZVS. With voltage match control strategy, the target is to achieve ZVS but the cost is neglected. Switches are turned on with ZVS but turned off at very high current. This causes high turn-off loss especially as the input voltage is low. The duty cycle of the LVS is determined by both side voltages and it is independent of the load. What is more, the clamp voltage is controlled to be constant and the LVS duty cycle range is limited. This leads to the unstable operation issue under very light load conditions.

For CF-SDAB topology, this paper proposes an optimized hybrid control strategy. Under very light load, output voltage operation stability is the first concern and thus the duty cycle is determined by the output voltage regulation and the clamp voltage becomes very low. This reduces the transmitted power and thus the light load operation can be stable. At medium or heavy load conditions, an instantaneous current tracking loop is used to determine the clamp voltage to make sure ZVS can be ensured with less loss. The clamp voltage value varies automatically as the input voltage and the output power.

6.1.1 Topology and Operation Principle

The converter topology is shown in Fig. 6.1. There are two interleaved boost circuits with a coupled inductor in the LVS. The coupled inductor is composed of two windings with the same turns so that the self-inductance of the two inductors, L_1 and L_2, are identical. In order to reduce current ripple and magnetic flux ripple, inverse coupled inductor is adopted and the mutual inductance value is negative.

The clamp voltage V_c depends on the input voltage V_{in} and the duty cycle of the bottom switches Q_1 and Q_2. In the HVS, the top switches of the full bridge are replaced by two diodes D_{r1} and D_{r2}, which define the power flow as unidirectional. The LVS and the HVS are connected with a high-frequency transformer, T. L_m and n are the magnetic inductance and turns ratio of the transformer, respectively. L_r is the combination of the leakage inductance and the external inductance.

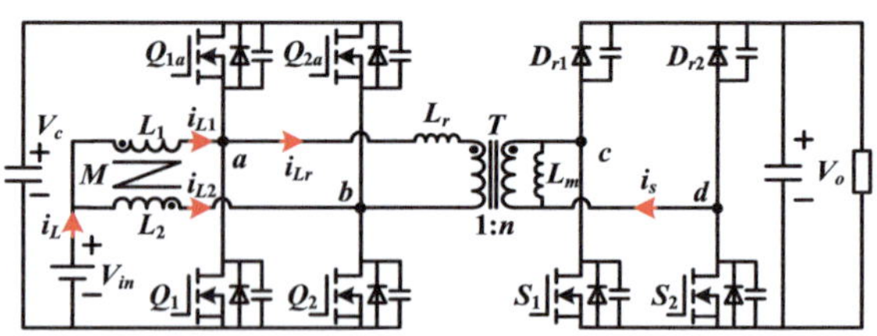

Fig. 6.1 CF-SDAB dc-dc converter with couple inductor

The steady-state waveforms with the proposed control strategy are shown in Fig. 6.2. D_1 and D_2 are the duty cycle of the LVS bottom switches and the HVS switches, respectively. d_1 is the effective duty cycle of v_{ab}. The phase-shift angle φ is defined as the ratio between the rising edge of v_{ab} from 0 to V_c and the rising edge of v_{cd} from 0 to V_o. The leakage inductor current expressions at different times are given in Table 6.1. According to the waveforms of the leakage inductor current, the operation principles can be divided into three modes. In Mode 1, both the clamp voltage and the duty cycle are very low, the converter operates under light load conditions. During t_1–t_2, the leakage inductor current i_{Lr} is decreased to 0. Then the output junction capacitor resonates with L_r until v_{ab} drops to 0. In Mode 2, i_{Lr} will not reach 0 during t_1–t_2. Unlike from Mode 1 and Mode 2, there is overlapping time when both v_{ab} and v_{cd} are 0 in Mode 3.

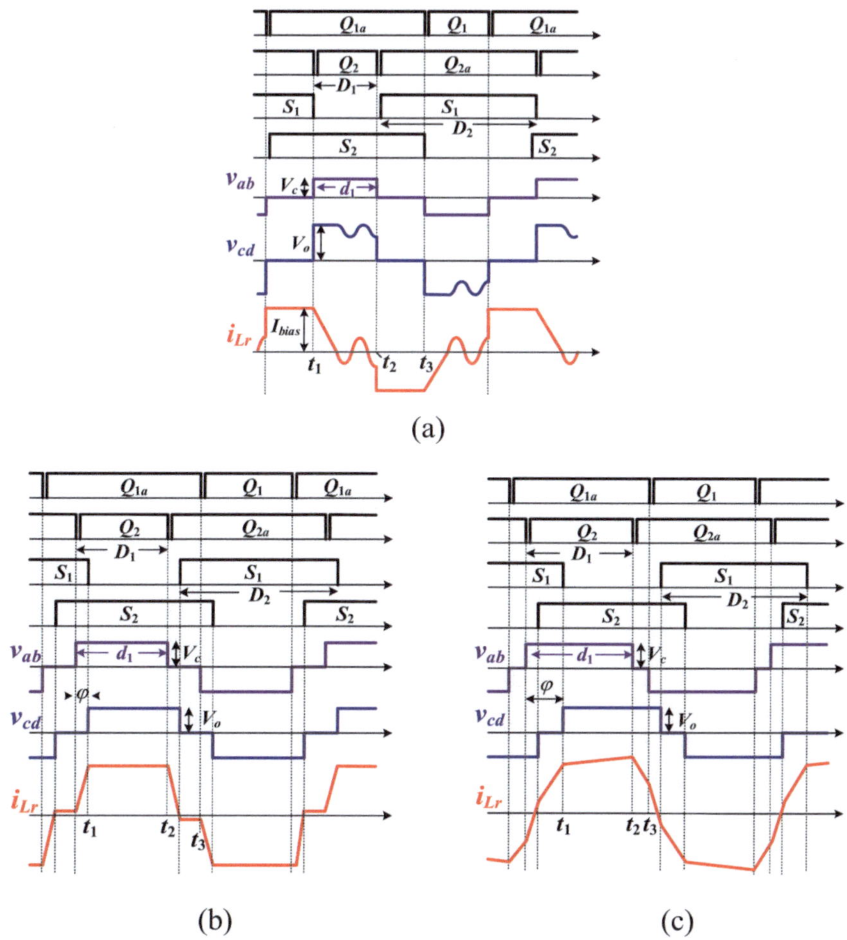

Fig. 6.2 Steady-state waveforms with the proposed control. **a** Mode 1. **b** Mode 2. **c** Mode 3

Table 6.1 Expressions of leakage inductor current

	Mode 1	Mode 2	Mode 3
$i_{Lr}(t_1)$	I_{bias}	$\frac{V_c\varphi}{L_r}T$	$\frac{V_c(2\varphi+0.5-d_1)}{3L_r}T$
$i_{Lr}(t_2)$	$-I_{\text{bias}}$	$\frac{V_cd_1-V_sd_1+V_s\varphi}{L_r}T$	$\frac{V_c(\varphi+d_1-0.5)+V_s(1-2d_1)}{3L_r}T$
$i_{Lr}(t_3)$	$-I_{\text{bias}}$	$-I_{\text{bias}}$	$\frac{(V_c+V_s)(\varphi+d_1-0.5)}{3L_r}T$

According to the volt-second balance, the duty cycle of the HVS is given in (6.1), in which $V_s = V_o/n$.

$$D_2 = \begin{cases} 1 - d_1 \text{ , Mode 1 and Mode 2} \\ 0.5 + \varphi - \dfrac{V_s\varphi + V_c d_1 - 0.5V_s}{2V_c + V_s} \text{ , Mode 3} \end{cases} \tag{6.1}$$

6.1.2 ZVS Conditions Analysis

The ZVS of LVS switches depends on the difference between the dc inductor current and the leakage inductor current. The dead time should be designed properly to reduce the conduction loss. Take Q_{1a} and Q_1 for example, the current flows through Q_1 when the ZVS resonant process begins is larger than Q_{1a}. Therefore, it is easier to achieve ZVS for the top switch Q_{1a} than the bottom switch Q_1. The soft switching process of Q_1 is analyzed in detail.

The ZVS of LVS switches depends on the difference between the dc inductor current and the leakage inductor current. The dead time should be designed properly to reduce the conduction loss. Take Q_{1a} and Q_1 for example, the current flows through Q_1 when the ZVS resonant process begins is larger than Q_{1a}. Therefore, it is easier to achieve ZVS for the top switch Q_{1a} than the bottom switch Q_1. The soft switching process of Q_1 is analyzed in detail.

Figure 6.3 shows the equivalent circuit during the ZVS resonant process of Q_1. Under very light load, the dc inductor current is negative enough and ZVS condition is not a concern. Under heavy load, the leakage inductor current is large enough so the ZVS is easy to be achieved, too. It is most difficult to achieve ZVS at the boundary between Mode 2 and Mode 3. v_{ds1} and v_{ds2} are the drain-source voltage of Q_1 and Q_{1a}, respectively. The circuit is so complicated that it's difficult to get the expression v_{ds1} during the resonant process. Reference [7] provides a charge-based ZVS analysis method of the bridge leg switches during commutation. As shown in Fig. 6.4, k_1 and k_2 represent the absolute value of the slope of the leakage inductor current when the resonant process begins and finishes, respectively. The slope of the leakage inductor current at different modes is given in Table 6.2. The waveforms of v_{ds1} and v_{ds2} intersect at $V_c/2$. The total charge needed to complete the commutation of a bridge leg, Q_{leg}, is divided into two equal parts by the intersection. For the sake of calculation, the leakage current in the two parts is linearized. The current flowing through the bridge leg as the LVS upper switch turns off is defined as i_{off}. The variation of the dc inductor current is ignored during the resonant process. Based on the principle of charge balance,

$$i_{\text{off}} t_{r1} - \frac{k_1 t_{r1}^2}{2} = (i_{\text{off}} - k_1 t_{r1}) t_{r2} - \frac{k_2 t_{r2}^2}{2} = \frac{Q_{\text{leg}}}{2} \tag{6.2}$$

If $k_1 = 0$, the relationship between dead time and the current turn-off current i_{off} is given by,

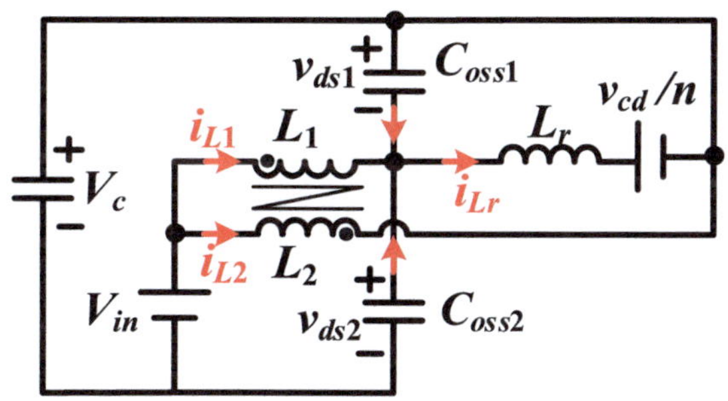

Fig. 6.3 The equivalent circuit during the ZVS resonant process of Q_1

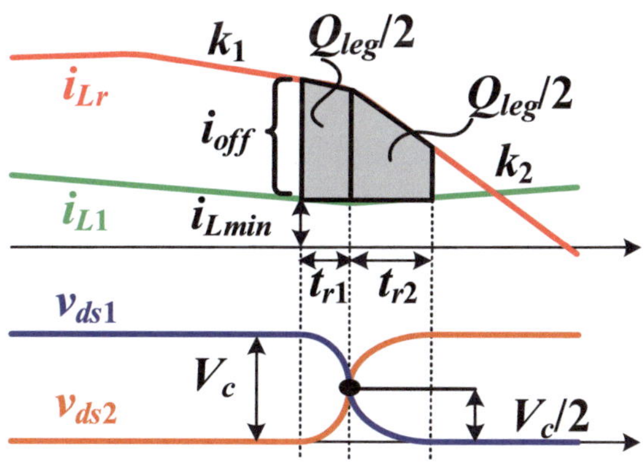

Fig. 6.4 The soft switching conditions for Q_1

Table 6.2 The slope expressions of the leakage inductor current

	$D_1 > 0.5$	$D_1 < 0.5$
Mode 1	/	$k_1 = 0, k_2 = \frac{V_c}{L_r}$
Mode 2	$k_1 = 0, k_2 = \frac{V_s}{L_r}$	$k_1 = 0, k_2 = \frac{V_c}{L_r}$
Mode 3	$k_1 = 0, k_2 = \frac{V_s}{L_r}$	$k_1 = \frac{V_s}{L_r}, k_2 = \frac{V_c+V_s}{L_r}$

$$t_d = \Delta t + \frac{Q_{\text{leg}}}{2i_{\text{off}}} + \frac{i_{\text{off}} - \sqrt{i_{\text{off}}^2 - k_2 Q_{\text{leg}}}}{k_2} \tag{6.3}$$

If $k_1 > 0$,

$$t_d = \Delta t + \frac{i_{\text{off}} - \sqrt{i_{\text{off}}^2 - k_1 Q_{\text{leg}}}}{k_1} + \frac{\sqrt{i_{\text{off}}^2 - k_1 Q_{\text{leg}}} - \sqrt{i_{\text{off}}^2 - k_1 Q_{\text{leg}} - k_2 Q_{\text{leg}}}}{k_2} \tag{6.4}$$

Δt is the time to compensate the difference between the rise time and fall time of the drive signal, based on the prototype. The minimum turn-off current $I_{\text{off\_min}}$ can be derived from (6.4),

$$I_{\text{off\_min}} = \sqrt{k_1 Q_{\text{leg}} + k_2 Q_{\text{leg}}} \tag{6.5}$$

Infineon OptiMOS IPT007N06N is used in the LVS, Q_{leg} is selected as 0.7 μC considering the nonlinearity of the junction capacitance and parasitic parameters. The dead time versus the clamp voltage and the turn-off current when $D_1 < 0.5$ is shown in Fig. 6.5.

Figure 6.6 depicts the ZVS resonant process of S_1. Once $i_s = 0$, the top diode will turn off with ZCS naturally. Then the output junction capacitor begins to be resonant with L_r. Even under very light load, a negative bias current will flow through S_1 after the resonance. The drain-source voltage will be clamped by the body diode and then ZVS can be achieved. The resonant process can be described by the following differential equations,

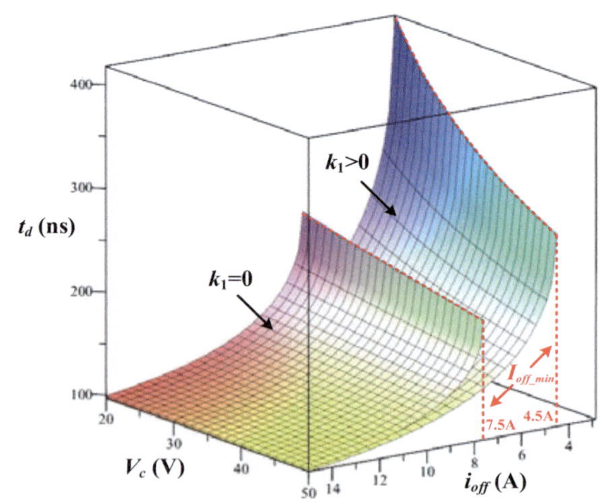

Fig. 6.5 The dead time versus the clamp voltage and the turn-off current when $D_1 < 0.5$

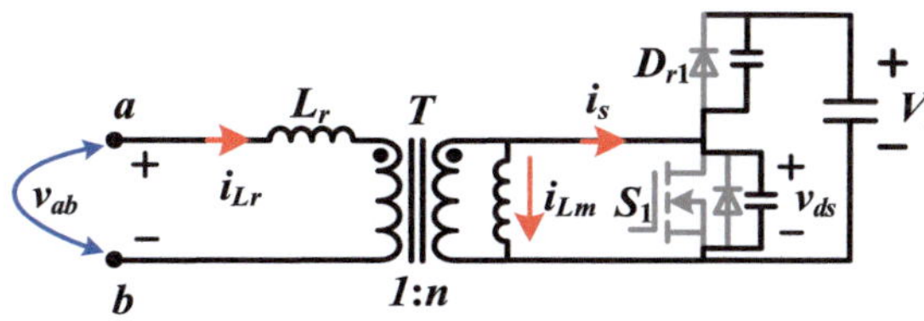

Fig. 6.6 The equivalent circuit during the resonant process of S_1

$$\begin{cases} v_{ab} - nv_{ds}(t) = L_r \dfrac{\mathrm{d}i_{Lr}(t)}{\mathrm{d}t} \\ i_{Lm} - ni_{Lr}(t) = C_{\mathrm{ossH}} \dfrac{\mathrm{d}v_{ds}(t)}{\mathrm{d}t} \end{cases} \tag{6.6}$$

in which $v_{ab} = 0$ in Mode 1 and Mode 2, $v_{ab} = V_c$ in Mode 3.

The initial values for the differential equations are

$$v_{ds}(0) = V_{\mathrm{o}},\ i_{Lr}(0) = ni_{Lm} \tag{6.7}$$

The expression of $v_{ds}(t)$ and $i_{Lr}(t)$ can be deduced from (6.6) and (6.7). In Mode 1 and Mode 2,

$$\begin{cases} v_{ds}(t) = V_{\mathrm{o}} \cos \omega t \\ i_{Lr}(t) = V_{\mathrm{o}} \sqrt{\dfrac{C_{\mathrm{ossH}}}{L_r}} \sin \omega t + ni_{Lm} \end{cases} \tag{6.8}$$

where $\omega = \frac{1}{n}\sqrt{\frac{1}{C_{\mathrm{ossH}} L_r}}$. After a quarter of a resonant period, v_{ds} drops to zero. The resonant time can be expressed as

$$t_r = \frac{\pi n}{2} \sqrt{C_{\mathrm{ossH}} L_r} \tag{6.9}$$

Similarly, in Mode 3,

$$t_r = n\sqrt{C_{\mathrm{ossH}} L_r} \cos^{-1}\left(\frac{V_c}{V_c + V_s}\right) \tag{6.10}$$

The resonant process ends when v_{ds2} drops to zero. After that, S_2 should be turned on as soon as possible to avoid additional loss caused by its body diode. Based on (6.9) and (6.10), the resonant time is 300 ns and 200 ns, respectively.

6.1.3 Implementation of Closed-Loop Control Strategy

The flowchart of the proposed closed-loop control strategy is shown in Fig. 6.7. The voltage tracking loop is composed of the outer voltage loop and inner current loop, while q is the output of the inner current loop and $q_1 = 1 - V_{in}/V_s$. For the outer voltage loop, the proportional and integration coefficients $k_{pv} = 0.8$, $k_{iv} = 0.01$. For the inner current loop, $k_{pi} = 0.1$, $k_{ii} = 0.01$. The parameters are determined by PSIM simulation and the stability is verified by PSIM frequency sweeping. Figure 6.8 shows the bode plot of the closed-loop control by PSIM frequency sweeping. Based on the gain curve and the phase curve, the phase margin is about 75°. Therefore, the simulation results show that the selected control parameters make the system stable.

An instantaneous turn-off current tracking loop is adopted to achieve ZVS of the LVS. The control parameters are: $k_p = 0.001$, $k_i = 0.0001$. The real-time turn-off

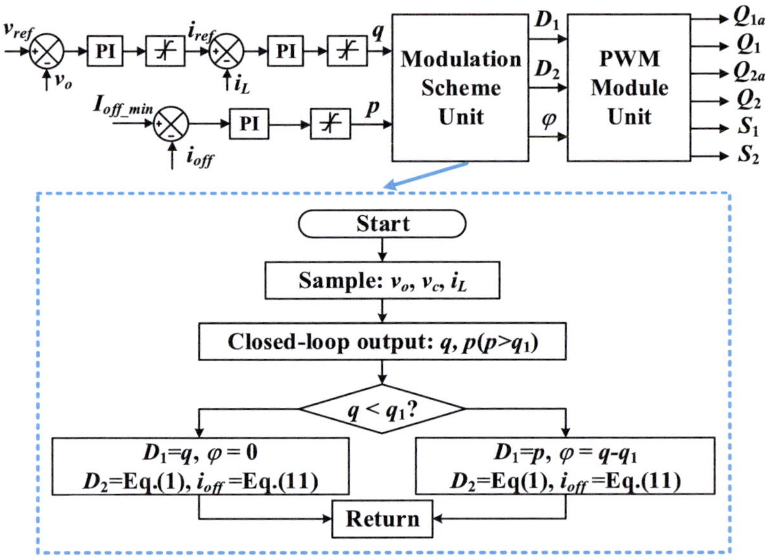

Fig. 6.7 The flowchart of the proposed closed-loop control

Fig. 6.8 Bode plot of the closed-loop control

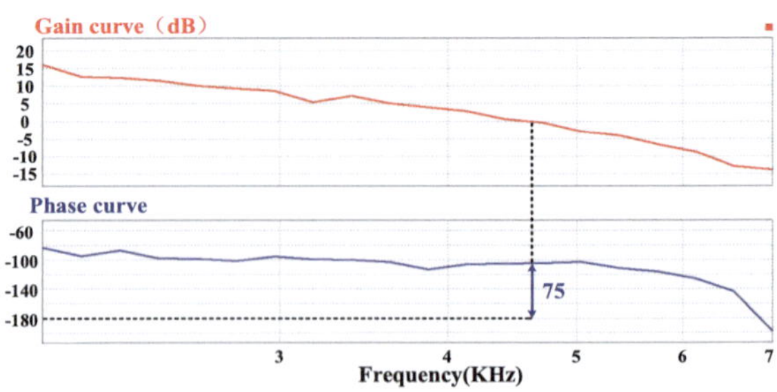

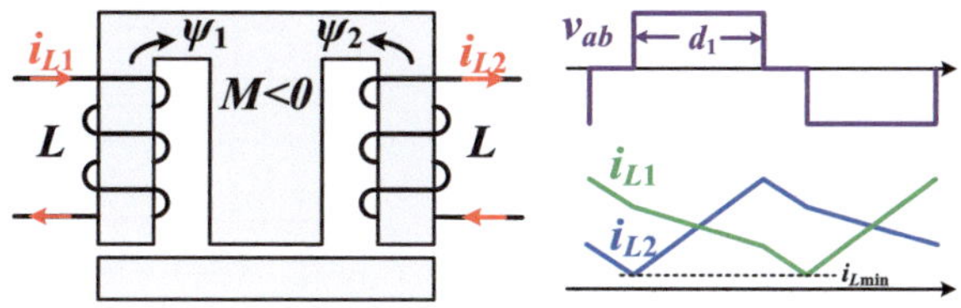

Fig. 6.9 The couple inductor model and its operating waveform

current, as the feedback of the closed-loop control, is calculated by DSP based on (6.11).

$$i_{\text{off}} = \begin{cases} i_{Lr}(t_3) - i_{L\min}, D_1 < 0.5 \\ i_{Lr}(t_2) - i_{L\min}, D_1 > 0.5 \end{cases} \tag{6.11}$$

Figure 6.9 shows the coupled inductor model and the waveforms of the coupled inductor current. The minimum value of the coupled inductor current can be given as,

$$i_{L\min} = I_L - (k_M V_c d_1^2 + k_L V_c d_1(1 - d_1)) \tag{6.12}$$

in which $k_M = \frac{TM}{2(L^2 - M^2)}$, $k_L = \frac{TL}{2(L^2 - M^2)}$. I_L is the average value of the input current, which can be obtained by the current sampling. L is the self-inductance of the couple inductor, M is the mutual inductance.

Under very light load conditions, the LVS duty cycle D_1 is determined by the output voltage tracking loop and the phase-shift angle is 0. As the load increases, the phase-shift control is to stabilize the output voltage V_o and D_1 will depend on the ZVS condition of the LVS switches. To avoid large leakage inductor current, the clamp voltage will be no less than the matched value even if $i_{\text{off}} > I_{\text{off\_min}}$, so $p > q_1$. If $i_{\text{off}} < I_{\text{off\_min}}$, i_{off} will be regulated by the turn-off current loop to track the reference value. With the proposed closed-loop control strategy, seamless transition among different modes can be achieved.

6.2 Comparison Between the Proposed Control and the Voltage Match Control Strategy

6.2.1 Light Load Stability

With voltage match control strategy in [5], the bias current of the leakage inductor is neglected when calculating the output power. Compared with the peak value of the leakage inductor current, the bias current is so small that can be neglected under heavy load conditions. However, the bias current cannot be neglected under very light load conditions. Figure 6.10 shows the simplified waveforms during a half period

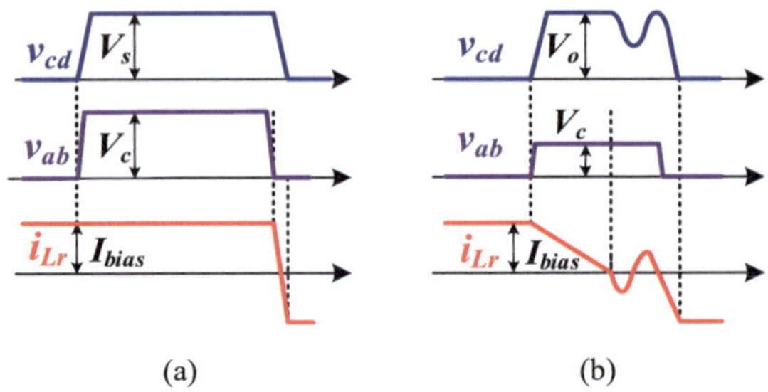

Fig. 6.10 Simplified waveforms when phase-shift angle is 0. **a** With voltage match control. **b** With the proposed control

under very light load. With voltage match control strategy, the converter delivers minimum power when the phase-shift angle reaches its minimum value, and the minimum output power $P_{\min}$ can be expressed as

$$P_{\min} = 2V_s I_{\text{bias}} d_1 \tag{6.13}$$

Due to the existence of the bias current, the minimum output voltage power $P_{\min}$ is always higher than 0. If the output power is lower than the minimum output power, the output voltage cannot be regulated to track the reference voltage. Figure 6.11a shows the simulation waveforms when the output power is 50 W with 400 V reference voltage. With the proposed control strategy, the converter works in Mode 1 and Mode 2 under very light load. Figure 6.11b shows the simulation waveforms with the proposed control strategy when the output power is 2 W (Consider the sampling loss). When φ reaches 0, the output power can be expressed as, in Mode 1,

$$P_1 = \frac{V_c I_{\text{bias}}^2 L_r}{(V_s - V_c)T} \tag{6.14}$$

in Mode 2,

$$P_2 = V_c(2I_{\text{bias}} d_1 - \frac{(V_s - V_c)d_1^2}{L_r} T) \tag{6.15}$$

Figure 6.12 shows the output power versus the clamp voltage and input voltage under very light load.

6.2.2 *The Cost of ZVS Achievement*

The ZVS achievement of the LVS switches depends on the leakage inductor current i_{Lr}, dc inductor current i_{L1} (or i_{L2}) and the dead time t_d. With voltage match control strategy, the dc inductance has to be designed small enough to achieve ZVS of the LVS bottom switches. It is most difficult for the LVS switches to achieve ZVS when

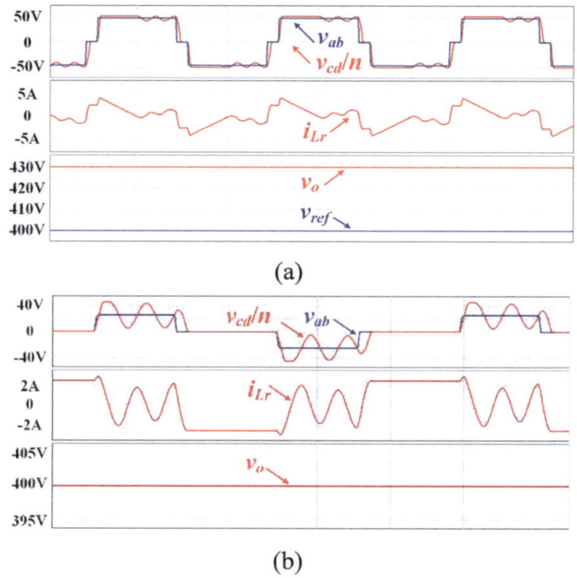

Fig. 6.11 Simulation waveforms under very light load conditions. **a** With voltage match control under 50 W. **b** With the proposed control under 2 W

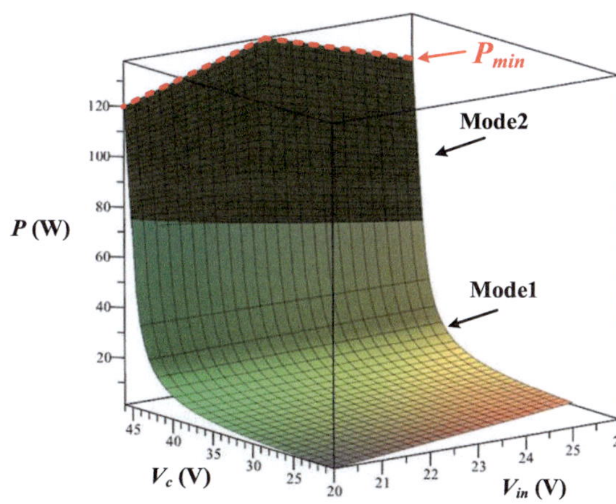

Fig. 6.12 Clamp voltage versus input voltage and output power under light load

the duty cycle is minimum. Figure 6.13 shows the soft switching conditions for Q_1 with voltage match control. t_s is the resonant time during the dead time. As shown in Fig. 6.13a, large current ripple is needed to achieve ZVS when $D_1 < 0.5$. In Fig. 6.13b, the ZVS condition is too sufficient when $D_1 > 0.5$. Though ZVS can be achieved, large current ripple, high turn-off loss and additional conduction loss are unnecessary.

The dc inductance is designed to achieve full range ZVS of the LVS switches. With the proposed control strategy, the dc inductance can be designed larger than the voltage match control strategy. The current flow through the LVS switches can be reduced. Figure 6.14 shows the turn-off current and RMS current of the LVS

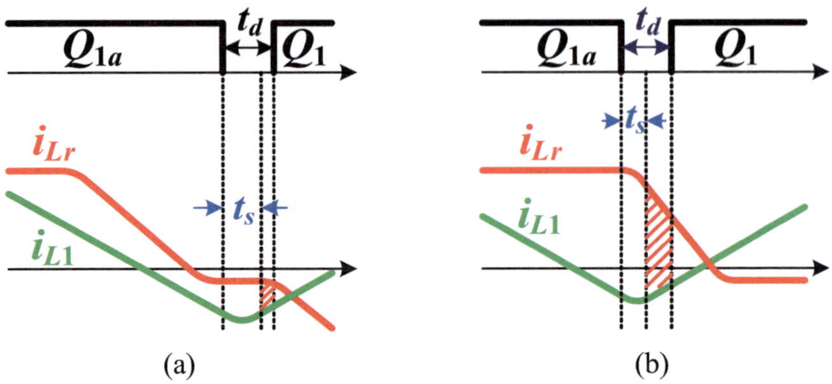

Fig. 6.13 Soft switching conditions for Q_1 with voltage match control. **a** $D < 0.5$. **b** $D > 0.5$

switches, which are obtained by simulation. Figure 6.15 shows the turn-off process of the switch, the turn-off loss can be expressed as,

$$P_{\text{off\_ loss}} = \frac{1}{T}\int_{0}^{t_f} v_{ds} i_{ds} \text{d}t \approx \frac{V_{ds} I_{ds} t_f}{6T} \tag{6.16}$$

The fall time t_f is 22 ns, the drain-source on-state resistance R_{ds} is 0.75 mΩ. The turn-off loss and the conduction loss of the LVS are shown in Fig. 6.16. With the proposed control strategy, the turn-off loss and conduction loss are reduced.

6.3 Experimental Results

6.3.1 Prototype

An experimental prototype employing TMS320F28335 DSP control has been fabricated. The prototype picture is shown in Fig. 6.17. The system specifications are listed in Table 6.3. To implement the closed-loop control, LV 25-P is selected as the voltage sensor and CHB-100A is selected as the current sensor.

6.3.2 Experimental Waveforms

Figure 6.18 shows the steady-state waveforms with the proposed control under different loads when $V_{\text{in}} = 20$ V. Figure 6.18a, b show the steady-state waveforms

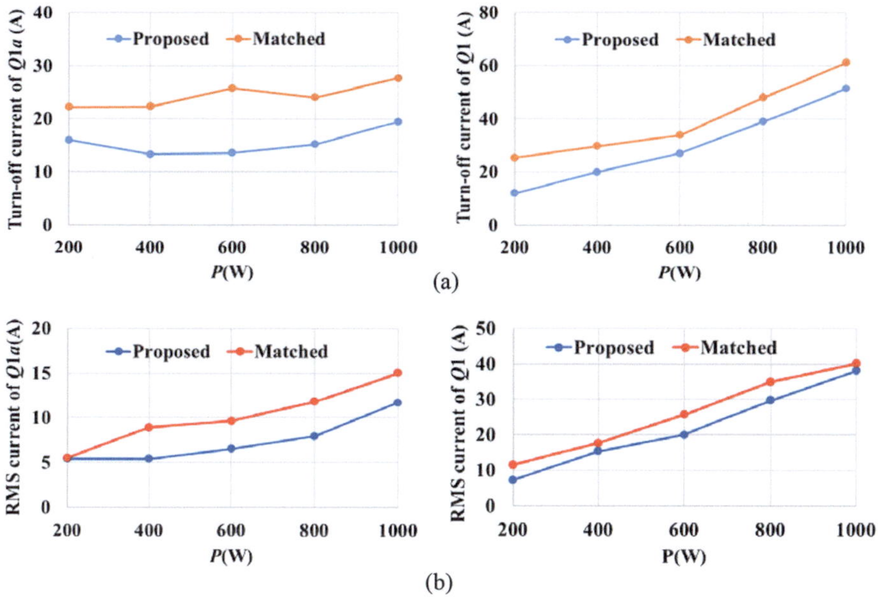

Fig. 6.14 Current comparison between the voltage match control and the proposed control strategy when $V_{\text{in}} = 20$ V

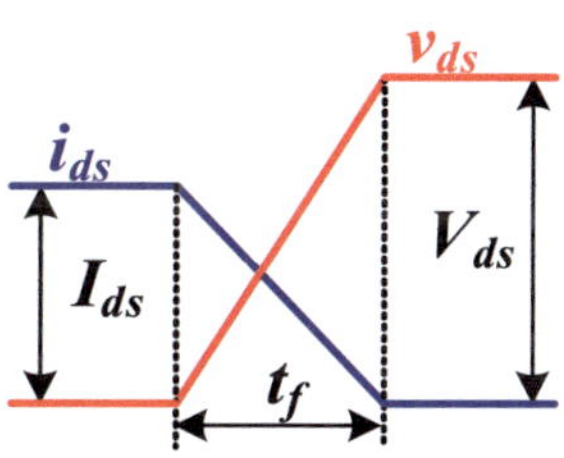

Fig. 6.15 Turn-off process

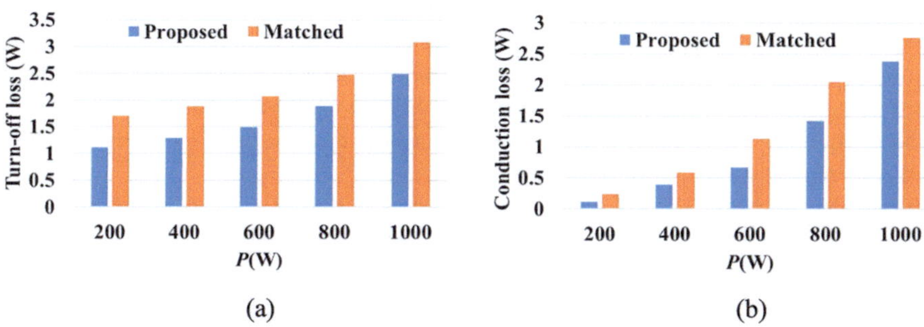

Fig. 6.16 Loss comparison of the LVS. **a** Turn-off loss. **b** Conduction loss

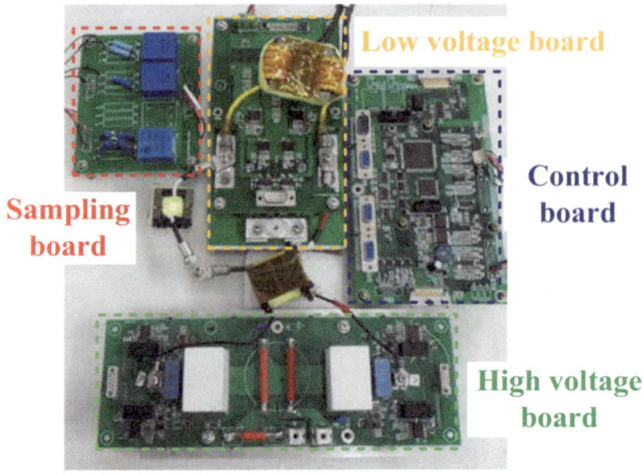

Fig. 6.17 The prototype of the CF-SDAB dc-dc converter

Table 6.3 System parameters

Symbol	Description	Value
V_{in}	Input voltage	20–26 V
V_o	Output voltage	400 V
L	Self-inductance	3.74 μH
L_r	Leakage inductance	1.72 μH
M	Mutual inductance	−0.76 μH
n	Turns ratio	35:4
P	Output power	1 kW
Q_{1a}–Q_2	LVS switches	IPT007N06N
S_1–S_2	HVS switches	C3M0045060J
D_{r1}–D_{r2}	Diodes	BYV30W-600PT2

under very light load. The LVS duty cycle D_1 is determined by the output voltage tracking loop and the phase-shift angle is 0. The converter can operate stably under light load. As the load increases, the clamp voltage will be increased to V_s. Then, the phase-shift angle will be increased as the load increases. Figure 6.18c and d show the steady-state waveforms in Mode 2 and Mode 3, respectively. The LVS duty cycle is larger than 50% and the turn-off current is enough to achieve ZVS. Therefore, the clamp voltage is matched with the output voltage.

Figure 6.19 shows the steady-state waveforms when $V_{in} = 26$ V. The turn-off current will be controlled by the instantaneous turn-off current tracking loop to meet the ZVS condition. The clamp voltage varies with the load automatically. When $P = 400$ W and $P = 600$ W, the clamp voltage exceeds the matched voltage so that i_{off} is big enough to achieve ZVS. When the load increases further, i_{off} is larger than I_{off_min} and ZVS can be achieved even though the clamp voltage matches with the output voltage.

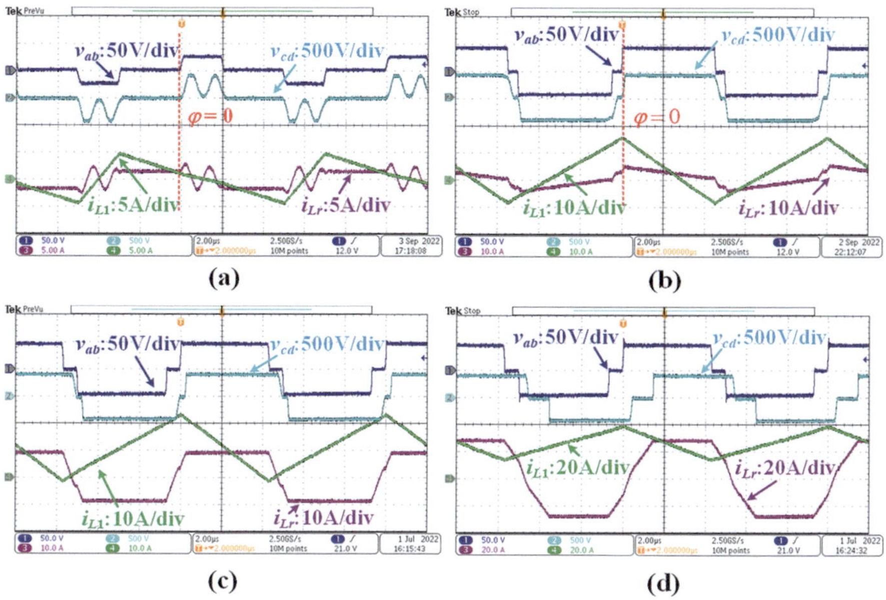

Fig. 6.18 Steady-state operating waveforms with the proposed control when $V_{\text{in}} = 20$ V. **a** When $P = 0$ W. **b** When $P = 100$ W. **c** When $P = 400$ W. **d** When $P = 1000$ W

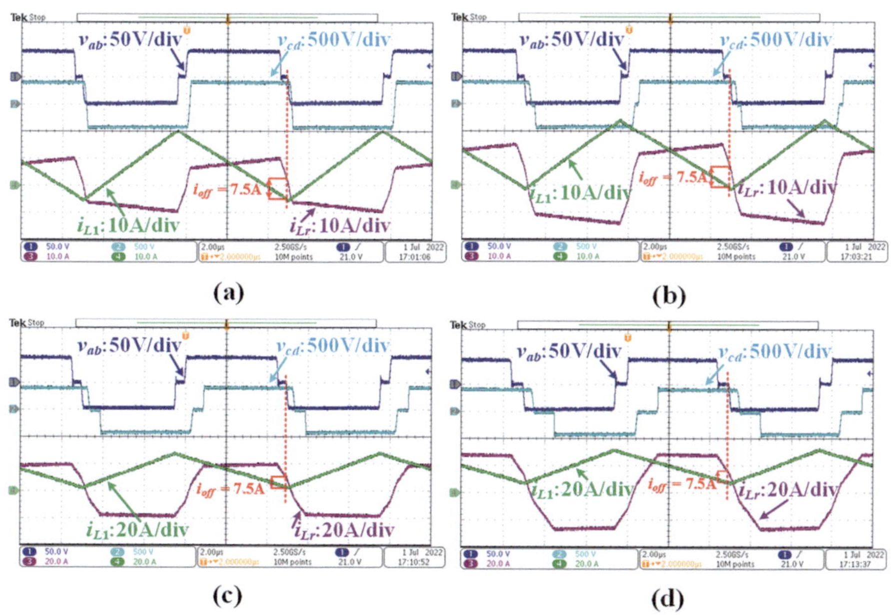

Fig. 6.19 Steady-state operating waveforms with the proposed control when $V_{\text{in}} = 26$ V. **a** $P =$ 400 W. **b** $P = 600$ W. **c** $P = 800$ W. **d** $P = 1000$ W

The ZVS conditions of Q_1 and S_2 under different loads are shown in Fig. 6.20. The voltage and current during the turn on process are demonstrated. The zero point of the switch current is artificially marked in Fig. 6.20. As seen in the figures, negative current flows through the switch before the turn-on gate-source voltage is imposed. The drain-source voltage decreases to 0 before the switch turns on, ZVS can be achieved in this way. The soft-switching performance will not be influenced during the wide load conditions. The light load ZVS achievement is shown in Fig. 6.21 when V_{in} is 20 V. As seen, even at no load conditions, ZVS for Q_1 under light load conditions can be ensured successfully.

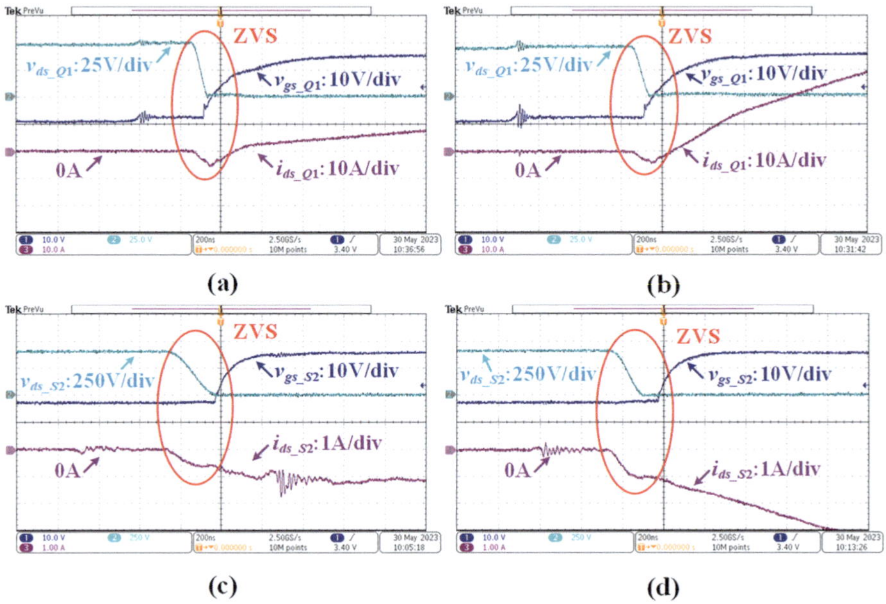

Fig. 6.20 The ZVS conditions. **a** For Q_1, $V_{\text{in}} = 26$ V, $P = 400$ W. **b** For Q_1, $V_{\text{in}} = 26$ V, $P = 1$ kW. **c** For S_2, $V_{\text{in}} = 20$ V, $P = 400$ W. **c** For S_2, $V_{\text{in}} = 20$ V, $P = 1$ kW

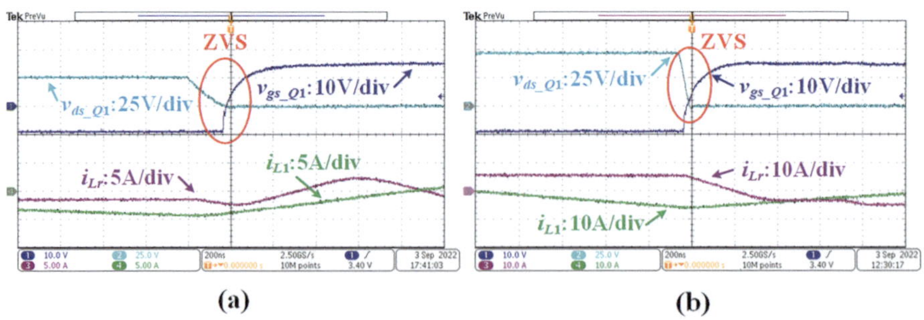

Fig. 6.21 ZVS conditions for Q_1 when $V_{\text{in}} = 20$ V. **a** $P = 0$ W. **b** $P = 200$ W

Figure 6.22 shows the ZVS conditions of S_2 when $V_{\text{in}} = 23$ V and $P = 1$ kW. With the matched control in [5], duty cycles of the LVS and the HVS are equal to be 05. After the drain-source voltage drops to 0, S_2 is turned on after a delay of 1200 ns in which the current flows through the body diode of S_2. This causes much conduction loss in its body diode. With the proposed control in this paper, the HVS duty cycle is calculated by DSP. In this case, $D_2 = 0.612$. The time that the current flowing through the body diode is reduced to 80 ns, the conduction loss over the body diode is significantly minimized.

Figure 6.23 shows the dynamic response of load transition with the proposed control strategy, and the input voltage is 26 V. It can be seen that the dynamic response is fast and the overshoot and the undershoot of the output voltage is relatively small.

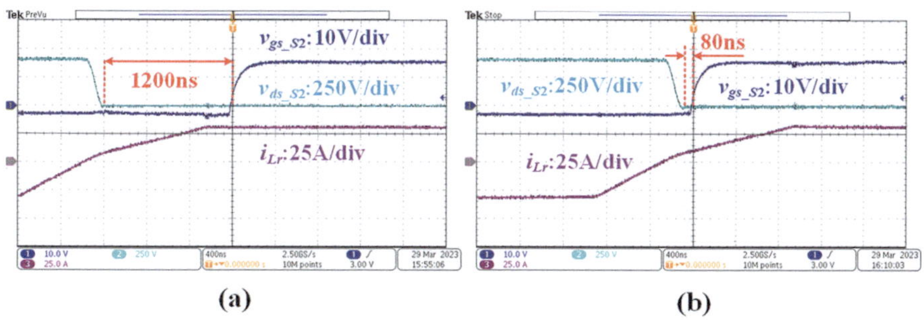

Fig. 6.22 ZVS conditions when $V_{\text{in}} = 23$ V, $P = 1$ kW. **a** In [5], $D_1 = D_2 = 0.5$. **b** Proposed, $D_1 = 0.5$, $D_2 = 0.612$

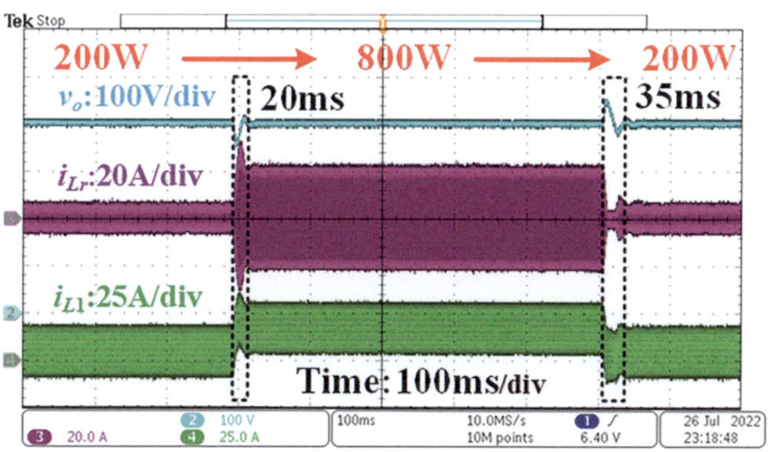

Fig. 6.23 Dynamic response of load transition

6.3.3 Conversion Efficiency and Comparison

Table 6.4 shows the comparison of different control strategies for CF-SDAB dc-dc converter. The clamp voltage V_c and HVS duty cycle D_2 are constant under different load in [5, 8, 9]. As a result, the current flows through the body diodes of the HVS switches for a much longer time, which causes additional conduction loss. In [5, 9], D_1 and D_2 are equal so the LVS switches cannot achieve ZVS under light load. In [8], $D_1 < D_2$ and ZVS can be achieved under light load. However, D_1 cannot be designed around 50% and high-voltage-rating devices have to be used in the LVS. The input current ripple depends on D_1 and the dc inductance. With the proposed control, all switches can achieve within full load range ZVS. Figure 6.24 shows the conversion efficiency of [5] and this paper. Obviously, the conversion efficiency has been improved in all cases with the proposed control strategy.

Table 6.4 Comparison of different control strategies

	Matched with DPS [5]	Matched with duty cycle compensation [8]	Mismatched [9]	Proposed
V_c	Constant	Constant	Constant	Variable
D_1	Around 50%	>50%	Around 50%	Around 50%
D_2	Constant	Constant	Constant	Variable
Conduction loss	Moderate	High	Moderate	Low
ZVS range	Wide range	Full range	Wide range	Full range
Current ripple	Moderate	High	Moderate	Low

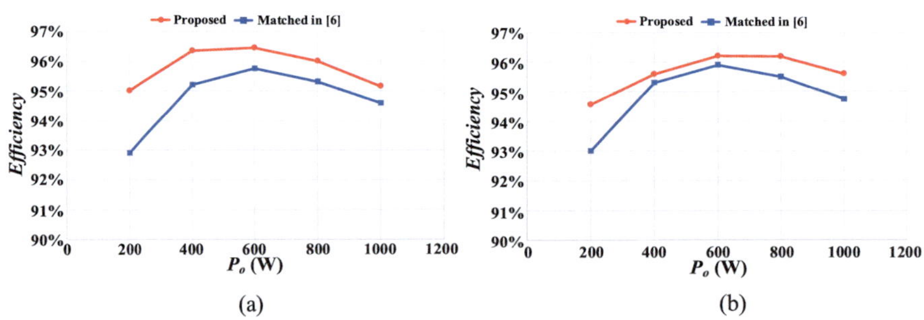

Fig. 6.24 Efficiency curve comparison with the proposed control and voltage match control. **a** $V_{in} = 20$ V. **b** $V_{in} = 26$ V

6.4 Conclusion

In this paper, an optimized hybrid control strategy is proposed for the CF-SDAB dc-dc converter. The inner current tracking loop is proposed so that the clamp voltage can change with the input voltage and load automatically. ZVS can be achieved in the full load range and the no load operating instability issue can be dealt with. The effectiveness of the proposed control has been verified by the experiment. With the proposed control strategy, the efficiency of the converter has been comprehensively improved compared with the voltage match control strategy.

References

1. Sha D, Zhang J, Xu Y (2017) Improved boundary operation for voltage-fed semi-DAB with ZVS achievement and nonactive power reduction. IEEE Trans Ind Electron 64(8):6179–6189
2. Lu M, Li X, Chen G (2020) A hybrid control of a semi dual-active bridge dc–dc converter with minimum current stress. IEEE Trans Power Electron 35(3):3085–3096
3. Wang L, Wang H, Fu M (2023) A three-port energy router for grid-tied PV generation systems with optimized control methods. IEEE Trans Power Electron 38(1):1218–1231
4. Zhang Y, Ding L, Hou N (2022) A dual-current-fed dual-active-bridge DC/DC converter with high-frequency current-ripple-friendly ports. IEEE Trans Power Electron 37(12):15084–15098
5. Sha D, You F, Wang X (2016) A high-efficiency current-fed semi-dual-active bridge dc–dc converter for low input voltage applications. IEEE Trans Ind Electron 63(4):2155–2164
6. Zhang J, Wu H, Qin X (2015) PWM plus secondary-side phase-shift controlled soft-switching full-bridge three-port converter for renewable power systems. IEEE Trans Ind Electron 62(11):7061–7072
7. Everts J, Krismer F, Van den Keybus J, Driesen J, Kolar JW (2014) Optimal ZVS modulation of single-phase single-stage bidirectional DAB AC-DC converter. IEEE Trans Power Electron 29(8):3954–3970
8. Sha D, Wang X, Chen D (2018) High-efficiency current-fed dual active bridge DC–DC converter with ZVS achievement throughout full range of load using optimized switching patterns. IEEE Trans Power Electron 33(2):1347–1357
9. Sha D, Wang X, Liu K, Chen D (2019) A current-fed dual-active-bridge DC–DC converter using extended duty cycle control and magnetic-integrated inductors with optimized voltage mismatching control. IEEE Trans Power Electron 34(1):462–473
10. Sha D, Zhang J, Sun T (2019) Multimode control strategy for SiC MOSFETS based semi-dual active bridge dc–dc converter. IEEE Trans Power Electron 34(6):5476–5486

Chapter 7
Current-Fed Three Phase Semi-DAB DC-DC Converter

Abstract In Chaps. 5 and 6, single phase current-fed semi-dual active bridge (1p-CF-SDAB) DC-DC converter and control strategy have been discussed, which is only suitable for the low power applications, not suitable for high power application due to the high current stress in switches. So in order to reduce the current stress, in this chapter, a high-frequency isolated unidirectional three phase current-fed semi-dual active bridge (3p-CF-SDAB) dc-dc converter is proposed for unidirectional power flow applications suitable for wide input and load variations with small input current ripple such as for fuel stacks and PV array. A hybrid operating mode control was proposed to handle the operating stability throughout the full load range especially the light load conditions. It is with which dual-PWM and phase shifted (DPPS) control with fixed clamping voltage was implemented over certain kind of load while varying the clamping voltage using only dual-PWM (D-PWM) control was used under light load conditions. The ideal power transferred expressions are derived based on the analysis of power transmission characteristics. The ideal RMS expressions of the leakage inductor current are ideally derived for leakage inductance value optimization or minimize the loss. And the ideal soft-switching conditions based on charges has been given. A 2 kW 3p-CF-SDAB prototype using SiC MOSFEs has been built to verify the validity of the proposed control strategy. With the proposed control, the steady state performance throughout the full load of range is pretty well. Seamless transition can be obtained between the two operating modes and the dynamic performance is good. The conversion efficiency is high and the loss distribution is given.

Keywords Three phase current fed semi dual active bridge DC-DC converter · High voltage interface · Wide input range · Small current ripple

D. Sha et al., *High Frequency Isolated Semi Dual Active Bridge DC-DC and AC-DC Converter*, CPSS Power Electronics Series,
https://doi.org/10.1007/978-981-95-1641-4_7

7.1 Mode Analysis with Proposed Control Strategy

A new concept of 1p-VF-SDAB is proposed in [1], whose output side top switches are replaced by diodes. Nevertheless, there is still room for further reduction of non-active power. A CF-SDAB converter introduced in [2] can significantly reduce the non-active power by applying dual-PWM and phase shift control (DPPS), but the current stress on the input side bottom switches is comparatively high. As aforementioned, 1p configuration is not suitable for high power applications. 1p-SDAB are therefore generalized to multiphase, 3p for example, to minimize current stress and output capacitor size. The multiphase converter has a simpler circuit and higher reliability than a simple series and parallel combination of multiple 1p converters. The comparison with the other topology is summarized in Table 7.1. However, the discussion on SDAB is not extensive, and the demonstration of multiphase SDAB lacks of standards. Moreover, in the existing literature, the light load conditions of the VF-SDAB and CF-SDAB is rarely discussed except [3, 4]. Their stability under light load conditions is also unknown.

In this paper, a hybrid operating mode control is proposed to meet wide voltage variations and full load range operation capability requirement. Its light load operation conditions have been envisaged. Dual-PWM control strategy has been proposed to completely deal with the issue of light load stable operation.

The proposed topology is shown in Fig. 7.1. For the primary side, L_A, L_B and L_C are DC inductors. S_{a1}–S_{a3} and S_1–S_3 are the power switches with body diodes and parasitic capacitors. C_c is the clamping capacitor. The AC inductors L_{kA}, L_{kB}, L_{kC} represent the sum of the secondary side referred transformer leakage inductance and the external AC inductance for each phase, respectively. In the secondary side, Q_1–Q_3 are the power switches. D_1–D_3 are the power diodes. The transformer is Y-Y connected. The current flow direction has been highlighted in Fig. 7.1. The ratio of the on-time of primary side bottom switch to the entire period is defined as the duty cycle D. The phase different between the driving signal of the secondary side switch and the primary side bottom switch in each phase is the phase shift angle. The ratio between the phase shift angle and 2π is defined as D_{ϕ}.

One of the main steady state waveforms of the proposed DC-DC converter during one complete switching period is shown in Fig. 7.2. Each phase is a 120 degree phase shifted with each other. The driving signals of each leg are complementary and each phase is phase shifted by 120 degrees. i_{LkA}, i_{LkB}, i_{LkC} represent the phase

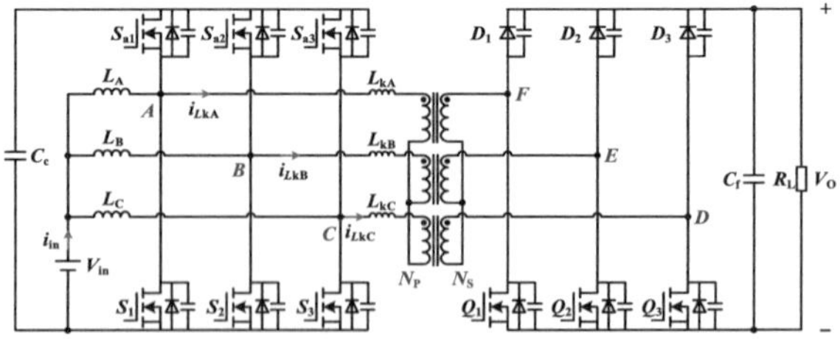

Fig. 7.1 Topology of 3p-CF-SDAB DC-DC converter

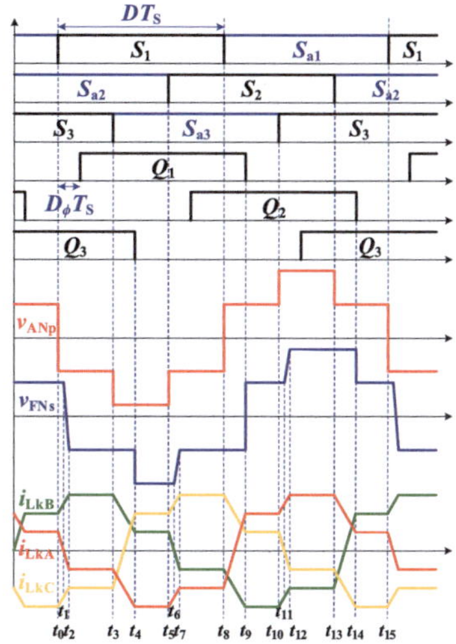

Fig. 7.2 Steady-state waveforms of the proposed DC-DC converter with DPPS control

to neutral current, namely the leakage inductor current of each phase. v_{ANp} is the primary side phase to neutral voltage of phase A and it depends on the primary side driving signals. v_{FNs} is the secondary side phase to neutral voltage of phase A which is not only determined by the secondary side driving signals but also the current zero-crossing moment in diodes. T_S is the switching period.

The typical operation modes are shown in Fig. 7.3. The detailed modes analysis are described as follows.

Stage 1 (before t_0): S_{a1}, S_{a2}, S_3, Q_3, D_1 and D_2 conduct, and v_{ANp}, v_{FNs}, i_{LkA}, i_{LkB} are positive, and i_{LkC} is negative. $v_{ANp} = V_{Cc}/3$ and $v_{FNs} = V_O/3$, where V_{Cc} is the clamping voltage of C_c and V_O is the output voltage. During this stage, the power flows from the primary side (V_{in}) to the secondary side (V_O).

Stage 2 ($t_0 - t_1$): At t_0, S_{a1} is turned off and S_1 is turned on. v_{ANp} becomes $-V_{Cc}/3$ and v_{FNs} does not change. i_{LkB}, i_{LkC} increase positively and i_{LkA} decreases negatively until i_{LkA} falls to 0.

Stage 3 ($t_1 - t_2$): When i_{LkA} falls to 0, D_1 is reverse biased, and v_{FNs} starts to gradually change from $V_O/3$ to $-V_O/3$. i_{LkA} will continue to decrease negatively and eventually become negative. At this time, the current on the secondary side of transformer phase-A simultaneously charges the junction capacitor of D_1 and discharges the junction capacitor of Q_1, waiting for Q_1 to turn on with ZVS.

Stage 4 ($t_2 - t_3$): Since the voltages of the primary and secondary sides match each other, v_{ANp}, v_{FNs}, i_{LkA}, i_{LkB} and i_{LkC} are all maintained at the previous values until the time instant of t_3.

Stage 5 ($t_3 - t_4$): At t_3, S_3 is turned off. The difference between i_{LC} and i_{LkC} begins to charge/discharge the junction capacitors of S_{a3} and S_3 simultaneously. Then S_{a3}

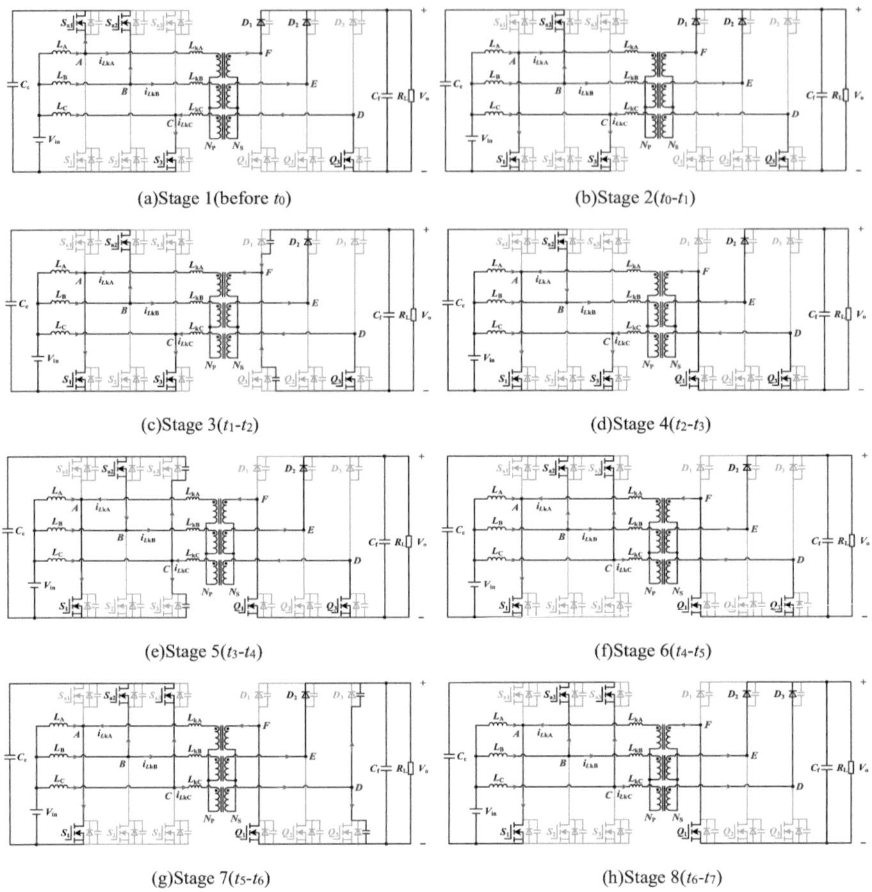

Fig. 7.3 Modes of the proposed topology during on third time

is turned on with ZVS. v_{ANp} becomes $-2V_{\mathrm{Cc}}/3$, and i_{LkC} increases positively whist i_{LkA} and i_{LkB} decrease negatively.

Stage 6 ($t_4 - t_5$): At t_4, Q_3 is turned off, the current on the secondary side of the transformer phase-C simultaneously charges the junction capacitance of Q_3 and discharges the junction capacitance of D_3. Then D_3 conducts and v_{FNs} becomes $2V_O/3$ consequently. After that, v_{ANp}, v_{FNs}, i_{LkA}, i_{LkB} and i_{LkC} remain unchanged. At t_5, S_{a2} is turned off while S_2 is turned on. The corresponding voltages and currents of phase A, B and C change alternately, and the first one-third of the period is reproduced until the end of the entire period. Thus they are omitted herein for the sake of brevity.

7.2 Performance Analysis and Discussion

7.2.1 Leakage Inductor Current and Output Power

As previously analyzed, v_{ANp} varies negatively once S_{a1} is turned off and S_1 is turned on since the leakage inductor suffers from reverse voltage. v_{FNs} begins to change its polarity when the polarity of the leakage inductor current changes completely. At this time, the junction capacitors of D_1 and Q_1 are charged and discharged by the leakage inductor current. Fully parsing this resonance process is complicated. Thence, this process can be approximated by a quadratic equation. Partial enlargement of it is shown in Fig. 7.4. The leakage inductor current expressions can be derived as (7.1).

$$i_{Lk}(t) = \begin{cases} -k \cdot k_1 \cdot (t - t_0) + i_{Lk}(t_0) & [t_0\,,\, t_1) \\ i_{Lk2}(t) & [t_1\,,\, t_2) \\ -k \cdot k_2 \cdot (t - t_2) + i_{Lk}(t_2) & [t_2\,,\, t_3) \\ k \cdot k_3 \cdot (t - t_3) + i_{Lk}(t_3) & [t_3\,,\, t_4) \\ -2k \cdot k_2 \cdot (t - t_4) + i_{Lk}(t_4) & [t_4\,,\, t_5) \\ -k \cdot k_4(t - t_5) + i_{Lk}(t_5) & [t_5\,,\, t_6) \\ i_{Lk7}(t) & [t_6\,,\, t_7) \\ -k \cdot k_2 \cdot (t - t_7) + i_{Lk}(t_7) & [t_7\,,\, t_8) \\ k \cdot k_1 \cdot (t - t_8) + i_{Lk}(t_8) & [t_8\,,\, t_9) \\ k \cdot k_2 \cdot (t - t_9) + i_{Lk}(t_9) & [t_9\,,\, t_{10}) \\ -k \cdot k_3 \cdot (t - t_{10}) + i_{Lk}(t_{10}) & [t_{10}\,,\, t_{11}) \\ i_{Lk12}(t) & [t_{11}\,,\, t_{12}) \\ 2k \cdot k_2 \cdot (t - t_{12}) + i_{Lk}(t_{12}) & [t_{12}\,,\, t_{13}) \\ k \cdot k_4 \cdot (t - t_{13}) + i_{Lk}(t_{13}) & [t_{13}\,,\, t_{14}) \\ k \cdot k_2 \cdot (t - t_{14}) + i_{Lk}(t_{14}) & [t_{14}\,,\, t_{15}] \end{cases} \tag{7.1}$$

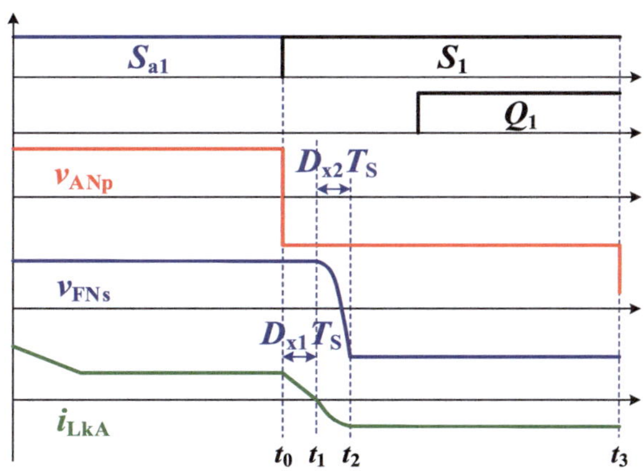

Fig. 7.4 Partial interval waveform enlargement

where n is the turn ratio of the transformer, $D_E = 1/3$, $k = D_E/L_{kA}$, $k_1 = V_{Cc} + nV_O$, $k_2 = V_{Cc} - nV_O$, $k_3 = nV_O - 2V_{Cc}$, $k_4 = V_{Cc} - 2nV_O$.

Suppose the expression of v_{FNs} within $t \in [t_1, t_2)$ (marking $\Delta_2 = t_2 - t_1$) is expressed as

$$v_{FNs}(t) = D_E V_O \left[1 - \frac{2}{\Delta_2^2}(t - t_1)^2 \right] \tag{7.2}$$

And then the corresponding leakage inductor current can be deduced as the following,

$$-D_E V_{Cc} - n v_{FNs}(t) = L \frac{di_{Lk2}(t)}{dt} \tag{7.3}$$

Integrating (7.3) and substituting the initial conditions yields the following,

$$i_{Lk2}(t) = \frac{2knV_O}{3D_{X2}^2 T_S^2}(t - t_1)^3 - k \cdot k_1 \cdot (t - t_1) \tag{7.4}$$

Similarly, the expressions of $i_{Lk7}(t)$ and $i_{Lk12}(t)$ also can be obtained accordingly, which are shown below.

$$i_{Lk7}(t) = -\frac{knV_O}{3D_{X2}^2 T_S^2}(t - t_6)^3 - k \cdot k_4 \cdot (t - t_6) + i_{Lk}(t_6) \tag{7.5}$$

$$i_{Lk12}(t) = -\frac{knV_O}{3D_{X2}^2 T_S^2}(t - t_{11})^3 - k \cdot k_3 \cdot (t - t_{11}) + i_{Lk}(t_{11}) \tag{7.6}$$

Theoretically, there is no DC component in the leakage inductor current. Namely, the average value of the leakage inductor current during a switching period is zero. D_{x1} can be obtained accordingly which is as follows.

$$D_{X1} = D_E - \frac{1 - 2D_{X2}}{3V_{Cc}} V_O n \tag{7.7}$$

Meanwhile, the junction capacitors of D_1 and Q_1 are simultaneously charged and discharged, respectively. According to the principle of conservation of charge, the following formula stands. Where C_D, C_Q are the junction capacitance value of D_1 and Q_1, respectively.

$$n \int_{t_1}^{t_2} i_{Lk}(t)dt = (C_D + C_Q)V_O \tag{7.8}$$

Simplifying and rearranging (7.8) yields,

$$D_{X2} = f_S \sqrt{\frac{6V_O(C_D + C_Q)}{nk(3V_{Cc} + 2nV_O)}} \tag{7.9}$$

The power expression can be obtained consequently which is written below,

$$\begin{aligned} P_O &= 3f_S \int_{t_0}^{t_{15}} i_{Lk}(t) v_{ANp}(t) dt \\ &= kT_S V_{Cc} V_O n \begin{bmatrix} (D_\phi + D_{X1}) + 2D_{X2}(D_E - D_{X1}) \\ -1.5(D_\phi^2 + D_{X1}^2) - 0.75D_{X2}^2 \end{bmatrix} \end{aligned} \tag{7.10}$$

The output power P_O versus the phase shift angle ratio D_ϕ is plotted in Fig. 7.5. The specific parameters are listed in Table 7.1. The dark blue solid line is the calculation result based on (7.10), and the red dashed line with red dots is the simulation result (based on PSIM). These two curves fit perfectly, with some deviations only at very light load conditions. Therefore, (7.10) has theoretical guiding significance for output power.

Meanwhile, it must be noted that the output power is not zero when D_ϕ is diminished to zero, which indicates that the output voltage is uncontrollable under light

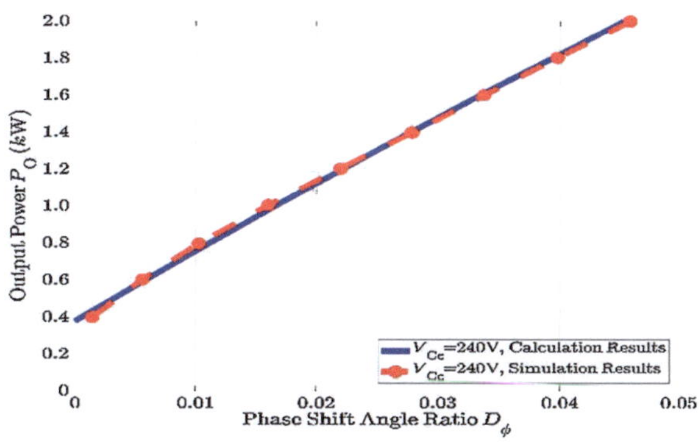

Fig. 7.5 Output power P_O versus phase shift angle ratio D_ϕ

Table 7.1 System parameter

Components	Parameters
Output voltage (V_O)	400 V
Input voltage (V_{in})	100–160 V
Clamping voltage (V_{Cc})	240 V
Output power (P_{omax})	2 kW
Switching frequency (f_S)	100 kHz
Turn ratio (n)	9:15
DC inductance (L)	40 μH
Leakage inductance (L_k)	5 μH

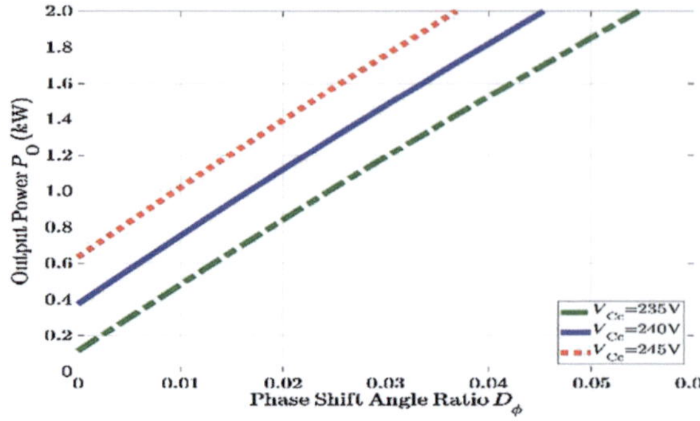

Fig. 7.6 P_O versus D_ϕ with different clamp voltages

load or even no load conditions only using DPPS control. Therefore, it is imperative to search an effective method to handle the light load issue.

Figure 7.6 shows P_O versus D_ϕ with different clamping voltages based on (7.10). As shown, the required D_ϕ is inversely related to the clamping voltage at the same power rating. Therefore, it is possible to suppress the slop of the leakage inductor current by lowering the clamping voltage, thereby achieving the condition of regulating the light load operation of the converter. This is a pure and extremely effective method that can completely solve the problem of uncontrollable light load conditions. This method is referred to herein as a dual-PWM (D-PWM) control. An exhaustive control block diagram will be explained later in detail.

7.2.2 Leakage Inductor Current Stress Analysis

The magnitude of the leakage inductance limits the power transferred capability and affects the RMS value of i_{Lk}. It is feasible to discuss the current stress of i_{Lk} to reduce the converter losses. The leakage inductor current equation in the whole period is given by (7.1) and the phase shift angle ratio can be obtained from (7.10). Then the leakage inductor current RMS value expression $i_{\mathrm{Lk(RMS)}}$ can be expressed by (7.11).

$$
\begin{aligned}
i_{Lk(\mathrm{RMS})} &= \sqrt{f_S \int_{t_0}^{t_{15}} i_{Lk}^2(t)dt} \\
&= kT_S V_C \sqrt{\left[\begin{array}{c} \frac{2}{3} D_\phi (D_\phi + 2D_{X1}) - D_\phi^3 \\ -D_{X1}^2 \left(\frac{302}{35} D_{X1} - \frac{8}{3} \right) \end{array} \right]}
\end{aligned}
\tag{7.11}
$$

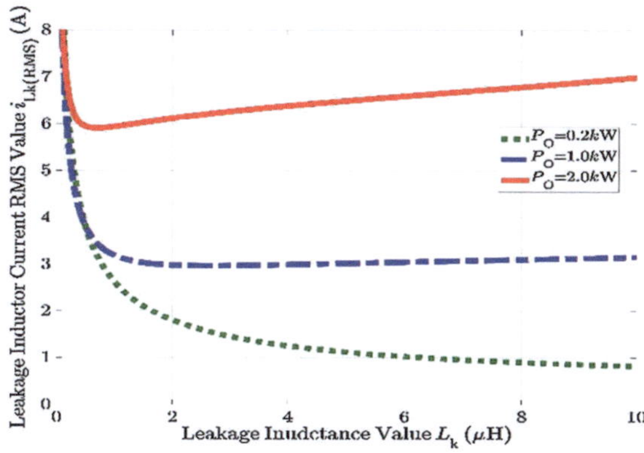

Fig. 7.7 L_k versus the leakage inductor current RMS value $i_{Lk(RMS)}$

The $i_{Lk(RMS)}$ versus L_k can be plotted in Fig. 7.7. The specific parameters used for plotting are listed in Table 7.1. There is a pole point in each of the $i_{Lk(RMS)}$ curves shown in Fig. 7.7. Each power rating has a different pole point coordinate. And $i_{Lk(RMS)}$ gets the minimum at these pole points.

Proverbially, larger leakage inductor requires larger phase shift angle at the same power rating. Meanwhile, the larger leakage inductance can alleviate the light load uncontrollable issue of SDAB dc-dc converter to a certain extent and broaden its stable working range. However, an excessive inductance will result in a large inductor size. In view of this, the effective total leakage inductance is designed around 5 μH in this paper.

7.2.3 Soft Switching Conditions

Since the other two phases are entirely chime with phase A, only the phase A is taken as an example for analysis. Figure 7.8 indicates the phase to neutral current of phase A. Furthermore, the impact of the dead time is also taken into account. The implementation of ZVS requires a large enough current to draw the charge on the MOSFETs before it is turned on. The exhaustive analysis will be developed below.

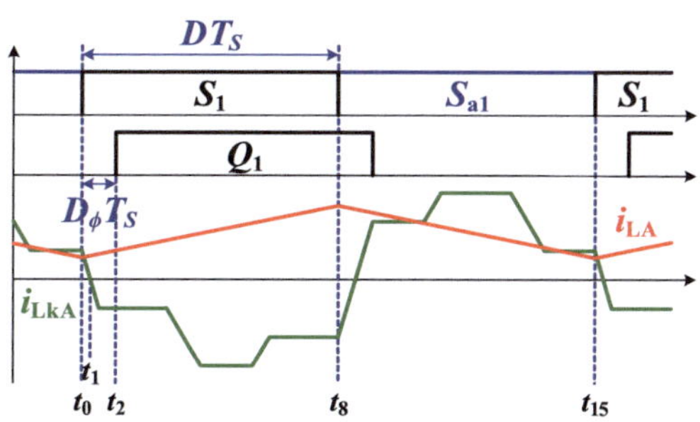

Fig. 7.8 Key waveforms of ZVS

(1) ***Primary side switches***

The maximum and minimum values of the DC inductor current of phase A are $I_{\mathrm{LAmax}} = I_{\mathrm{LA}} + 0.5\Delta i_{\mathrm{LA}}$ and $I_{\mathrm{LAmin}} = I_{\mathrm{LA}} - 0.5\Delta i_{\mathrm{LA}}$, respectively. I_{LA} is the average current of DC inductor of phase A and $I_{\mathrm{LA}} = P_{\mathrm{O}}/(3V_{\mathrm{in}})$. Δi_{LA} is the current ripple of the DC inductor of phase A and $\Delta i_{\mathrm{LA}} = V_{\mathrm{in}} D T_{\mathrm{S}}/L_{\mathrm{A}}$.

Thereafter, the expression of the DC inductor current in a switching period can be expressed as follows.

$$i_{LA}(t) = \begin{cases} I_{LA\,\min} + \frac{V_{in}}{L_A}(t - t_0) & t \in [t_0,\, t_8) \\ I_{LA\,\max} + \frac{V_{in} - V_{Cc}}{L_A}(t - t_8) & t \in [t_8,\, t_{15}] \end{cases} \tag{7.12}$$

Before S_{a1} is turned on, the difference between i_{LA} and i_{LkA} must thoroughly charge/discharge the junction capacitors of S_1 and S_{a1} within the dead time. The ZVS condition of S_{a1} can be formed consequently as the following,

$$\int_{t_8}^{t_8 + T_{dP}} [i_{LA}(t) - i_{LkA}(t)]dt \geq 2C_{\mathrm{oss}\cdot P} V_{Cc} \tag{7.13}$$

where T_{dP} is the primary side dead time, $C_{\mathrm{oss}\cdot\mathrm{P}}$ is the junction capacitance of the primary side MOSFETs. Analogously, the ZVS condition of S_1 can be established as follow,

$$\int_{t_0}^{t_0 + T_{dP}} [i_{LkA}(t) - i_{LA}(t)]dt \geq 2C_{\mathrm{oss}\cdot P} V_{Cc} \tag{7.14}$$

Actually, Fig. 7.8 intuitively indicates that the ZVS condition of S_{a1} is more easily satisfied than that of S_1. It can be proved that (7.13) is always satisfied regardless of the input voltage and the output power discussed in this paper in the case of $L < 160\ \mu\mathrm{H}$. Therefore, (7.14) is the main consideration. After reorganization and simplification (7.14), the ZVS condition of the primary side, namely the selection of the DC inductance, can be obtained. The result is as follows,

$$L \leq \frac{3L_k T_{dP} V_{in}^2 (DT_S - T_{dP})}{2\left[L_k P_O T_d + \left(6C_{\mathrm{oss}\cdot P} L_k + T_{dP}^2 - 2D_{X1} T_S T_{dP}\right) V_{in} V_{Cc}\right]} \tag{7.15}$$

Substituting the necessary parameters listed in Table 7.1 and (7.10) into (7.15), the relationship between the ratio of $M = L/L_{\mathrm{k}}$ and D_{ϕ} can be obtained and the corresponding picture with different D values is plotted in Fig. 7.9.

Figure 7.9 illustrates that the higher the input voltage, the wider the selection range of DC inductance values at the same power rating. And the ZVS range of S_1 becomes narrower as the power rating increases. This indicates that it requires more current ripple. In order to achieve the ZVS of primary side bottom switches in heavy load, $M = 8$ is taken in this paper.

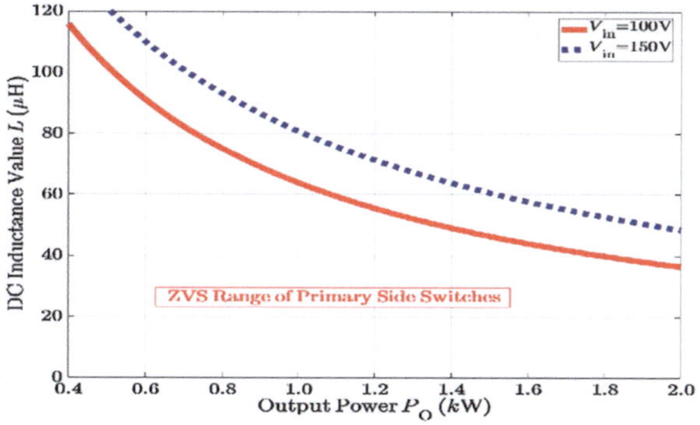

Fig. 7.9 ZVS range of primary side bottom switches with different duty cycles

(2) ***Secondary side switches***

As described above, the leakage inductor reflected to the secondary side resonates with the junction capacitors of secondary side MOSFETs and diodes. Obviously, ZVS of Q_1 can be fully guaranteed only if $D_{X2} < D_\phi$. Thus the ZVS condition of Q_1 is summarized as (7.16).

$$P_O \geq T_S V_{Cc}^2 k \left(D_{X1} + 5D_E D_{X2} - \frac{3}{2} D_{X1}^2 - \frac{9}{4} D_{X2}^2 - 2D_{X1} D_{X2} \right) \tag{7.16}$$

The right side of the inequality is a constant value. Substituting the parameters listed in Table 7.1 into (7.16) and calculating, this constant value is approximately equal to 660 W. That is to say, when the converter discussed in this paper operates above 660 W, the ZVS of Q_1 can be guaranteed. It should be emphasized here that the ZVS conditions discussed above were made with $D_\phi > 0$. The case where $D_\phi =$ 0 and the reference value of V_{Cc} is regulated has not been discussed. As can be seen from Fig. 7.5, the proportion of the second case is very small and it can only occur under very light load conditions. Therefore, ZVS of that situation is not discussed herein.

7.2.4 *Control Strategy*

As aforementioned that the transferred power is not zero when D_ϕ is zero if DPPS control is implemented by using the phase shift angle ratio defined in this paper. This means that if the minimum of D_ϕ is designed to be 0 then the output voltage will be out of control at light load conditions when using DPPS control. Of course, this issue can be alleviated rather than thoroughly solved by extending the D_ϕ to be negative. However, the monotonic interval of the output power with respect to D_ϕ is not fixed, and it varies with the loads and input voltages. Its turning point is hard to find. Therefore, the output of the limiter needs to be adjusted with the input and

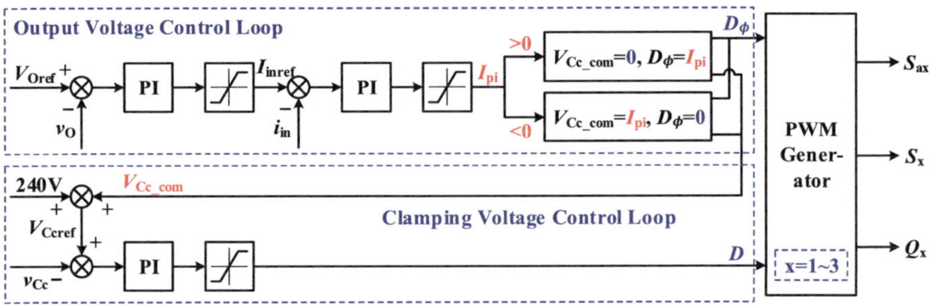

Fig. 7.10 Control block diagram

load variations. It is troublesome and uncertain. Especially at no load conditions, the output voltage will increase sharply, posing a hazard. Moreover, the circulating current or reverse power is very high if $D_\phi < 0$. It must cause a very low conversion efficiency. Therefore, it is not suitable to regulate the output voltage only using DPPS by allowing minus D_ϕ at light load conditions. Figure 7.6 has already illustrates that when D_ϕ is diminished to zero, the output power can be suppressed by lowering the clamping voltage. A varying control strategy is consequently proposed in this paper based on this pure idea to thoroughly handle this issue. The control block diagram of the proposed method is shown in Fig. 7.10.

As seen in Fig. 7.10, v_{O}, v_{Cc} are the feedback information of the output and the clamping voltage, respectively. V_{Oref}, V_{Ccref} are their corresponding reference. The input of the PI compensator is the difference between V_{Oref} and v_{O}, and its output is restricted by the limiter and used as the reference I_{inref} for the DC inductor current inner loop. The difference between I_{inref} and i_{in} serves as the input of another PI compensator, and the output of this PI compensator is also limited by the limiter. When the output of the second limiter is positive, the V_{Ccref} is set to 240 V and remains constant, and the output power is only controlled by D_ϕ. In comparison, when its output is negative, D_ϕ is set to zero and the V_{Ccref} is regulated negatively to restrict the output power. The clamping voltage control loop has a similar working process. What's more, D is always regulated automatically.

Figure 7.11 illustrates the simulation results by using different control strategies when $V_{\text{in}} = 120$ V and $P_O = 0.2$ kW. Figure 7.11a is under the DPPS control by allowing the minus D_ϕ while Fig. 7.11b is applied by D-PWM control. The top of each figure are phase to neutral voltage v_{ANp} and v_{FNs} of phase A. The phase to neutral current of phase A is shown in the middle of each figure. And the bottom one shows the instantaneous reverse power calculated by v_{ANp} times i_{LkA} and only shows the minus value. Obviously, the peak current in Fig. 7.11a is much higher than that in Fig. 7.11b. So is the reverse power. The average reverse power in Fig. 7.11a is about 319.33 W while only 4.68 W in Fig. 7.11b. These are vastly different.

Consequently, the efficiency by using DPPS control with minus D_ϕ at light load condition is much lower than that of D-PWM control.

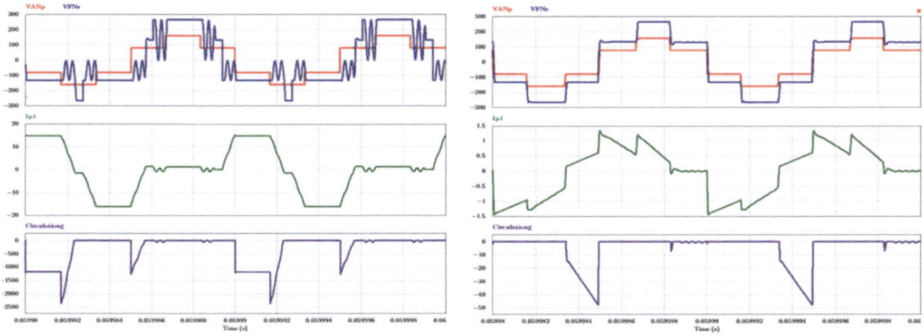

(a) DPPS control with minus D_ϕ (b) D-PWM control

Fig. 7.11 Steady state waveforms when $V_{in} = 120$ V and $P_O = 0.2$ kW

7.3 Experimental Results

7.3.1 Prototype

A 2 kW experimental prototype has been built to verify the effectiveness of the 3p-CF-SDAB dc-dc converter utilizing the proposed control. The prototype photograph is shown in Fig. 7.12. The parameters of the converter system is listed in Table 7.1 and all device patterns are summarized in Table 7.2.

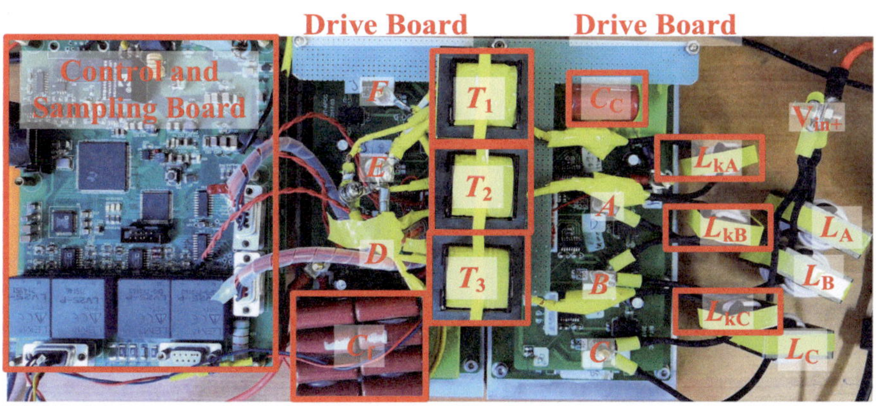

Fig. 7.12 Photograph of the test prototype

Table 7.2 Device patterns

Devices	Patterns
Control chip	TMS320F28335
Drive chip	Si8233
SiC MOSFETs	C3M0065090J
Diodes	C3D10060G
Transformer core	DMR95 EE42A
DC inductor core	DMR95 EE42A
Leakage inductor core	DMR95 EI30

7.3.2 *Steady State Experimental Results*

Figures 7.13 and 7.14 show the experimental waveforms of the proposed converter under different input voltages and different loads with the proposed control strategy. It can be seen that the phase to neutral voltage waveforms (v_{ANp} and v_{FNs}) in Figs. 7.13 and 7.14 are not so regular, which are different from the ideal analysis. This is due to the fact that the neutral point of the primary and secondary side is floating. The asymmetry of each phase parameters result in a very irregular neutral point voltage waveform. This affects the phase to neutral voltage waveforms. Meanwhile, the constituent components such as the transformers, inductors and power devices are not ideal, and the effect of dead time is also taken into account. In comparison, the experimental current waveforms is relatively regular and pretty similar to their theoretical waveforms. What's more, D_ϕ is positive as the output power is 1.6 kW (Fig. 7.13) while the D_ϕ is zero when the output power is 0.2 kW (Fig. 7.14).

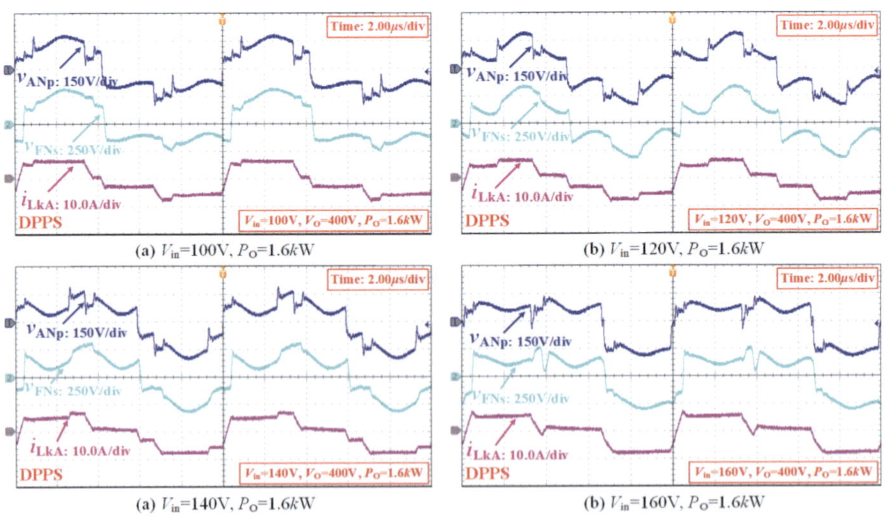

Fig. 7.13 Steady state experimental waveforms in 1.6 kW with different input voltages

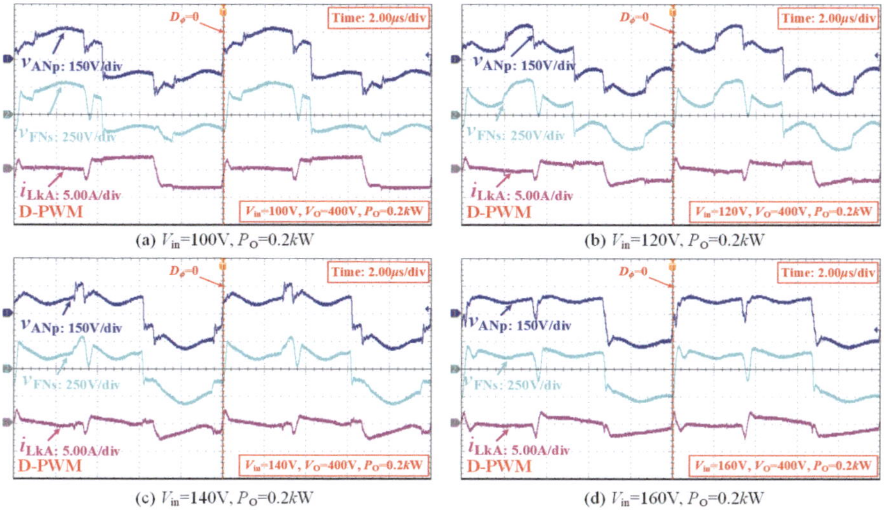

(a) V_{in}=100V, P_O=0.2kW (b) V_{in}=120V, P_O=0.2kW (c) V_{in}=140V, P_O=0.2kW (d) V_{in}=160V, P_O=0.2kW

Fig. 7.14 Steady state experimental waveforms in 0.2 kW with different input voltages

Figure 7.15 shows the primary and secondary side line to line voltage ($v_{AB} = v_{ANp} - v_{BNp}$, $v_{FE} = v_{FNs} - v_{ENs}$) and phase to neutral current waveforms. It can be seen that the primary side line to line voltage is very regular and smooth. This is because the primary side line to line voltages are only determined by switching logic of the primary side switches rather than the leakage inductor current. What's more, the effect of the neutral point waveform is eliminated on the line to line voltage waveforms. The secondary side line to line voltage waveforms in Fig. 7.15 is also very clean except Fig. 7.15d. The spikes appear in Fig. 7.15d happen exactly at the natural zero crossing of the phase to neutral or line to line current.

At this time, the diode is reverse biased and the leakage inductor resonates with the junction capacitors of the power devices.

Figure 7.16 shows the no-load conditions experimental waveforms. The proposed converter can work very stably at no-load conditions by utilizing the proposed control strategy, which is enough to prove the effectiveness of this method.

Figure 7.17 shows the total input current and input current of phase-A of the proposed converter. It illustrates that the current ripple of each phase is large even the DC inductance value is large enough due to the high input voltage level. However, the ripple of the total input current is quite small thanks to the 3p interleaving structure. The smaller current ripple is beneficial to extend the lifetime of input source such as fuel cell stack and the PV array. Figure 7.18 shows the ZVS achievement situations of different active switches when $P_O = 1.6$kW.

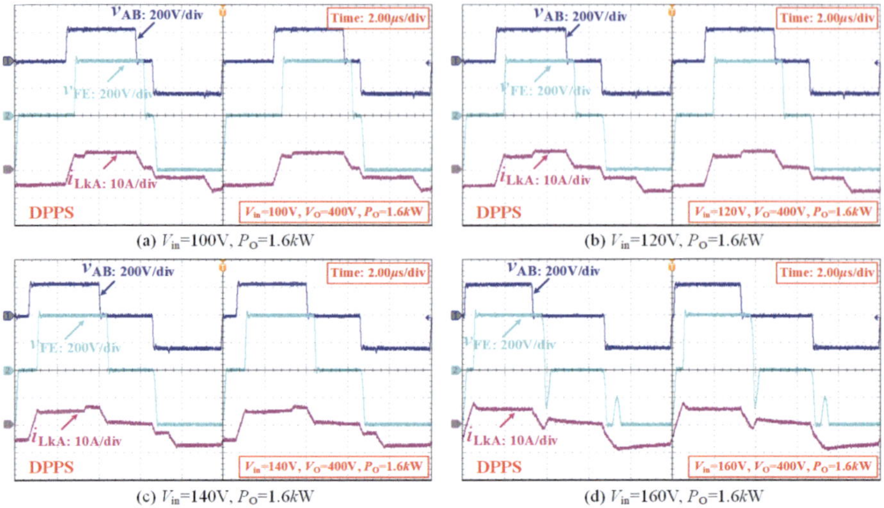

Fig. 7.15 Primary and secondary side line to line voltage and phase to neutral current waveforms

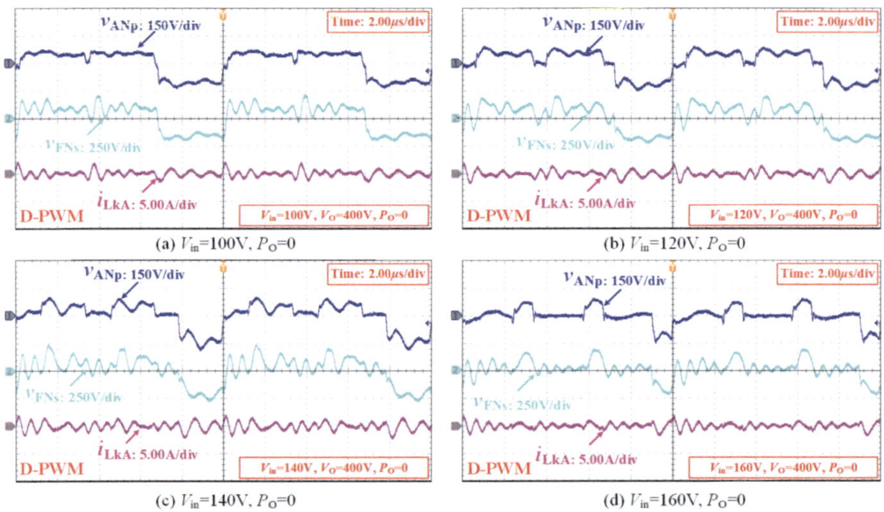

Fig. 7.16 Steady state experimental waveforms in 0.0 kW with different input voltages

7.3.3 Dynamic Response Experimental Results

Figure 7.19 shows the dynamic response only using DPPS control without minus D_ϕ. The load switches back and forth between 0.8 and 0.2 kW when the $V_{in} = 120$ V. When the converter is working normally, $D_\phi > 0$ under 0.8 kW condition and $D_\phi = 0$ under 0.2 kW condition. It can be seen from the experimental results that the

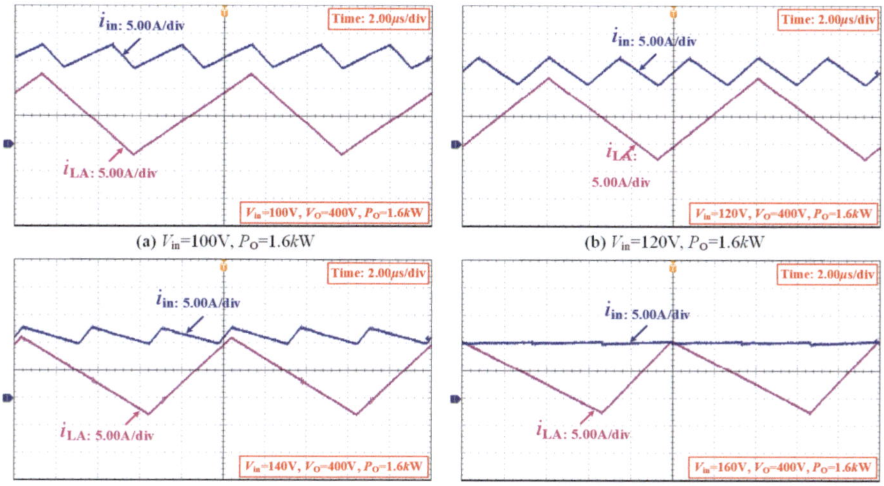

(a) V_{in}=100V, P_O=1.6kW (b) V_{in}=120V, P_O=1.6kW

(c) V_{in}=140V, P_O=1.6kW (d) V_{in}=160V, P_O=1.6kW

Fig. 7.17 Input current ripple waveforms in 1.6 kW with different input voltages

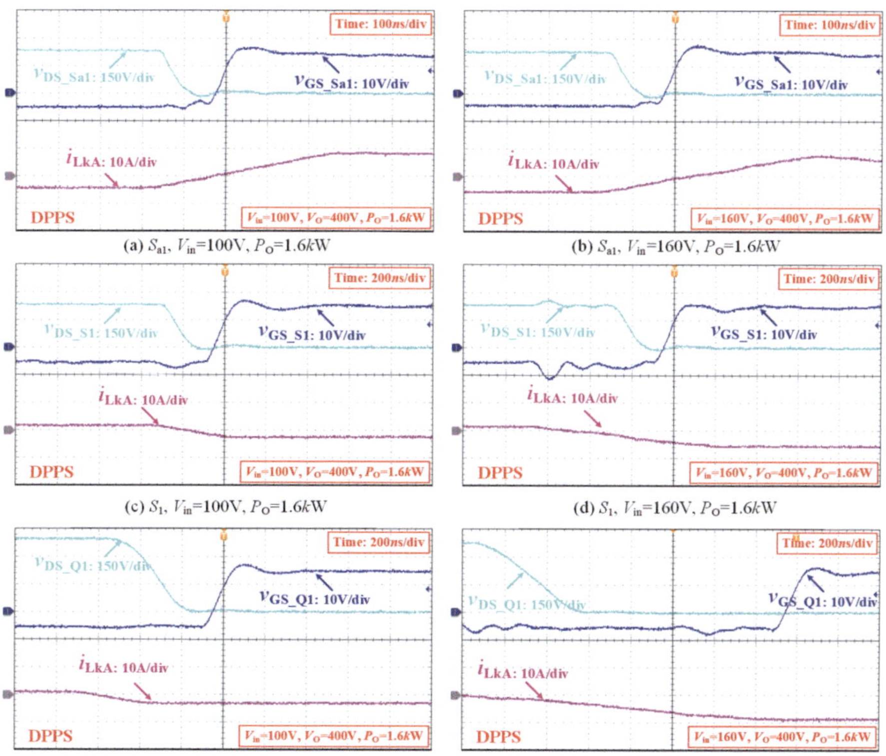

(a) S_{a1}, V_{in}=100V, P_O=1.6kW (b) S_{a1}, V_{in}=160V, P_O=1.6kW

(c) S_1, V_{in}=100V, P_O=1.6kW (d) S_1, V_{in}=160V, P_O=1.6kW

(e) Q_1, V_{in}=100V, P_O=1.6kW (f) Q_1, V_{in}=160V, P_O=1.6kW

Fig. 7.18 ZVS operating conditions of active switches when $P_O = 1.6$ kW

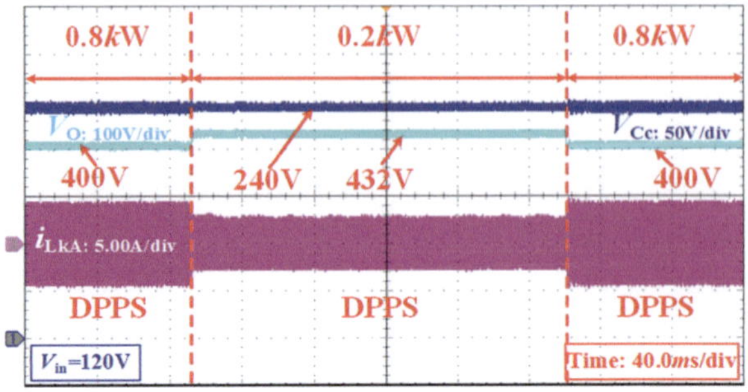

Fig. 7.19 Dynamic response only using DPPS control without minus D_ϕ

output voltage is uncontrollable when the load is very light. At this moment the regulator fails and the converter operate in an open loop. The output voltage will rise to 432 V under the operating conditions shown in Fig. 7.19. When the load is further reduced, the output voltage will increase rapidly, which is very dangerous. As aforementioned, the stable operating range of the converter can be broadened to some extent by allowing the minus D_ϕ theoretically. But this issue cannot be completely solved. The reason has been described in detail above (Fig. 7.18).

Load switching experiments between 0.2 kW (D-PWM control) and 0.8 kW (DPPS control) were implemented to demonstrate the smoothness of the control strategies switching between D-PWM control and DPPS control. Figure 7.20 illustrates the dynamic responses experimental results when $V_{\text{in}} = 120$ V. Figure 7.20a shows the macro view, and it was divided into 3 areas. Figure 7.20b–d show the detail of area 1 to area 3, respectively. Although there is a small drop or bulge in the output voltage at the transient of load switching, it quickly returns to the ideal value. Switching between these two modes does not cause a large transient of voltage or current. It indicates that the proposed method enables the converter to have a fast dynamic response speed and a smooth modal transition. Load switching experiments at other input voltages were also implemented, and the experimental results have the same conclusion. For brevity, it is omitted herein.

7.3.4 Conversion Efficiency and Loss Breakdown

Figure 7.21 illustrates the measured efficiency curves under different input voltages and different loads. The high conversion efficiency is guaranteed over a wide variations of input voltage and load conditions. Peak efficiency always occurs around $P_O = 2.0$ kW regardless of the input voltage.

Moreover, the efficiency is positively correlated with the input voltage due to the reduction of core loss of the magnetic components. The maximum efficiency is about 97.29%.

Figure 7.22 illustrates the loss distribution. As can be seen from Fig. 7.22, the core loss of the leakage (AC) inductors and the copper loss of each magnetic element

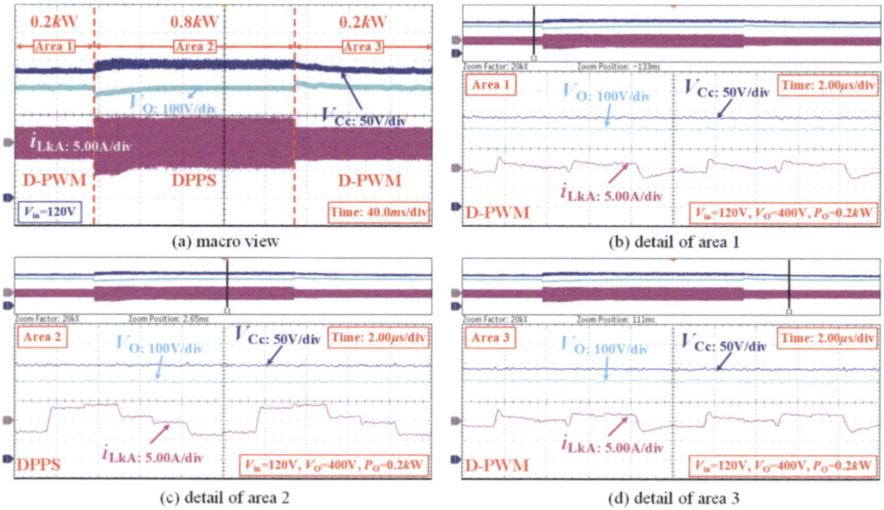

Fig. 7.20 Dynamic response under different loads

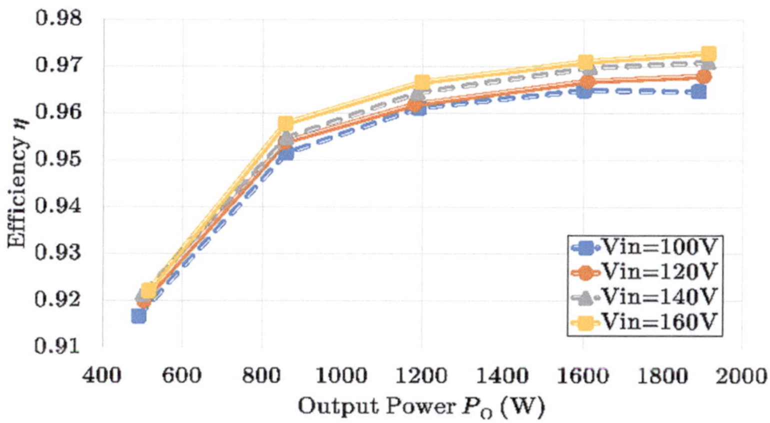

Fig. 7.21 Power conversion efficiency

are very small and can be ignored. According to the operating mode of the topology, the volt-second area of the transformers is independent of the input voltage and load, so the core loss of the transformer does not change. And it occupies a considerable proportion. Due to the voltage clamping of the clamp capacitor, the duty cycle does not change with load at the same input voltage. In the D-PWM mode, the duty cycle changes a little because the clamping voltage changes less. Therefore, the core loss of the DC inductor can also be regarded as a fixed loss under different loads. The achievement of ZVS of primary side bottom switches (S_1, S_2, S_3) needs a large DC current ripple. The difference between the DC inductor current and the leakage inductor current flows into each switch on the primary side. Therefore, the current RMS value and the turn off current in the switches, especially the primary side bottom switches, increase as the load increases. The switching losses of primary side bottom switches account for more than half of the total switching losses. But this proportion has decreased at light loads. The sum of the conduction losses of the primary side bottom switches and secondary side diodes occupies 90% of the total conduction loss.

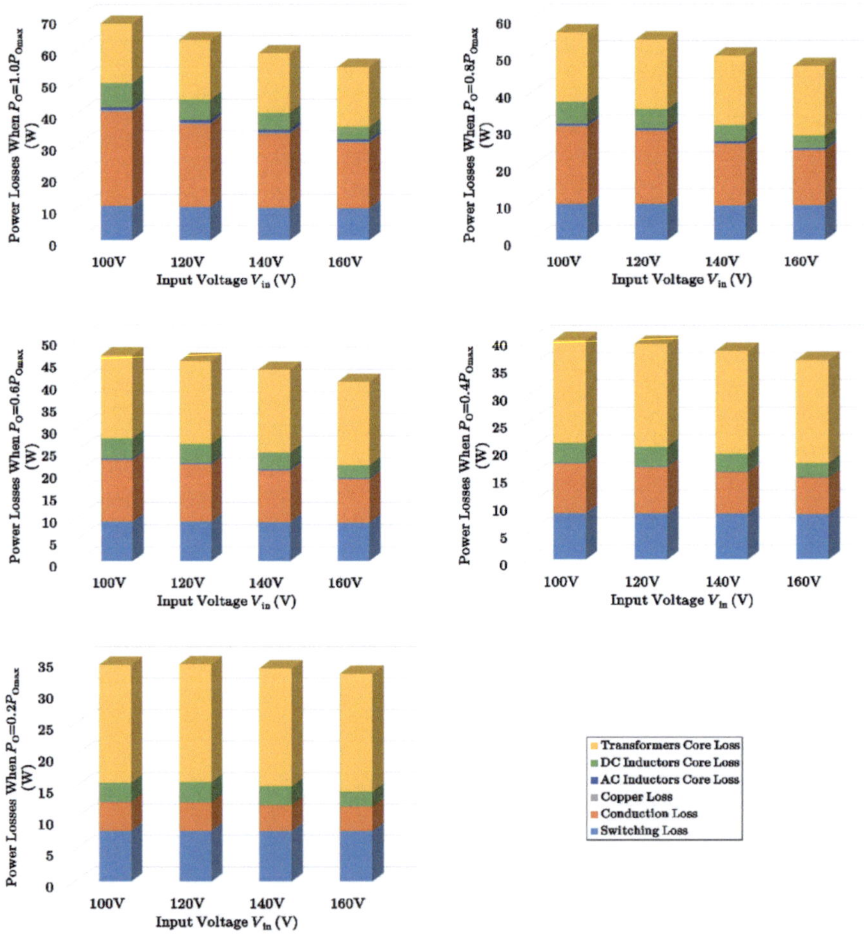

Fig. 7.22 Loss distribution

7.4 Conclusion

A high-frequency isolated 3p-CF-SDAB dc-dc converter is proposed for unidirectional power flow applications with wide input voltage range. The ideal power transferred expressions have been obtained for analyzing power transmission characteristics. The ideal RMS expressions of the leakage inductor current are ideally derived for the optimal leakage inductance value design. The ideal soft-switching conditions based on charges have been given. In order to deal with the issue of that the 3p-CF-SDAB is uncontrollable when it operates under light load or even no load condition, a hybrid operating mode was proposed. Over medium or heavy load conditions, only DPPS is implemented. In comparison, under light load conditions, the uncontrollable issue for CF-SDAB brought by using DPPS control can be solved with DPWM control by regulating the clamp voltage instantaneously. The validity of the proposed converter and the proposed control strategy has been verified by simulation and experimental results of a 2 kW prototype. The converter system has good steady state and dynamic performance even between operating mode transition. The converter with the proposed control has high conversion efficiency and works stably throughout the full load range. The loss distribution analysis is given.

References

1. Kulasekaran S, Ayyanar R (2014) Analysis, design, and experimental results of the semidual-active-bridge converter. IEEE Trans Power Electron 29(10):5136–5147
2. Sha D, You F, Wang X (2016) A high-efficiency current-fed semi-dual active bridge DC–DC converter for low input voltage applications. IEEE Trans Ind Electron 63(4):2155–2164
3. Sha D, Chen D, Kham S, Guo Z (2019) Voltage fed three phase semi dual active bridge DC-DC converter utilizing varying operating mode with high conversion efficiency. IEEE Trans Power Electron 34(10):9447–9458
4. Sha D, Zhang J, Sun T (2019) Multi-mode control strategy for SiC MOSFETs based semi dual active bridge DC-DC converter. IEEE Trans Power Electron 34(6):5476–5486

Chapter 8
Single-Stage Semi-DAB AC-DC Converter with Basic Modulation Strategy

Abstract According to the previous introduction for semi dual-active-bridge (SDAB) DC-DC converter topology, modulation and control strategy, the voltage-fed semi-DAB is suitable for extremely wide voltage conversion gain range. So a novel single-phase single-stage AC-DC converter based on semi-DAB DC-DC converter is proposed for unidirectional power flow applications in this chapter. Dual-phase-shift (DPS) and Single-phase-shift (SPS) modes are adopted individually according to its operation. Only two control variables which are the primary side phase shift and the phase shift between the two bridges are used. The optimal voltage and current waveforms in different operation modes are analyzed. The detailed transition process between the DPS mode and SPS mode are given, and the mode transition is seamless. The proposed converter can achieve high power factor correction without inner current tracking loop. In addition, the switches can achieve zero-voltage-switching (ZVS) within a wide operating range and the rectifier diode can obtain zero-current-switching (ZCS) off. The power transfer capability, mode transition mechanism, soft switching range, mode operation range and parameter design are analyzed. Besides, a 1 kW experimental prototype was built to verify the effectiveness of the proposed control strategy.

Keywords Semi dual-active-bridge (SDAB) · Zero voltage switching · Seamless mode transition · Single stage AC-DC converter

8.1 Circuit Topology and Equivalent Circuits

8.1.1 *Power Circuit Topology*

Single-phase isolated AC-DC converters with high power factor correction (PFC) are widely used in many fields such as electrical vehicles, power electronic transformers, telecom power supplies and future data centers. The traditional two-stage conversion solution are widely used, such as three phase Vienna rectifier followed by the LLC DC-DC converter, where the front stage converts the grid voltage to a stable DC bus

D. Sha et al., *High Frequency Isolated Semi Dual Active Bridge DC-DC and AC-DC Converter*, CPSS Power Electronics Series,
https://doi.org/10.1007/978-981-95-1641-4_8

with power factor correction (PFC), the second stage is generally a high-frequency isolated DC-DC converter to achieve galvanic isolation and DC voltage regulation. However, the two-stage AC-DC converters have some disadvantages, such as the lower power density because of the large size electrolytic capacitor needed to smooth the DC bus voltage. And the conversion efficiency is not high and the control is more complex due to the requirement of individual control systems for each stage. In comparison, the single-stage AC-DC converter can realize galvanic isolation, high power density and high effectiveness employing only one stage [1, 2].

No doubt, DAB DC-DC converter can be used in single stage AC-DC converters where the power flow is unidirectional [3–6]. Actually, DAB converter has the bidirectional power flow capability feature. However, for unidirectional power flow applications, there are many switches and this causes more gating signals and drivers and it is not cost effective. Compared with traditional DAB employing eight switches, semi-DAB is simplified by replacing some switches with diodes [7]. Thus the power circuit and the driver circuits can be simplified, it becomes cost effective and more reliable by using diodes.

The power circuit topology of the single-stage single-phase isolated AC-DC converter based on semi-DAB is shown in Fig. 8.1. For the grid side, the uncontrolled rectifier composed of four diodes (D_{r1}–D_{r4}) is followed by a semi-DAB DC-DC converter. The AC grid voltage v_{ac} is converted to DC voltage v_{in} through the diode rectifier and its output DC voltage is fed to the semi-DAB DC-DC converter. Its primary side is composed of four MOSFETs (Q_1–Q_4) converting v_{in} into high frequency AC voltage v_{ab}. The secondary side is composed of two MOSFETs (Q_5–Q_6) and two diodes (D_{o1}–D_{o2}) converting high frequency AC voltage v_{cd} into DC voltage v_o. The HF capacitors C_{in} with very low capacitance is used to bypass the high frequency current. The high frequency transformer T turns ratio is n:1. L_r represents its leakage inductance and L_m is its magnetizing inductance. M is the voltage gain conversion ratio and it is defined as $M = nv_o/|v_{\mathrm{ac}}|$. It should be noted that as the grid voltage magnitude approaches zero, the minimum value of the $|v_{\mathrm{ac}}|$ is set to be no less than 3 V in this paper. This means if the sampling AC input voltage $|v_{\mathrm{ac}}| \leq$ 3 V, then $|v_{\mathrm{ac}}|$ is set to be 3 V.

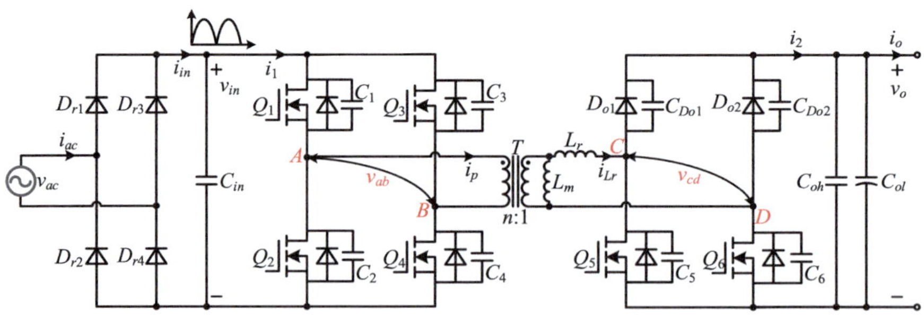

Fig. 8.1 Single-stage single-phase AC-DC topology based on semi-DAB

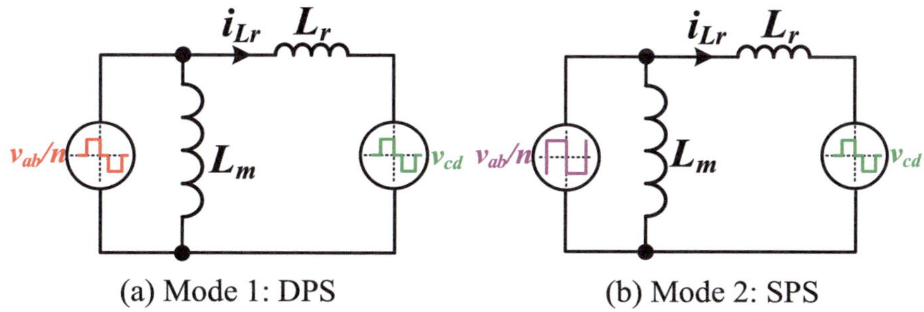

Fig. 8.2 Equivalent circuit model

8.1.2 Equivalent Circuit Model

The equivalent circuit model of the semi-DAB converter is shown in Fig. 8.2. By referring the primary side voltage to the secondary side, a simplified equivalent model of the semi-DAB converter can be obtained.

8.2 Control Strategy

8.2.1 Operating Modes and Circuit Operating Principle

In [8], three basic modulation modes and the boundary conditions between them are given, the optimal modulation modes are obtained by minimizing the peak value of the transformer leakage inductor current. The optimal modulation modes under the optimal control strategy are shown in Fig. 8.3a, b. $D_1 T$ represents the time length Q_1 leading Q_6, $D_2 T$ represents the time duration length in which v_{AB} and v_{CD} have the same non-zero polarity. It is worth noting that $D_1 + D_2 <= 1$. T represents the half switching period and the total switching period is $T_S = 2T$.

The optimal control trajectories are expressed in (8.1),

$$\begin{cases} D_2 = \frac{D_1}{M-1} & \text{(Mode 1)} \\ D_2 = 1 - D_1 & \text{(Mode 2)} \end{cases} \tag{8.1}$$

The intersection of two optimal control trajectories can be obtained through (8.1). The intersection is,

$$\begin{cases} D_1 = 1 - \frac{1}{M} \\ D_2 = \frac{1}{M} \end{cases} \tag{8.2}$$

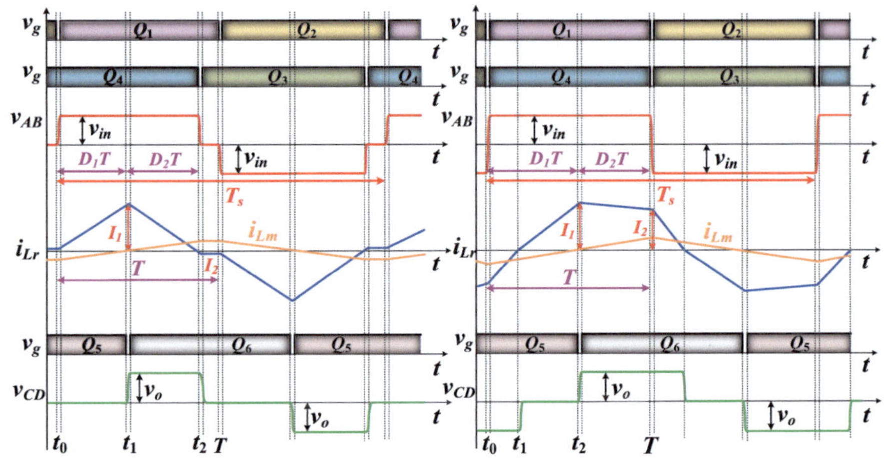

(a) Mode 1: DPS (b) Mode 2: SPS

Fig. 8.3 The optimal modulation modes

The maximum value of the transfer power under different conversion gains occur at mode 2 where,

$$\begin{cases} p_{o,\max} = \frac{v_o^2 T}{2L_r}\cdot\frac{M+1}{M(M^2+2M+2)} \\ D_{1,\max}(M) = \frac{M^2+M+1}{M^2+2M+2} \\ D_{2,\max}(M) = \frac{M+1}{M^2+2M+2} \end{cases} \tag{8.3}$$

The primary side current expressions, transformer leakage inductance current expressions at various moments in the two optimal modes are listed in Tables 8.1 and 8.2 respectively, which are used in the ZVS range and mode operation region analysis later. C_{oss} and C_D represent the junction capacitance value of switches Q_5–Q_6 and diodes D_{o1}–D_{o2} respectively.

Table 8.1 Expressions of transformer primary side currents

	Mode 1	Mode 2
$i_p(t_0)$	$-\frac{1}{n}\cdot\left[\frac{\frac{\lvert v_{ac}\rvert}{n\cdot L_m}\cdot(D_1+D_2)T}{2}-v_O\sqrt{\frac{C_{oss}+C_D}{L_r}}\right]$	$-\frac{1}{n}\cdot\left[\frac{v_oT}{L_r}\cdot\frac{(1+M)(1-M+MD_1)}{M(M+2)}+\frac{\lvert v_{ac}\rvert}{n\cdot L_m}\times\frac{T}{2}\right]$
$i_p(t_1)$	$\frac{1}{n}\left[v_O\sqrt{\frac{C_{oss}+C_D}{L_r}}+\frac{v_oT}{L_r}\cdot(\frac{D_1}{M})+\frac{\lvert v_{ac}\rvert(D_1-D_2)T}{2nL_m}\right]$	$\frac{1}{n}\left[\frac{\lvert v_{ac}\rvert}{2nL_m}\frac{M(2D_1-3)T}{M+2}\right]$
$i_p(t_2)$	$\frac{1}{n}\cdot\left[\frac{\frac{\lvert v_{ac}\rvert}{n\cdot L_m}\cdot(D_1+D_2)T}{2}-v_O\sqrt{\frac{C_{oss}+C_D}{L_r}}\right]$	$\frac{1}{n}\left[\frac{v_oT}{L_r}\cdot\frac{2D_1+M-1}{M(2+M)}+\frac{\lvert v_{ac}\rvert(2D_1-1)T}{2nL_m}\right]$
$i_p(T)$	$\frac{1}{n}\cdot\left[\frac{\frac{\lvert v_{ac}\rvert}{n\cdot L_m}\cdot(D_1+D_2)T}{2}-v_O\sqrt{\frac{C_{oss}+C_D}{L_r}}\right]$	$\frac{1}{n}\cdot\left[\frac{v_oT}{L_r}\cdot\frac{(1+M)(1-M+MD_1)}{M(M+2)}+\frac{\lvert v_{ac}\rvert}{n\cdot L_m}\times\frac{T}{2}\right]$

Table 8.2 Expressions of transformer leakage inductance currents

	Mode 1	Mode 2
$i_{Lr}(t_0)$	$v_o\sqrt{\frac{C_{oss}+C_D}{L_r}}$	$-\frac{v_oT}{L_r}\cdot\frac{(1+M)(1-M+MD_1)}{M(2+M)}$
$i_{Lr}(t_1)$	$v_o\sqrt{\frac{C_{oss}+C_D}{L_r}}+\frac{v_oT}{L_r}\cdot(\frac{D_1}{M})$	0
$i_{Lr}(t_2)$	$-v_o\sqrt{\frac{C_{oss}+C_D}{L_r}}$	$\frac{v_oT}{L_r}\cdot\frac{2D_1+M-1}{M(2+M)}$
$i_{Lr}(T)$	$-v_o\sqrt{\frac{C_{oss}+C_D}{L_r}}$	$\frac{v_oT}{L_r}\cdot\frac{(1+M)(1-M+MD_1)}{M(2+M)}$

The semi-DAB converter is needed to adapt extremely wide voltage gain variation because the input voltage is a folded sinusoidal wave. The maximum and minimum value for the voltage conversion gain M lie at the minimum value $|v_{ac}|_{min}$ and the maximum value $|v_{ac}|_{max}$ respectively. Thus the optimal control trajectory needs to consider the gain decreasing from infinity to M_{min}. Three-dimensional optimal control trajectory is shown in Fig. 8.4a, the light blue surface indicates the set of optimal control trajectories of mode 1 (DPS) at different voltage conversion gains, and the deep blue surface represents the set of optimal control trajectories of mode 2 (SPS) at different conversion gains as shown in (8.1). For example, the red surface and orange surface indicate the two section plane at $M = 5$ and $M = 2$ respectively, the intersection line of the red surface and two blue surfaces is the optimal control trajectory under $M = 5$ as the green solid line, and the turning point is the mode transition point between the DPS mode and SPS mode. The same case is similar for $M = 2$. The mode transition point moves along the pink dotted line as the gain decreases and optimal control trajectory also changes all the time. In order to explain more clearly, the two-dimensional optimal control trajectories under $M = 5$ and $M = 2$ are given in Fig. 8.4b, c.

8.2.2 Power Transfer Analysis

The power transmitted in a switching period can be calculated through the following expression:

$$p = \frac{\int_{t_o}^{T}[i_{Lr}(t)\cdot v_{AB}(t)/n]dt}{T} \tag{8.4}$$

In mode 1, the power transfer expression with the optimal control trajectory is shown as the following,

$$p = \frac{v_o^2T}{2L_r}\cdot\frac{D_1^2}{M(M-1)} \tag{8.5}$$

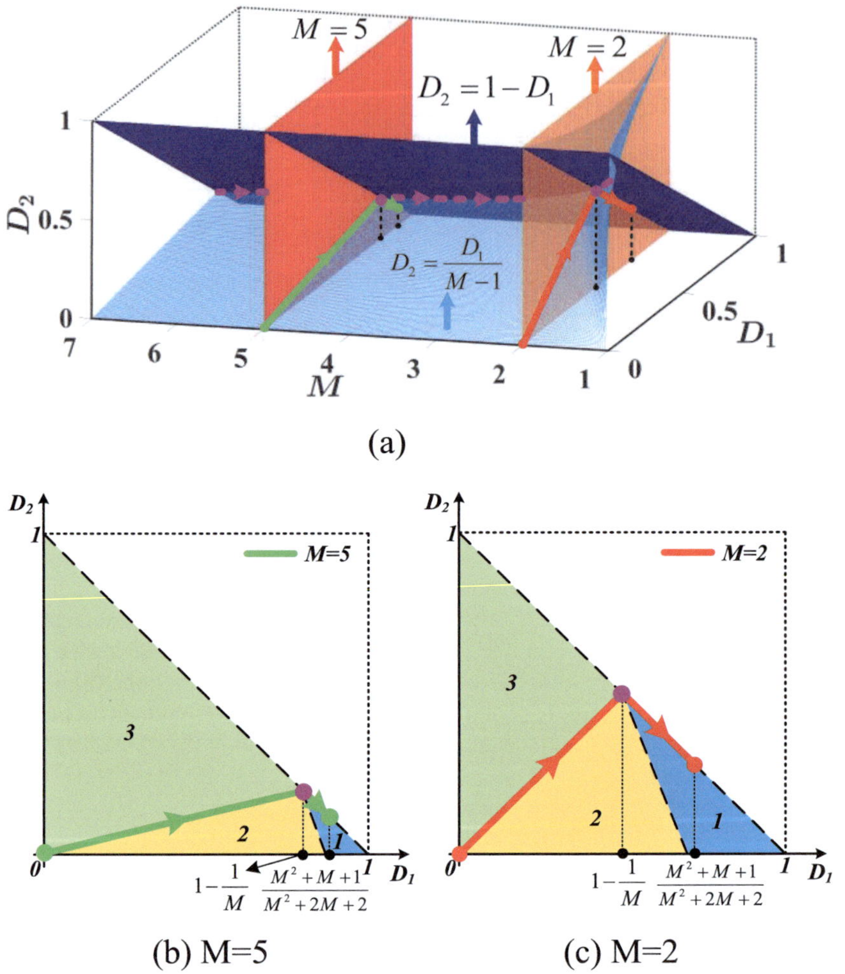

Fig. 8.4 Control trajectories of the optimal control strategy

The three-dimensional power transfer surface of mode 1 with control variable D_1 and conversion gain M is shown in Fig. 8.5, the maximum value of D_1 under different gains is shown in (8.2), and the three-dimensional surface of $D_{1,\max}(M)$ is also shown in Fig. 8.5.

Substituting (8.2) and system parameters (Table 8.3) into (8.5) yields the maximum power transfer ability curve of mode 1 as the intersection line of two surfaces in Fig. 8.5.

In mode 2, the power transfer expression with the optimal control trajectory is shown in (8.6),

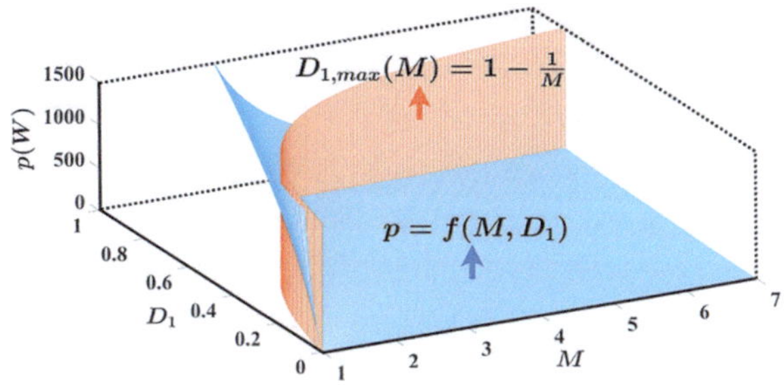

Fig. 8.5 The power transfer capability of mode 1

Table 8.3 System Parameters

Symbol	Value	Symbol	Value
v_{ac}	220V$_{rms}$	C_{ol}	10,000 μF
f_{ac}	50 Hz	Q_1, Q_2, Q_3, Q_4	C3M0045065K
v_o	400 V	Q_5, Q_6	C3M0065100K
C_{in}	2 μF	D_{o1}, D_{o2}	RURG80100
L_r	64.8 μH	$D_{r1}, D_{r2}, D_{r3}, D_{r4}$	MDS100B
L_m	230 μH	f_s	100 kHz
n	1/1.2: 1	t_{dead}	200 ns
C_{oh}	20 μF	P_{rated}	1 kW

$$p = \frac{v_o^2 T}{2L_r} \cdot \left[\frac{4D_1M^2 - 4D_1^2M - 4D_1^2 - 2D_1^2M^2}{M(M+2)^2} + \frac{4D_1M + 4D_1 - 2M^2 + M + 1}{M(M+2)^2} \right] \tag{8.6}$$

The three-dimensional power transfer surface of mode 2 versus control variable D_1 and conversion gain M is shown in Fig. 8.6, the maximum value of D_1 under different gains is shown in (8.7), and the three-dimensional surface of $D_{1,\max}(M)$ is also shown in Fig. 8.6.

$$D_{1,\max}(M) = \frac{M^2 + M + 1}{M^2 + 2M + 2} \tag{8.7}$$

Substituting (8.7) and system parameters (Table 8.3) into (8.6) yields the maximum power transfer ability cure of mode 2 as the intersection line of two surfaces in Fig. 8.6.

The following two-dimensional power curve can be obtained through the power transfer analysis under different modes. The red and blue solid lines indicate the maximum power transfer capacity of the mode 1 and mode 2 at different conversion gains respectively. The four dashed lines indicate the instantaneous power curves

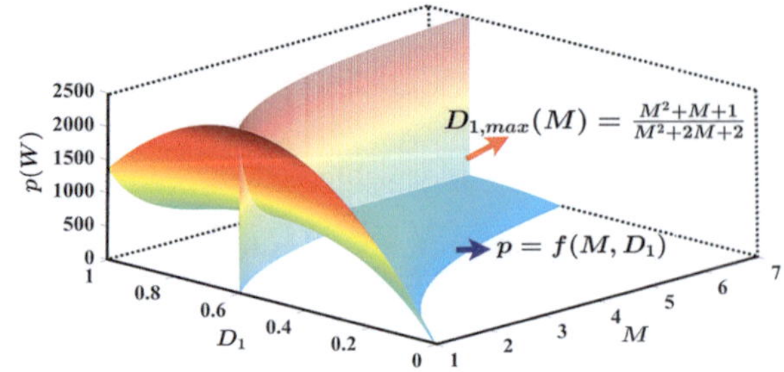

Fig. 8.6 The power transfer capability of mode 2

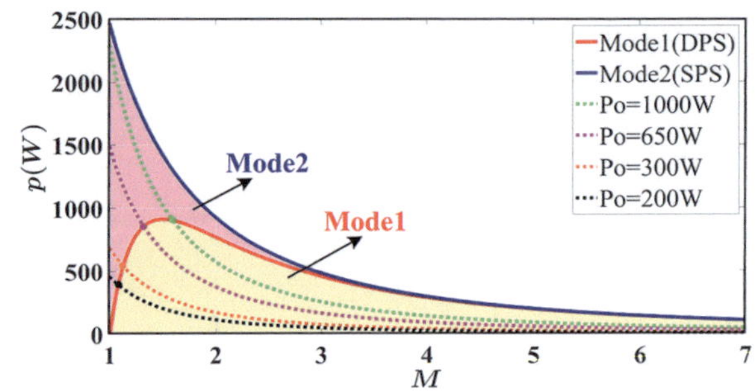

Fig. 8.7 The maximum power transfer capability of two modes and required power from grid side

transferred by the converter as the load power is 1000, 650, 300, and 200 W individually, which can be got through the required instantaneous power expression as (8.8). From Fig. 8.7 it is seen that the required instantaneous power p_{require} from the grid increases with the decrease in the conversion M, the maximum power transfer ability of mode 1 can satisfy the requirement when the dashed lines are located below the red solid line. However, the dashed lines will locate above the red solid line with the further decrease of the conversion gain M, the maximum power transfer ability of mode 1 cannot satisfy the requirement, so the required instantaneous power is provided by mode 2. The zone of mode 1 lies between the horizontal axis and the red solid line, and mode 2 stays between the red and blue solid lines.

$$\begin{cases} p_{\text{require}} = |v_{\text{ac}} \cdot i_{\text{ac}}| = 2P_o \sin(\omega t)^2 \\ M = \frac{nv_o}{|v_{\text{ac}}|} = \frac{nv_o}{V_m|\sin(\omega t)|} \end{cases} \tag{8.8}$$

8.2.3 Control Strategy and Mode Transition Mechanism

The main idea of the control strategy in this paper is that the average input current value of the semi-DAB varies as the required sinusoidal current of the power grid. So the average input current value of the DAB converter needs to be calculated under different modes. P.U. (Per Unit) values are used to simplify the analysis and calculation. The base value of the current is given by:

$$I_{\text{base}} = \frac{v_o T}{2nL_r} \tag{8.9}$$

In mode 1, the average input current value of semi-DAB and the P.U. value of input reference current are shown in (8.10), the control variable can be obtained as shown in (8.11).

$$\begin{cases} i^*_{in(av)} = \frac{D_1^2}{M-1} \\ i^*_{ref} = \frac{A}{I_{base}} \cdot \sin(\omega t) = A^* \cdot \sin(\omega t) = \frac{\frac{2P_O}{V_m}}{\frac{v_O T}{2nL_r}} \cdot \sin(\omega t) \\ i^*_{in(av)} = i^*_{ref} \end{cases} \tag{8.10}$$

$$\begin{cases} D_1 = \sqrt{i^*_{ref} \cdot (M-1)} = \sqrt{A^* \cdot \sin(\omega t) \cdot (M-1)} \\ D_2 = \frac{D_1}{M-1} \end{cases} \tag{8.11}$$

In mode 2, the average input current value of semi-DAB and the P.U. value of input reference current are shown in (8.12), the control variable can be obtained as shown in (8.13).

$$\begin{cases} i^*_{in(av)} = \frac{4D_1M^2-4D_1^2M-4D_1^2-2D_1^2M^2+4D_1M+4D_1-2M^2+M+1}{(M+2)^2} \\ i^*_{ref} = \frac{A}{I_{base}} \cdot \sin(\omega t) = A^* \cdot \sin(\omega t) = \frac{\frac{2P_O}{V_m}}{\frac{v_O T}{2nL_r}} \cdot \sin(\omega t) \\ i^*_{in(av)} = i^*_{ref} \end{cases} \tag{8.12}$$

$$\begin{cases} D_1 = \frac{(M+2)^2}{2(M^2+2M+2)} \cdot \left[\frac{2(M^2+M+1)}{(M+2)^2} - \frac{\sqrt{2} \cdot \sqrt{M - 2 \cdot A^* \sin(\omega t) - M^2 \cdot A^* \sin(\omega t) - 2M \cdot A^* \sin(\omega t) + 1}}{(M+2)} \right] \\ D_2 = 1 - D_1 \end{cases} \tag{8.13}$$

The mode transition mechanism can be obtained through the boundary condition between mode 1 and mode 2 as shown in the following,

$$\begin{cases} 0 < i^*_{ref} = A^* \sin(\omega t) < \frac{M-1}{M^2} \quad \text{(Mode 1)} \\ \frac{M-1}{M^2} < i^*_{ref} = A^* \sin(\omega t) < \frac{\frac{2P_{O\max}}{V_m}}{\frac{v_O T}{2nL_r}} \quad \text{(Mode 2)} \end{cases} \tag{8.14}$$

The operating zone of different modes and the mode transition mechanism with the proposed control strategy are shown in Fig. 8.8.

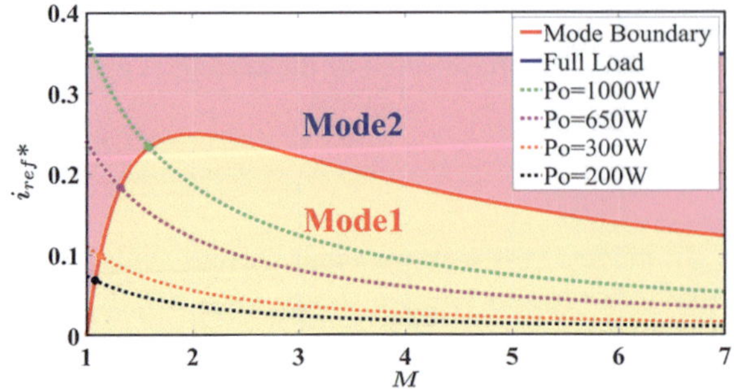

Fig. 8.8 The mode transition mechanism

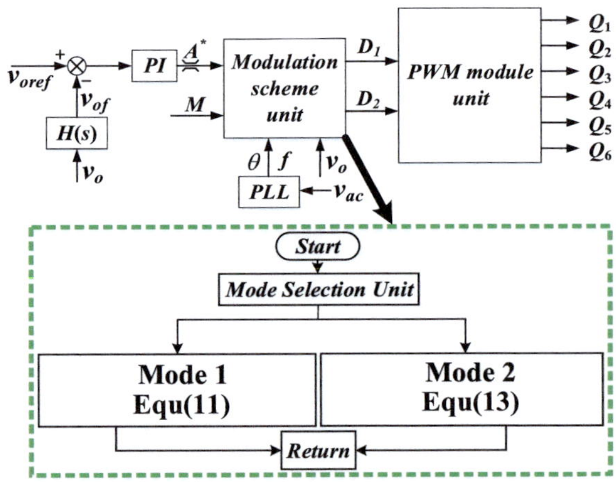

Fig. 8.9 Closed loop control diagram

The proposed closed loop diagram is shown in Fig. 8.9. The output voltage is controlled by the voltage loop through the PI controller, which outputs the P.U. magnitude value A^* for the input current reference. The phase and frequency of the power grid voltage can be obtained through the PLL module. The value A^*, the conversion gain M and the PLL output are used to calculate the control variables D_1 and D_2 under different modes according to mode transition principle and criteria. The PWM module generates all the gate drive signals for all the switches based on the values of D_1 and D_2.

8.2.4 Transition Mode Between Mode 1 and Mode 2

There is a transition mode between mode 1 and mode 2 due to the existence of the leakage inductance bias current in mode 1, the detailed transition process from mode 1 to mode 2 is given in Fig. 8.10 where Fig. 8.10a is the mode1, Fig. 8.10b is the boundary between mode 1 and the transition mode, Fig. 8.10c is the transition mode

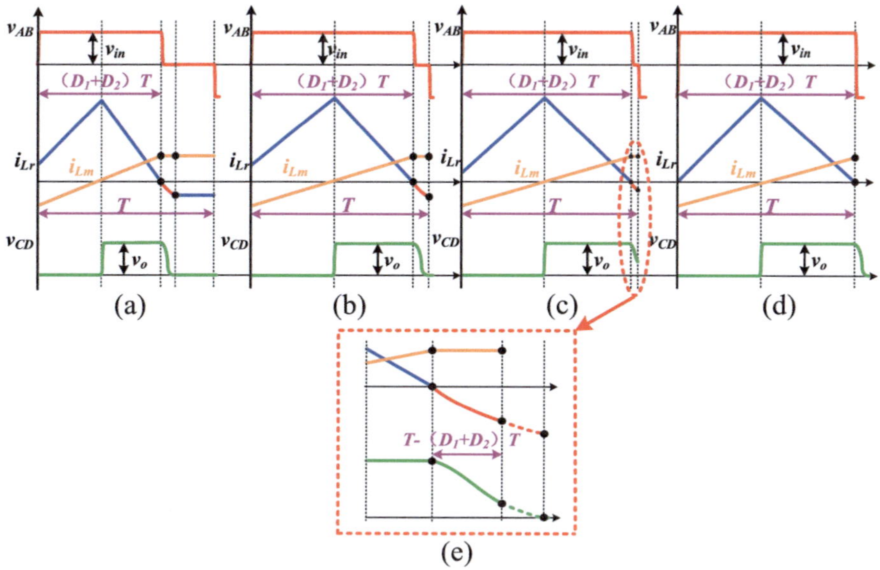

Fig. 8.10 Transition process from mode 1 to mode 2

and Fig. 8.10d is the boundary between the transition mode and mode 2. In order to show the mode transition process more clearly, the enlarged view of Fig. 8.10c is given in Fig. 8.10e.

The equivalent circuit to describe the resonant process for Q_5 and D_{o1} in transition mode is shown in Fig. 8.11, in which the resonant process can be described by the differential Eq. (8.15), where combining the initial condition yields analytical solutions as (8.17). The sign of the completion of the resonant process is that v_{ds5} drops to zero. From (8.17), the quarter of the resonant process time (the complete charging and discharging time for junction capacitor of Q_5 and D_{o1}) can be written as (8.18). The resonant current in the leakage inductor of the transformer under the transition mode is shown in (8.19).

$$\begin{cases} L_r \frac{di_{Lr}}{dt} = v_{ds5} \\ i_{Lr} = C_{Do1}\frac{d(v_o - v_{ds5})}{dt} - C_{Q5}\frac{dv_{ds5}}{dt} = -(C_{\text{oss}} + C_D)\frac{dv_{ds5}}{dt} \end{cases} \tag{8.15}$$

$$\begin{cases} i_{Lr}(0) = 0 \\ v_{ds5}(0) = v_o \end{cases} \tag{8.16}$$

$$\begin{cases} i_{Lr}(t) = v_o\sqrt{\frac{C_{\text{oss}}+C_D}{L_r}}\sin\left(\frac{t}{\sqrt{L_r(C_{\text{oss}}+C_D)}}\right) \\ v_{ds5}(t) = v_o\cos\left(\frac{t}{\sqrt{L_r(C_{\text{oss}}+C_D)}}\right) \end{cases} \tag{8.17}$$

$$i_{Lr}(T_r/4) = v_o\sqrt{\frac{C_{\text{oss}} + C_D}{L_r}} \tag{8.18}$$

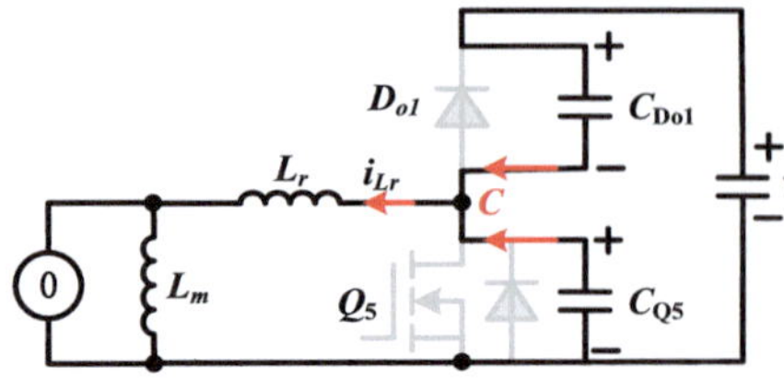

Fig. 8.11 Resonant process of Q_5 and D_{o1} in transition mode

$$i_{\text{resonant}} = v_o \sqrt{\frac{C_{\text{oss}} + C_D}{L_r}} \sin\left(\frac{T - (D_1 + D_2)T}{\sqrt{L_r(C_{\text{oss}} + C_D)}}\right) \tag{8.19}$$

The aforementioned analysis related to the transition mode is useful for the soft-switching range analysis.

8.3 Converter Design and Analysis

8.3.1 Leakage Inductance

The leakage inductor determines the power transmission capability. The maximum instantaneous power $p_{o,\max}$ with this control strategy lies in mode 2 at the maximum absolute value for the gird voltage. The leakage inductance can be designed as the following,

$$\begin{cases} p_{o,\max} > 2P_{\text{rated}} \\ p = \frac{v_o^2 T}{2L_r} \cdot \left[\frac{4D_1M^2 - 4D_1^2M - 4D_1^2 - 2D_1^2M^2}{M(M+2)^2} + \frac{4D_1M + 4D_1 - 2M^2 + M + 1}{M(M+2)^2}\right] \\ D_{1,\max}(M) = \frac{M^2+M+1}{M^2+2M+2} \end{cases} \tag{8.20}$$

Substituting the system parameters (Table 8.3) into (8.20) yields the maximum value of L_r as the following expression.

$$L_r < \frac{v_o^2 T(M_{\min} + 1)}{2P_{\text{rated}}(2M_{\min}^3 + 4M_{\min}^2 + 4M_{\min})} = 73\,\mu\text{H} \tag{8.21}$$

8.3.2 ZVS Range for Primary Side Switches Analysis

To achieve ZVS for the switches, the energy stored in the transformer leakage inductor should be sufficient to charge or discharge the parasitic capacitance of the power switches completely within the dead time.

The ZVS condition of primary side switches Q_1–Q_4 is given as follows,

$$I_{p.\mathrm{ZVS}} = \frac{|v_{\mathrm{ac}}|}{\sqrt{(n^2 L_r)/C_{\mathrm{eq.ac}}}} \tag{8.22}$$

where $I_{p.\mathrm{ZVS}}$ is the minimum current to achieve ZVS for the primary side switches, and $C_{\mathrm{eq.ac}}$ is the equivalent capacitance of a bridge leg on the AC side.

The primary side current $|i_p(T)|$ of transformer at time T determines the ZVS condition for Q_1–Q_4, so $|i_p(T)|$ in different modes needs to be calculated respectively. The corresponding switches can realize ZVS on when $|i_p(T)| > I_{p.\mathrm{ZVS}}$.

In mode 1:

$$\begin{cases} p = 2P_o \cdot \left(\frac{|v_{\mathrm{ac}}|}{V_m}\right)^2 \\ p = \frac{v_o^2 T}{2L_r} \cdot \frac{D_1^2}{M(M-1)} \end{cases} \tag{8.23}$$

$$D_1 = f_1(|v_{\mathrm{ac}}|, P_o) \tag{8.24}$$

$$\left|i_p(T)\right| = \frac{1}{n} \cdot \left[\frac{\frac{|v_{ac}|}{n \cdot L_m} \cdot (D_1 + D_2)T}{2} - v_o \sqrt{\frac{C_{\mathrm{oss}} + C_D}{L_r}} \right] \tag{8.25}$$

In transition mode:

The primary side current $|i_p(T)|$ of transformer at T under the transition mode can be derived from (8.19).

$$\left|i_p(T)\right| = \frac{1}{n} \cdot \left[\frac{\frac{|v_{ac}|}{n \cdot L_m} \cdot (D_1 + D_2)T}{2} - v_o \sqrt{\frac{C_{\mathrm{oss}} + C_D}{L_r}} \sin\left(\frac{T - (D_1 + D_2)T}{\sqrt{L_r (C_{\mathrm{oss}} + C_D)}} \right) \right] \tag{8.26}$$

In mode 2:

$$\begin{cases} p = 2P_o \cdot \left(\frac{|v_{ac}|}{V_m}\right)^2 \\ p = \frac{v_o^2 T}{2L_r} \cdot \left[\frac{4D_1 M^2 - 4D_1^2 M - 4D_1^2 - 2D_1^2 M^2}{M(M+2)^2} + \frac{4D_1 M + 4D_1 - 2M^2 + M + 1}{M(M+2)^2} \right] \end{cases} \tag{8.27}$$

$$D_1 = f_2(|v_{ac}|, P_o) \tag{8.28}$$

$$\left|i_p(T)\right| = \frac{1}{n} \cdot \left[\frac{v_o T}{L_r} \cdot \frac{(1 + M)(1 - M + MD_1)}{M(M + 2)} + \frac{|v_{ac}|}{n \cdot L_m} \times \frac{T}{2} \right] \tag{8.29}$$

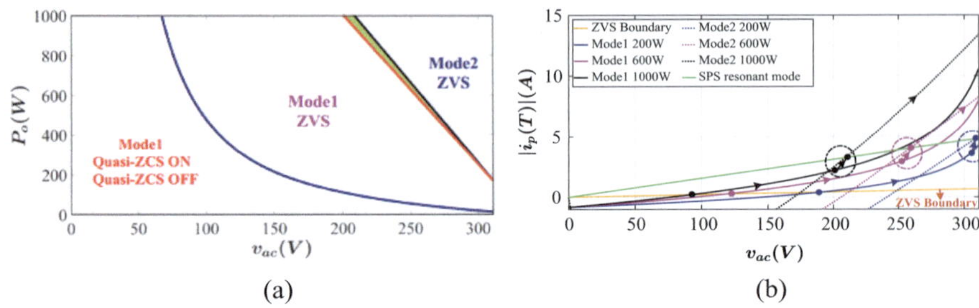

Fig. 8.12 Soft switching range of Q_1–Q_4

The ZVS range of primary side switches is given in Fig. 8.12. The blue curve indicates the boundary of the quasi-ZCS on/off region and ZVS region within mode 1, the red line indicates the boundary of mode 1 and the transition mode, and the black line represents the transition mode and mode 2. The region between the red line and black line is the operation region of the transition mode. As seen in Fig. 8.12a, Q_1–Q_4 can realize ZVS on at the right side of the blue curve in mode1, the transition mode and mode 2, and Q_1–Q_4 can realize quasi-ZCS on and quasi-ZCS off at the left side of the blue curve in mode 1 due to the very low turn on/off current. Figure 8.12a also reflects the mode operation region. As seen, the proportion of mode 2 increases as the load increases. The converter operates at mode 1 only when the load is very low.

Figure 8.12b also shows the ZVS range and the mode transition process, especially the transition mode (inside the dotted circles) between mode 1 and mode 2. The converter operates along the direction indicated by the arrow, working at mode 1 shown in the solid lines firstly, then turning to the transition mode shown in the dotted circles, finally operating at mode 2 shown in the dotted lines.

8.3.3 ZVS Range for Secondary Side Switches Analysis

The ZVS condition of secondary side switches Q_5–Q_6 is shown as follows,

$$I_{\text{s.ZVS}} = \frac{v_o}{\sqrt{L_r/C_{\text{eq.dc}}}} \tag{8.30}$$

where $I_{s.\text{ZVS}}$ is the minimum current to achieve ZVS for the secondary side switches, and $C_{\text{eq.dc}}$ is the equivalent capacitance of a bridge leg on the DC side.

The leakage inductor peak current I_1 determines ZVS condition of Q_5–Q_6, so I_1 in different modes needs to be calculated respectively. The corresponding switches can realize ZVS on if $I_1 > I_{s.\text{ZVS}}$.

In mode 1:

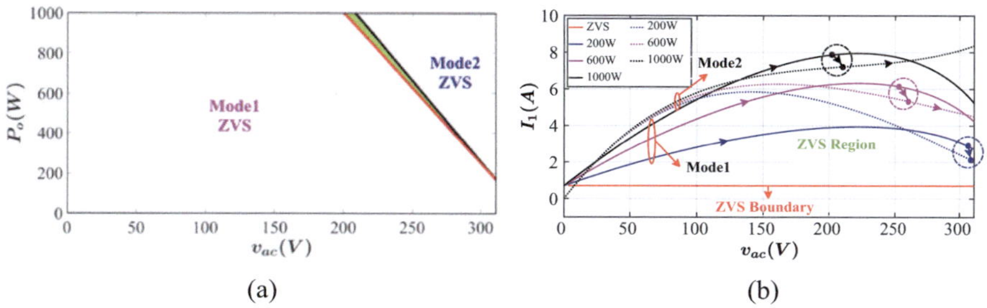

Fig. 8.13 Soft switching range of Q_5–Q_6

$$I_1 = v_o\sqrt{\frac{C_{\text{oss}} + C_D}{L_r}} + \frac{v_oT}{L_r} \cdot \frac{D_1}{M} \tag{8.31}$$

In transition mode:

$$I_1 = v_o\sqrt{\frac{C_{\text{oss}} + C_D}{L_r}} \sin\left(\frac{T - (D_1 + D_2)T}{\sqrt{L_r(C_{\text{oss}} + C_D)}}\right) + \frac{v_oT}{L_r} \cdot \frac{D_1}{M} \tag{8.32}$$

In mode 2:

$$I_1 = \frac{v_oT}{L_r} \cdot \frac{2D_1 + M - 1}{M(M + 2)} \tag{8.33}$$

Similarly, the ZVS range of secondary side switches is given in Fig. 8.13. The red line indicates the boundary of mode 1 and the transition mode, and the black line represents mode 2 and the transition mode. The region between the red line and black line is the operation region of the transition mode. As seen in Fig. 8.13a, Q_5–Q_6 can realize ZVS in the entire AC line voltage period within full load range. The ZVS range is so wide that it is helpful to improve the conversion efficiency.

Figure 8.13b shows the ZVS range of secondary side switches and the mode transition process, especially the transition mode (inside the dotted circles) between mode 1 and mode 2. The converter operates along the direction indicated by the arrow, which works at mode 1 shown in the solid lines firstly, then turning to the transition mode shown in the dotted circles, finally operates at mode 2 shown in the dotted lines.

As shown in Fig. 8.14, ZVS range of primary side switches is extended as the magnetizing inductance value is decreased. However, lower magnetizing inductance will cause higher $i_p(T)$ and this increases magnetic loss and current stress. Thus in order to make a tradeoff, the value of magnetizing inductance is designed as 230 μH.

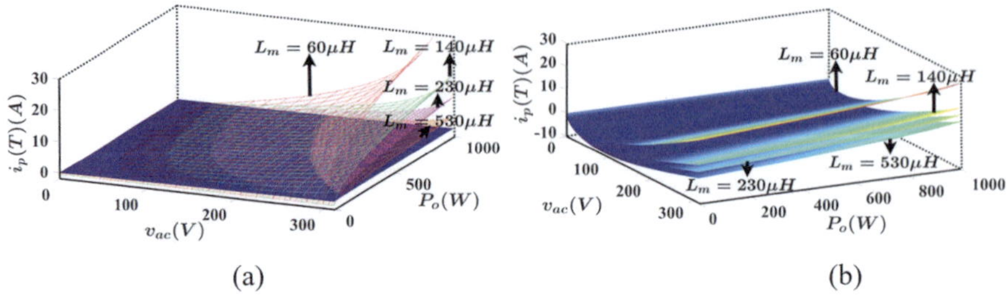

Fig. 8.14 Design of magnetizing inductance

8.4 Experimental Results

8.4.1 Experimental Parameters

A 1 kW experimental prototype as Fig. 8.15 was made to verify the effectiveness of the control strategy, the system parameters are listed in Table 8.3. The control and modulation strategy are implemented by DSP TMS320F28335.

Fig. 8.15 Experimental prototype

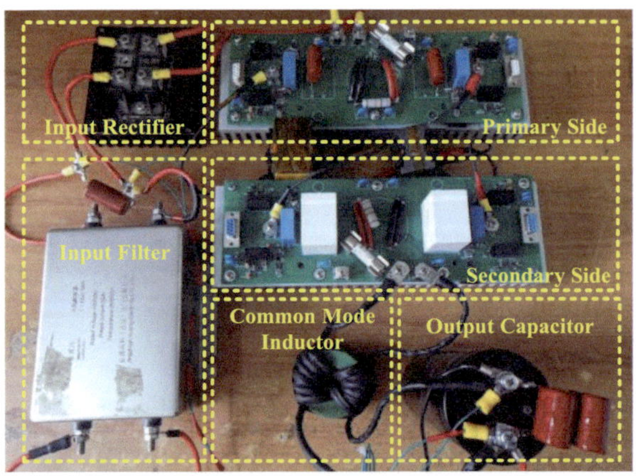

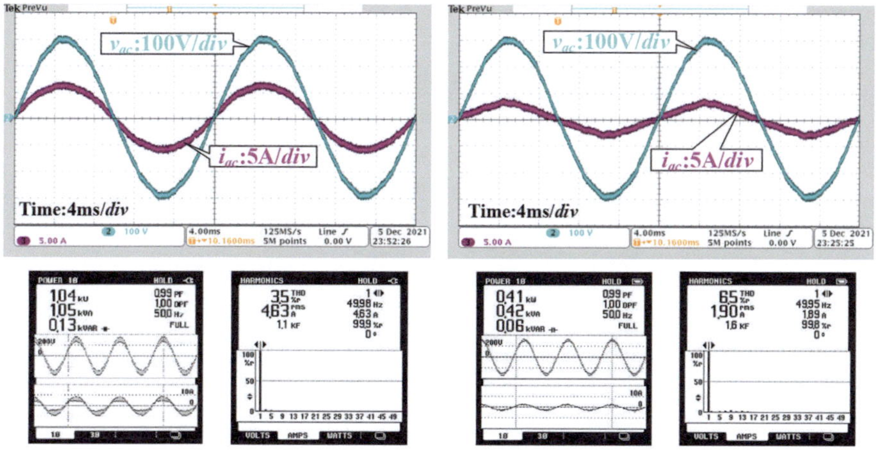

(a) (b)

Fig. 8.16 Grid side current voltage at steady state: **a** 1000 W, **b** 400 W

8.4.2 Steady-State Operating Waveforms

Figure 8.16 shows the steady state grid voltage and current waveforms at 1000 W and 400 W respectively. The PF value of the converter is 0.99 and 0.99 respectively. The THD is 3.5% and 6.5% respectively. High power quality can be obtained by using the proposed control.

8.4.3 Mode Operation and ZVS/ZCS Situations

Figure 8.17 shows the waveforms of v_{ab}, v_{cd}, i_{Lk}, i_p and different operation modes as the load is 1000 W and 400 W respectively. As shown in Fig. 8.17, the experiment results agree well with the previous theoretical analysis for the mode operating zone shown in Figs. 8.12a and 8.13a.

The peak current value of theoretical calculation, simulation and experiment under different voltage conversion gains when load power is 1000 W and 400 W are listed in Tables 8.4 and 8.5 respectively.

Figure 8.18 shows ZVS situation of different switches at different voltage conversion gains as the output power is 1000 W and 400 W. Only Q_2, Q_6 and D_{o2} are given to show the symmetry between the upper switch and the bottom switch in a half bridge. As illustrated in Fig. 8.18, the three figures in the first row show the switching state

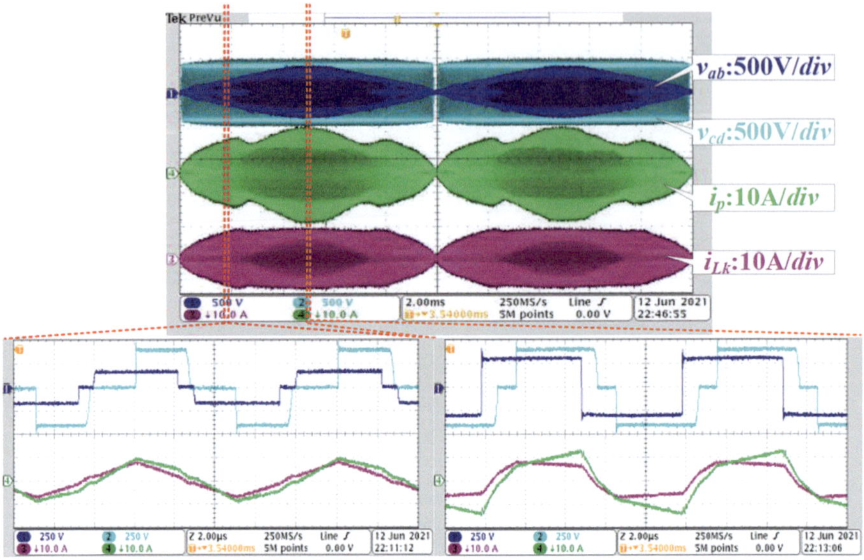

(a)

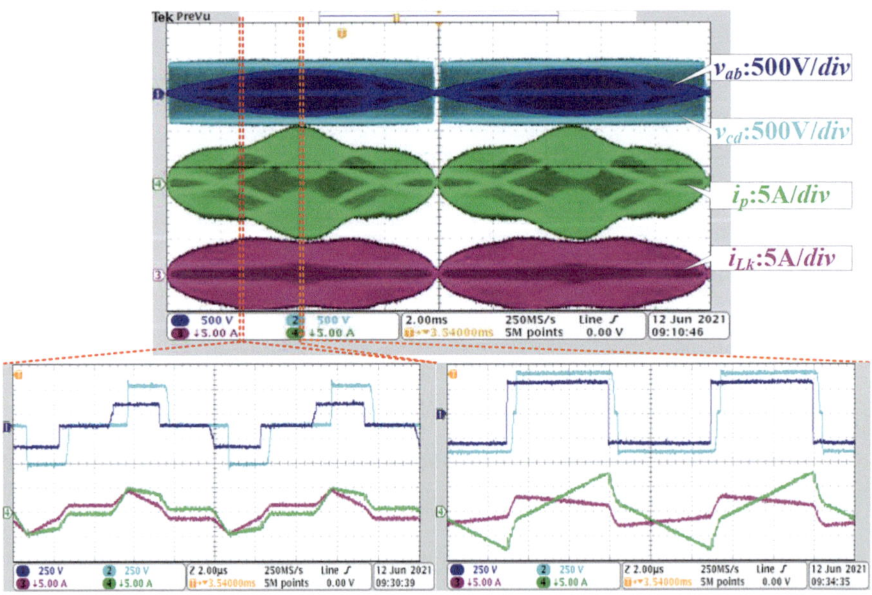

(b)

Fig. 8.17 Mode situation **a** 1000 W, **b** 400 W

Table 8.4 Peak current value of different conversion gain ($P_o = 1000$ W)

M	5	3	2	1.5	1.07
Calculated i_{peak} (A)	4.02	5.83	7.34	7.28	8.36
Simulated i_{peak} (A)	3.81	5.53	7.28	7.11	8.23
Experimental i_{peak} (A)	4.1	5.6	7.2	7.1	8.1

Table 8.5 Peak current value of different conversion gain ($P_o = 400$ W)

M	5	3	2	1.5	1.07
Calculated i_{peak} (A)	2.62	3.76	4.83	5.12	4.03
Simulated i_{peak} (A)	2.43	3.58	4.59	4.94	3.84
Experimental i_{peak} (A)	2.5	3.6	4.6	5.0	4.1

for Q_2 when v_{ac} is 48 V, 183 V, 311 V (M is 6.94, 1.82, 1.07) respectively, the three figures in the second row show the switching state for Q_6 with different v_{ac} values, the three figures in the third row represent the turn off waveform for the diode D_{o2} as v_{ac} is 48 V, 183 V, 311 V (M is 6.94, 1.82, 1.07) respectively. It can be found that Q_2 Can achieve ZVS when v_{ac} is 183 V and 311 V. Although ZVS on is lost for Q_2 as v_{ac} is 48 V, it can be viewed as almost ZCS on. Q_6 can achieve ZVS on in the entire AC voltage period within the full load range while D_{o2} can obtain ZCS off throughout the entire AC voltage period within the full load range. The experiment results of ZVS on for switches and ZCS off for diodes agree well with the previous theoretical analysis as shown in Figs. 8.12a and 8.13a.

8.4.4 Dynamic Response and Conversion Efficiency

The dynamic response performance of this converter is shown in Fig. 8.19 as the load is switched from 400 to 800 W. The output voltage drops a little in facing the load transient. Figure 8.20 shows the conversion efficiency curve, the peak value is 95.9%.

8.4.5 Comparison with Other Works

Table 8.6 represents the comparison with similar solutions. It can be found that, compared with other works, the proposed topology has only six switches and operates with higher constant switching frequency due to the use of SiC MOSFETs. The

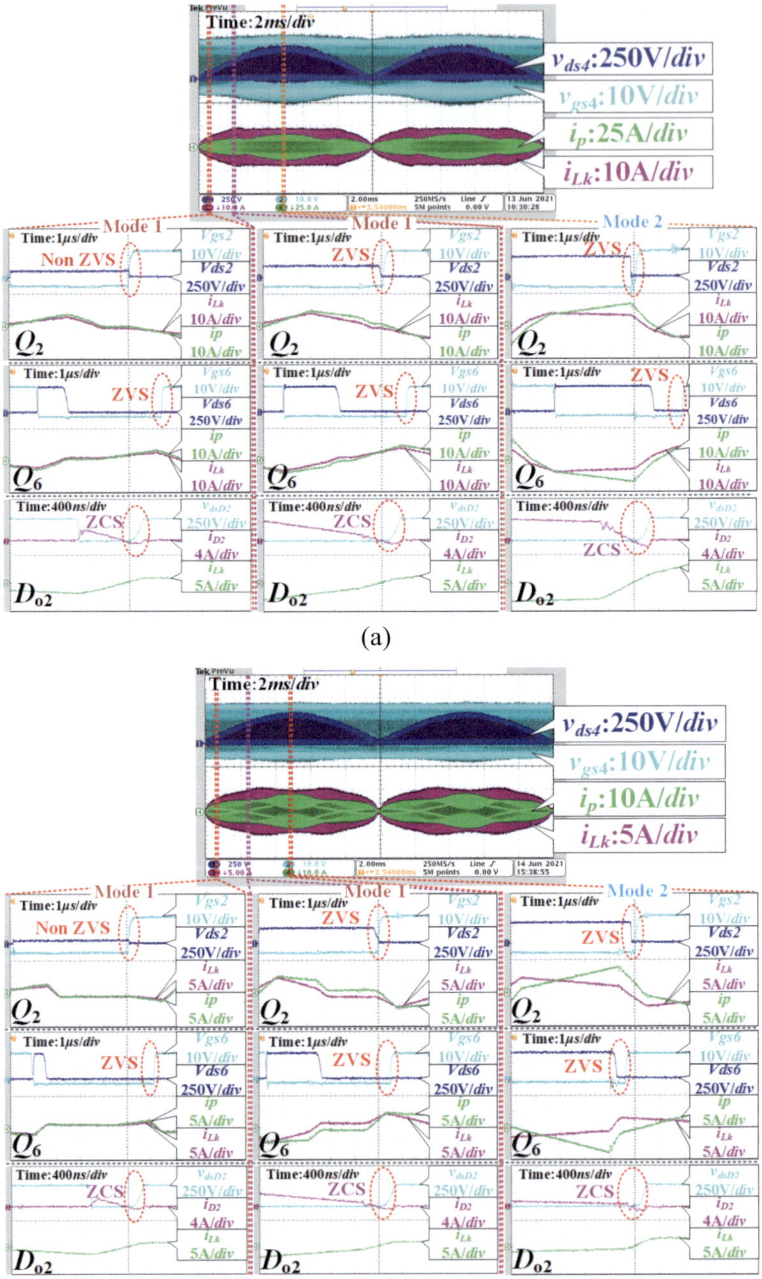

Fig. 8.18 ZVS/ZCS situation **a** 1000 W, **b** 400 W

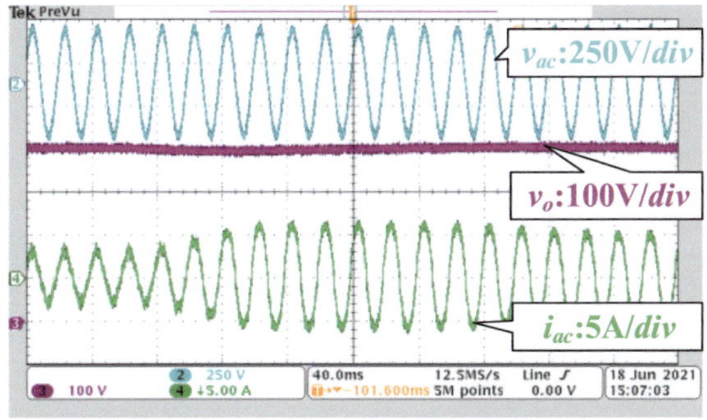

Fig. 8.19 Dynamic response of load transient from 400 to 800 W

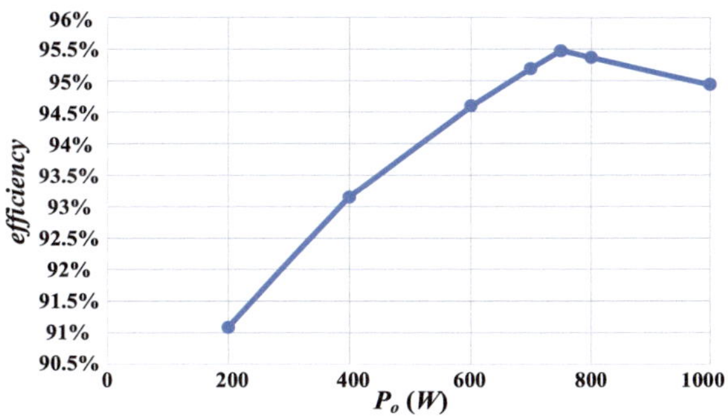

Fig. 8.20 Efficiency curve

control strategy is simple compared with other works without using inner current loops and look up tables. The AC interface to the grid is the active bridge with lower forward voltage drop in [3], while the diode rectifier used in this paper has higher conduction loss, and thus the peak efficiency 95.5% is a little less than that in Ref. [3] but still higher than most of the existing solutions.

Table 8.6 Comparison with other works

	DAB AC/DC [9]	DAB AC/DC [3]	DAB AC/DC [10]	Proposed one
Switch Numbers	12	12	12	6
Maximum power (kW)	1	3.7	1.5	1
Control strategy	Constant-switching frequency multimode	Variable- frequency with look-up tables	Constant-switching frequency multimode	Constant-switching frequency multimode
Switching frequency (kHz)	50	75–125	50	100
Transformer turns ratio	1:2	1:1	1:2	1:1.2
Operating modes	Two modes	Two modes	Two modes	Two modes
AC interface	Active bridge	Active bridge	Active rectifier	Diode rectifier
Current loop	/	√	/	/
Control complexity	Medium	Complex	Medium	Simple
Maximum efficiency (%)	92.4	96.2	94.5	95.5

8.5 Conclusion

A novel single-stage AC-DC rectifier based on the semi-DAB DC-DC converter is presented. The contribution can be summarized as the following,

1. The semi DAB topology is firstly proposed to be interfaced with folded grid AC voltage and the conversion gain may reach infinite theoretically.
2. The detailed transition process between the operating modes are analyzed, and the mode transition is seamless without needing to judge the output power.
3. For the semi DAB, the proposed control strategy can achieve high power factor correction without inner current tracking loop automatically.
4. With the proposed control, the primary side switches can achieve ZVS on or quasi-ZCS on/off within a wide operating range, the secondary side switches can achieve ZVS on and diodes can obtain ZCS off in the entire ac voltage period throughout the entire load.

The effectiveness of the control strategy is verified by a 1 kW experimental prototype.

References

1. Wang M, Guo S, Huang Q, Yu W, Huang AQ (2017) An isolated bidirectional single-stage DC–AC converter using wide-band-gap devices with a novel carrier-based unipolar modulation technique under synchronous rectification. IEEE Trans Power Electron 32(3):1832–1843
2. Wang M, Huang Q, Guo S, Yu X, Yu W, Huang AQ (2018) Soft-switched modulation techniques for an isolated bidirectional DC–AC. IEEE Trans Power Electron 33(1):137–150

3. Everts J, Krismer F, Van den Keybus J, Kolar JW (2014) Optimal ZVS modulation of single-phase single-stage bidirectional DAB AC–DC converters. IEEE Trans Power Electron 29(8):3954–3970
4. Jauch F, Biela J (2016) Combined phase-shift and frequency modulation of a dual-active-bridge AC-DC converter with PFC. IEEE Trans Power Electron 31(12):8387–8397
5. Sha D, Wang S (2021) A single-stage natural power factor corrector based on dual active bridge DC–DC converter without inner current tracking loop. IEEE Trans Power Electron 36(1):342–352
6. Zhang J, Sha D, Ma P (2022) A dual active bridge DC-DC based single stage AC-DC converter with seamless mode transition and high power factor. IEEE Trans Ind Electron 69(2):1411–1421
7. Kulasekaran S, Ayyanar R (2014) Analysis, design, and experimental results of the semidual-active-bridge converter. IEEE Trans Power Electron 29(10):5136–5147
8. Li F, Li Y, You X (2018) Optimal dual-phase-shift control strategy of an isolated buck-boost converter with a clamped inductor. IEEE Trans Power Electron 33(6):5374–5385
9. Li X, Wu F, Yang G, Liu H (2020) Improved modulation strategy for single-phase isolated quasi-single-stage AC-DC converter to improve current characteristics. IEEE Trans Power Electron 35(4):4296–4308
10. Luo S, Wu F, Wang G (2022) Single-stage hybrid three-level type resonant AC-DC converter. IEEE Trans Transp Electrif 8(1):799–807

Chapter 9
Single-Stage Semi-DAB AC-DC Converter with Decoupled Inductor Current Modulation Strategy

Abstract For AC-DC converter to charge a battery whose voltage range is wide, in the last chapter, an optimal peak current modulation strategy is proposed for the semi-DAB based single stage high frequency isolated AC-DC converter, but the output voltage is fixed, and there are only two modes as $M > 1$ within entire line frequency cycle, the turns ratio of the high frequency transformer is not optimal. The effect of the modulation strategies on THD of i_{ac} and the mechanism are not considered. The backflow power of the converter is relatively large. So in this chapter, a novel decoupled inductor current modulation strategy for a single-stage AC-DC converter based on semi-DAB is proposed, which helps to minimize the total harmonics distortion (THD) and improve the conversion efficiency. The transformer turns ratio is designed in an optimized method for higher conversion efficiency. Only two control variables are used. And the converter could achieve wider output voltage range and high power factor without mandatory current tracking loop and look-up tables. In addition, losses, THD and backflow power comparisons between the decoupled inductor current and the optimal peak current modulation are presented. The power transfer capability, mode transition mechanism, soft switching range and mode operation region and parameter design are analyzed. Switches can achieve zero-voltage-switching (ZVS) within a wide operating range and rectifier diodes can achieve zero-current-switching (ZCS) off in all operation situations. Besides, a 1 kW experimental prototype was built to verify the effectiveness of the proposed control strategy.

Keywords Semi dual-active-bridge (SDAB) · Single-stage AC-DC converter · Wider output voltage range · Total harmonic distortion (THD) · Decoupled inductor current modulation strategy

D. Sha et al., *High Frequency Isolated Semi Dual Active Bridge DC-DC and AC-DC Converter*, CPSS Power Electronics Series,
https://doi.org/10.1007/978-981-95-1641-4_9

9.1 Circuit Topology and Control Strategy

9.1.1 Power Circuit Topology

A single-stage isolated AC-DC converter composed of a synchronous rectifier followed by DAB is proposed in [1, 2]. A more accurate ZVS condition (current-dependent charge-based) is presented, wherein this control strategy requires lookup tables for real-time closed-loop control and extra inductor current will cause more magnetic loss. An improved hybrid current modulation scheme is proposed in [3], which solves the current distortion problem without increasing the root mean square (RMS) value of current, while the large switching loss reduces the efficiency. A novel control strategy is proposed in [4], which adopts the TPS modulation scheme and does not need inner current loop and lookup tables, but ZVS range becomes narrow as the load increases. To extend ZVS range, a multi-mode control with seamless transition is presented [5], but the peak or RMS value of the transformer leakage inductor current is not optimal. The nonlinear parasitic output capacitance is analyzed for the AC-DC converter [6]. The nonidealities that existed in the circuit will cause the error of theoretical analyses. A modeling approach of the DAB AC-DC converter is presented in [7] to reduce the modeling error. In addition, some single-stage AC-DC converter based resonant topologies have also been proposed.

The effect of most modulation strategies on THD of i_{ac} and the mechanism are seldom considered in previous works. And the backflow power of the converter is seldom optimized in most literatures. In [8], only two modes of $M > 1$ exist in the entire a line frequency period, the turns ratio is not optimal, and the output voltage is fixed. A novel modulation strategy has been proposed in this paper, the input voltage is 220 V (RMS) and the output voltage is 270–400 V (DC), and is preferred to be used in the OBC, which could overcome the above shortcomings, lower the THD of i_{ac}, and improve the conversion efficiency. The backflow power can be minimized further.

The single-stage isolated AC-DC converter based semi-DAB is shown in Fig. 9.1. The grid voltage v_{ac} is converted to DC voltage v_{in} through the diode rectifier and is fed to the semi-DAB DC-DC converter. Its primary side is composed of four MOSFETs (Q_1–Q_4). The secondary side is composed of two MOSFETs (Q_5–Q_6) and two diodes (D_{o1}–D_{o2}). The high frequency (HF) capacitor C_{in} with very low capacitance is used to bypass the high frequency current. The turns ratio of transformer is n:1. L_r represents the leakage inductance and L_m is the magnetizing inductance. M is the voltage conversion gain and is defined as $M = nv_o/|v_{ac}|$.

9.1.2 Operating Modes and Circuit Operating Principle

In [9], three basic modulation modes under $M > 1$and three basic modulation modes under $M <= 1$ are given respectively. Based on the boundary condition between these

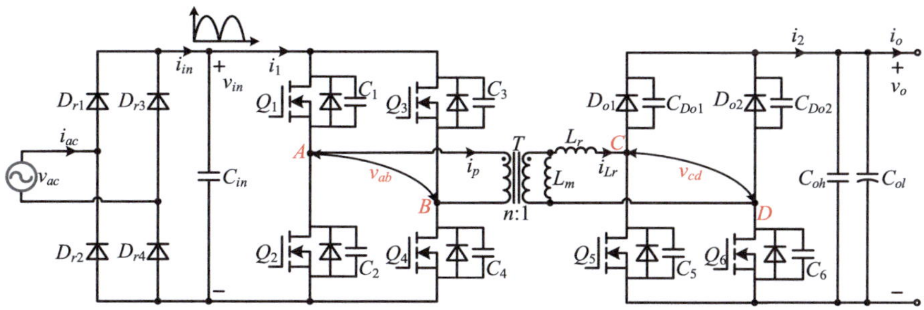

Fig. 9.1 Single-stage single-phase AC-DC topology based on semi-DAB

modes, the decoupled inductor current modulation strategy (which will be explained in detail in the Sect. 9.3.2) as proposed in Fig. 9.2. Figure 9.2a, b are modes under $M > 1$, while Fig. 9.2c, d are modes under $M <= 1$. D_1T represents the time length of Q_1 leading Q_6, D_2T represents the time duration length in which v_{AB} and v_{CD} have the same non zero polarity. It is worth noting that $D_1 + D_2 <= 1$. T represents the half switching period and the entire switching period is $T_S = 2T$.

Control trajectories or constraints between D_1, D_2 of four modes and the intersection points between mode 1 and mode 2, mode 3 and mode 4 are listed in the following Table 9.1.

The maximum instantaneous power transferred by the semi-DAB converter within the entire a line frequency period occurs in mode 4, where,

$$\begin{cases} p_{o,\max} = \frac{v_o^2 T}{2L_r} \cdot \frac{1}{M^2+M+1} \\ D_{1,\max}(M) = \frac{M^2}{M^2+M+1} \\ D_{2,\max}(M) = \frac{M}{M^2+M+1} \end{cases} \quad (9.1)$$

Two-dimensional control trajectories under $M > 1$ and $M < 1$ according to Table 9.1 are shown in Fig. 9.3 respectively. Six regions in Fig. 9.3 represent six basic modulation modes in [16], and four different color lines indicate four modes respectively. The semi-DAB needs to adapt extremely wide voltage gain variation because the input voltage is a sinusoidal wave. The maximum and minimum voltage gain M lie at the minimum and maximum value of $|v_{ac}|$ respectively. So the control trajectory of the decoupled inductor current strategy need to be considered when the voltage gain decreases from infinity to $M_{\min}$. Three-dimensional control trajectories under $M > 1$ and $M < 1$ are shown in Figs. 9.4 and 9.5 respectively. They are a set of two-dimensional control trajectories under different gains in fact.

In Fig. 9.4, the light blue surface refers to the set of control trajectories of mode 1 at different voltage gains, and the orange surface represents the set of control trajectories of mode 2 at different voltage gains. For example, the intersection line of the purple section plane, light blue and orange surface is the control trajectory under $M = 2$, which agrees well with Fig. 9.3a. The same case is similar for $M = 5$. The

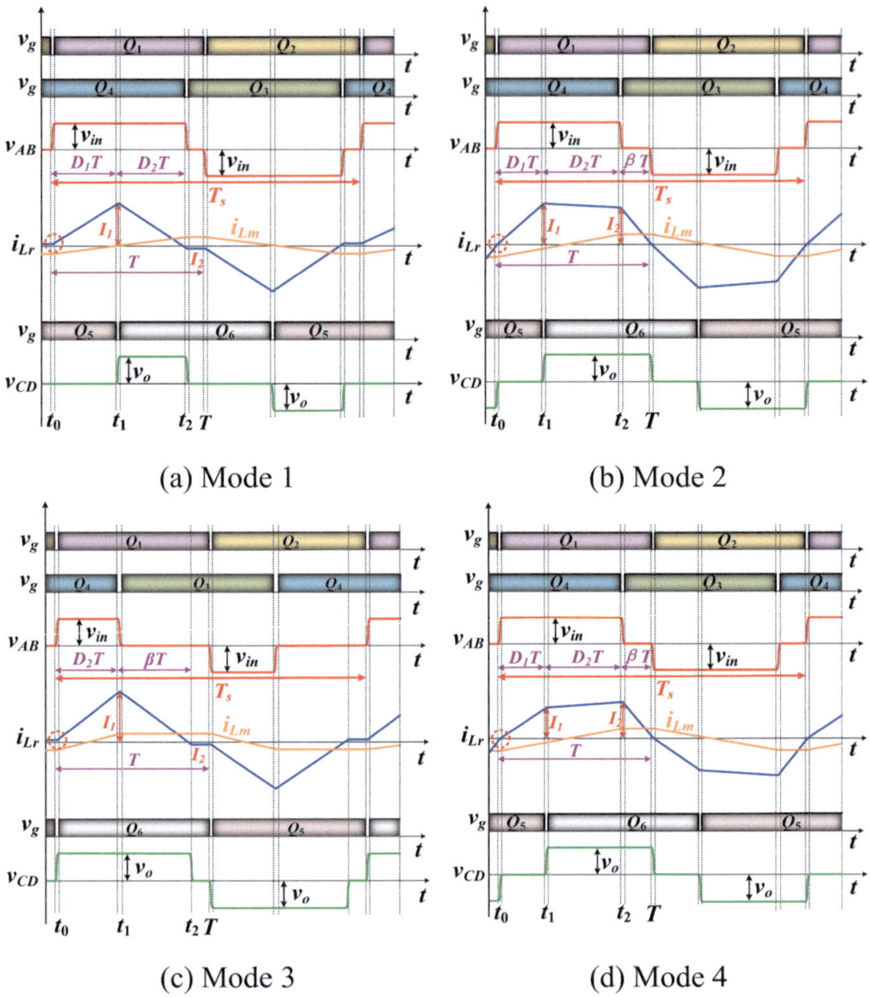

Fig. 9.2 The decoupled inductor current modulation strategy under M > 1 and M < 1

Table 9.1 Control trajectory and intersection point of different modes

Gain	Control trajectory	Intersection point
$M > 1$	$\begin{cases} D_2 = \frac{D_1}{M-1} & \text{(Mode 1)} \\ D_2 = M - (1+M)D_1 & \text{(Mode 2)} \end{cases}$	$\begin{cases} D_1 = 1 - \frac{1}{M} \\ D_2 = \frac{1}{M} \end{cases}$
$M < 1$	$\begin{cases} D_1 = 0 & \text{(Mode 3)} \\ D_2 = M - (1+M)D_1 & \text{(Mode 4)} \end{cases}$	$\begin{cases} D_1 = 0 \\ D_2 = M \end{cases}$

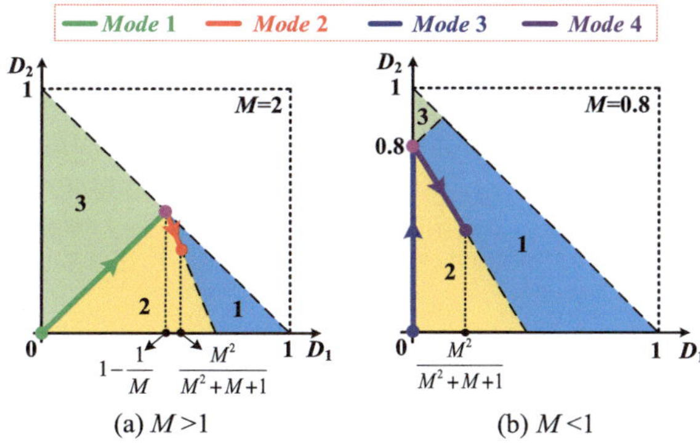

Fig. 9.3 Control trajectories of the decoupled inductor current modulation

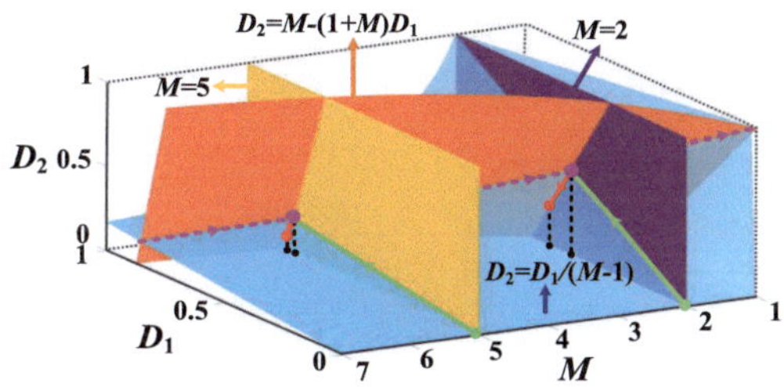

Fig. 9.4 Control trajectories of the decoupled inductor current strategy of M > 1

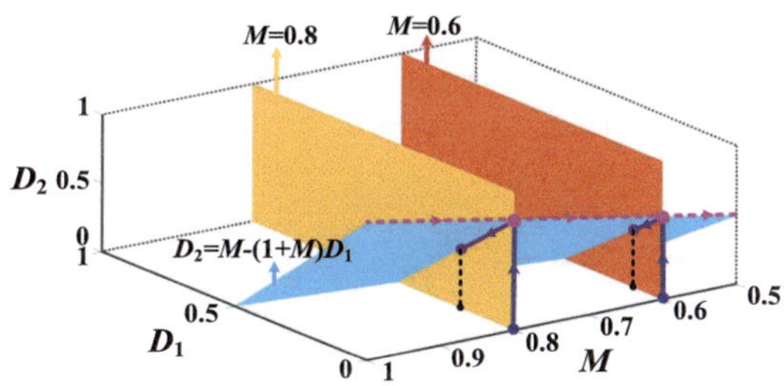

Fig. 9.5 Control trajectories of the decoupled inductor current strategy of M < 1

mode transition point moves along the pink dotted line as the voltage gain decreases. The analysis for the case of $M < 1$ is similar as shown in Fig. 9.5.

9.1.3 Power Transfer Analysis

The power transmitted within a switching period could be calculated by the following expression:

$$p = \frac{\int_{t_o}^{T} [i_{Lr}(t) \cdot v_{AB}(t)/n]dt}{T} \quad (9.2)$$

The power transfer expression under the corresponding control trajectory (Table 9.1) and the maximum value of the control variable of different modes are listed in Table 9.2 respectively. The three-dimensional power transfer surface of mode 1 with D_1 and M is shown in Fig. 9.6, the three-dimensional surface of $D_{1,\max}$(M) is also shown in Fig. 9.6. The intersection line of above two surfaces is the maximum power transfer capability curve of mode 1. The principle and analysis of mode 2, mode 3 and mode 4 are similar to mode 1, so are not shown again here.

Table 9.2 Power transfer expression of different modes

Mode	Power transfer expressions	The maximum of D_1 (D_2)
Mode 1	$p=\frac{v_o^2T}{2L_r}\cdot\frac{D_1^2}{M(M-1)}$	$D_{1,\max}(M)=1-\frac{1}{M}$
Mode 2	$p=-\frac{v_o^2T}{2L_r}\cdot[\frac{D_1^2M^2+D_1^2M}{M}+\frac{D_1^2-2D_1M^2+M^2-M}{M}]$	$D_{1,\max}(M)=\frac{M^2}{M^2+M+1}$
Mode 3	$p=\frac{v_o^2T}{2L_r}\cdot\frac{D_2^2(1-M)}{M^2}$	$D_{2,\max}(M)=M$
Mode 4	$p=-\frac{v_o^2T}{2L_r}\cdot[\frac{D_1^2M^2+D_1^2M}{M}+\frac{D_1^2-2D_1M^2+M^2-M}{M}]$	$D_{1,\max}(M)=\frac{M^2}{M^2+M+1}$

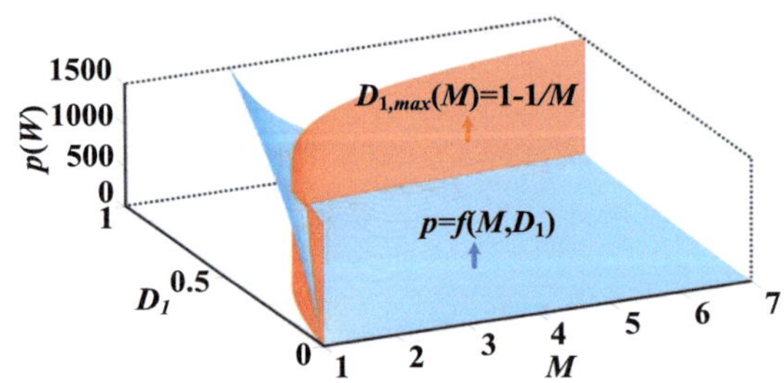

Fig. 9.6 The power transfer capability of mode 1

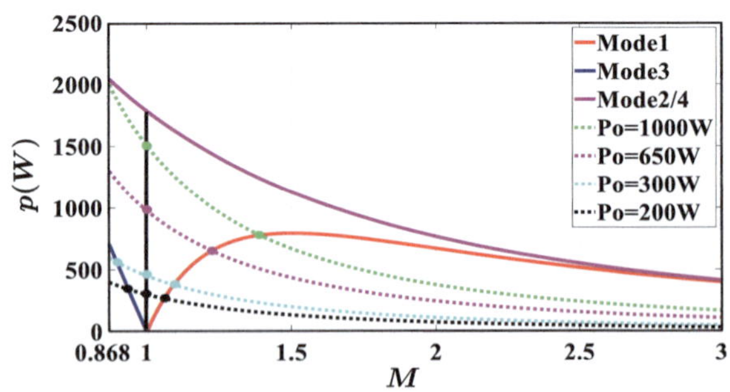

Fig. 9.7 Maximum power transfer capability of four modes and required instantaneous power from the grid side under different load powers

The following two-dimensional power transfer capability curve can be obtained through the above power transfer capability analysis of different modes. The red, blue and pink solid line refer to the maximum power transfer capability of the mode 1, mode 3 and mode 2/4 at different voltage gains respectively. The black solid line represents the boundary line ($M = 1$) of $M > 1$ and $M < 1$. Four dotted lines are required for instantaneous power curves transferred by the converter as the load power is 1000, 650, 300, and 200 W individually. As seen in Fig. 9.4, the required instantaneous power p_{require} from the grid increases as the voltage gain decreases. When $M > 1$, the maximum power transfer capability of mode 1 can satisfy the requirement when the dotted lines are located below the red solid line. However, the dotted line will locate above the red solid line and with further decrease of the voltage gain M, the maximum power transfer capability of mode 1 cannot satisfy the requirement, so the required instantaneous power is provided by mode 2. And the working principle is similar under $M < 1$.

9.1.4 Control Strategy and Mode Transition Mechanism

The control strategy is to make the average input current of the semi DAB equal to the required sinusoidal current of the power grid. So the average input current of semi-DAB can be calculated for different modes in terms of P.U. (Per Unit) values as the Table 9.3. The base value of the current is,

$$I_{\text{base}} = \frac{v_o T}{2nL_r} \tag{9.3}$$

The P.U. value of input reference current is shown in (9.4), two control variables can be obtained at any different electrical angle as shown in Table 9.4. Then power factor correction (PFC) can be realized naturally as long as the control is implemented by the real-time control variables D_1 and D_2.

$$i^*_{\text{ref}} = \frac{A}{I_{\text{base}}} \cdot \sin(\omega t) = A^* \cdot \sin(\omega t) = \left(\frac{2P_o}{V_m} / \frac{v_o T}{2nL_r} \right) \cdot \sin(\omega t) \tag{9.4}$$

Table 9.3 Average input current of different operating modes

Mode 1	$i^*_{\text{in(av)}} = \frac{D_1^2}{M-1}$
Mode 2	$i^*_{\text{in(av)}} = -(D_1^2 M^2 + D_1^2 M + D_1^2 - 2D_1 M^2 + M^2 - M)$
Mode 3	$i^*_{\text{in(av)}} = \frac{D_2^2(1-M)}{M}$
Mode 4	$i^*_{\text{in(av)}} = -(D_1^2 M^2 + D_1^2 M + D_1^2 - 2D_1 M^2 + M^2 - M)$

Table 9.4 Control variables calculation of different operating modes

Mode 1	$\begin{cases} D_1 = \sqrt{i_{\text{ref}}^* \cdot (M-1)} = \sqrt{A^* \sin(\omega t) \cdot (M-1)} \\ D_2 = \frac{D_1}{M-1} \end{cases}$
Mode 2	$\begin{cases} D_1 = \frac{M^2 - \sqrt{M - A^* \sin(\omega t) - M^2 A^* \sin(\omega t) - MA^* \sin(\omega t)}}{M^2+M+1}] \\ D_2 = M - (1+M)D_1 \end{cases}$
Mode 3	$\begin{cases} D_1 = 0 \\ D_2 = \sqrt{\frac{i_{\text{ref}}^* \cdot M}{1-M}} = \sqrt{\frac{A^* \sin(\omega t) \cdot M}{1-M}} \end{cases}$
Mode 4	$\begin{cases} D_1 = \frac{M^2 - \sqrt{M - A^* \sin(\omega t) - M^2 A^* \sin(\omega t) - MA^* \sin(\omega t)}}{M^2+M+1}] \\ D_2 = M - (1+M)D_1 \end{cases}$

Table 9.5 The mode transition mechanism

$M > 1$	$\begin{cases} 0 < i_{\text{ref}}^* = A^* \sin(\omega t) < \frac{M-1}{M^2} \quad \text{(Mode 1)} \\ \frac{M-1}{M^2} < i_{\text{ref}}^* = A^* \sin(\omega t) < \frac{\frac{2P_{o,\max}}{V_m}}{\frac{v_O T}{2nL_r}} \quad \text{(Mode 2)} \end{cases}$
$M < 1$	$\begin{cases} 0 < i_{\text{ref}}^* = A^* \sin(\omega t) < M(1-M) \quad \text{(Mode 3)} \\ M(1-M) < i_{\text{ref}}^* = A^* \sin(\omega t) < \frac{\frac{2P_{o,\max}}{V_m}}{\frac{v_O T}{2nL_r}} \quad \text{(Mode 4)} \end{cases}$

The mode transition criterion can be obtained through the boundary condition between mode 1 and mode 2 when $M > 1$, the boundary condition between mode 3 and mode 4 when $M < 1$ as shown in the following Table 9.5. Furthermore, the operating region of each mode and the mode transition mechanism with the proposed control strategy are shown in Fig. 9.8. The proposed closed loop diagram is shown in Fig. 9.9.

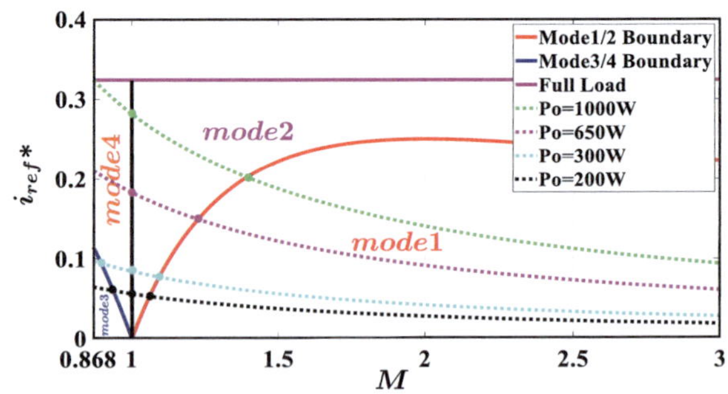

Fig. 9.8 The mode transition mechanism

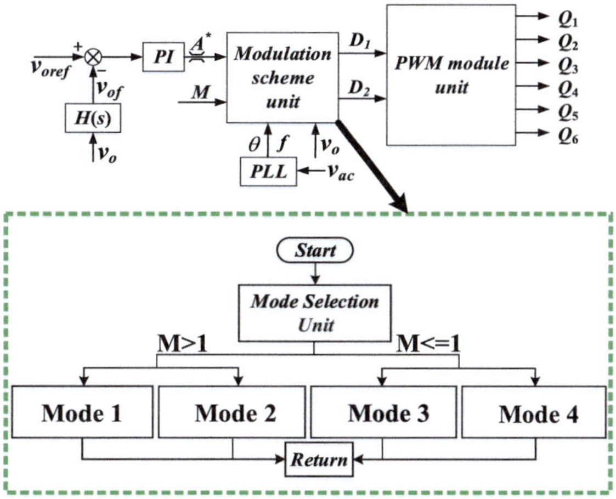

Fig. 9.9 Closed loop control diagram

9.2 Converter Design and Analysis

9.2.1 Leakage Inductance Design

The leakage inductance determines the power transmission capability. The maximum instantaneous power $p_{o,\max}$ with this control strategy lies in mode 4 at the maximum absolute value for the gird voltage. The leakage inductance can be designed as the following,

$$\begin{cases} p_{o,\max} > 2P_{\text{rated}} \\ p = -\frac{v_o^2 T}{2L_r} \cdot \left[\frac{D_1^2 M^2 + D_1^2 M + D_1^2 - 2D_1 M^2 + M^2 - M}{M}\right] \\ D_{1,\max}(M) = \frac{M^2}{M^2 + M + 1} \end{cases} \tag{9.5}$$

Substituting system parameters (Table 9.6) into (9.5) yields the maximum value of L_r as the following expression.

$$L_r < \frac{v_o^2 T}{4P_{\text{rated}}\left(M_{\min}^2 + M_{\min} + 1\right)} = 34.76\,\mu\text{H} \tag{9.6}$$

Table 9.6 System parameters

Symbol	Value	Symbol	Value
v_{ac}	220 V_{rms}	C_{ol}	10,000 μF
f_{ac}	50 Hz	Q_1, Q_2, Q_3, Q_4	C3M0045065K
v_o	270–400 V	Q_5, Q_6	C3M0065100K
C_{in}	2 μF	D_{o1}, D_{o2}	RURG80100
L_r	34 μH	$D_{r1}, D_{r2}, D_{r3}, D_{r4}$	KBPC2510
L_m	210 μH	f_s	100 kHz
n	1: 1	t_{dead}	200 ns
C_{oh}	20 μF	P_{rated}	1 kW

9.2.2 *Optimal Design of the Transformer Turns Ratio*

Different transformer turns ratio will affect the root mean square value and peak value of the primary and secondary side currents, which are related to the conduction loss and switching loss of switches, the core loss and copper loss of the magnetic components. So in order to reduce the loss and improve the conversion efficiency, the transformer turns ratio need to be optimal designed.

In [9], the turns ratio is 1:1.2, the input and output voltage are 220 V (AC) and 400 V (DC) respectively. In this case, the converter works in boost mode ($M > 1$) in the entire ac line period. While the RMS value and peak value of transformer currents are the smallest, the soft switching range is the widest, the total loss is the smallest and the conversion efficiency is the highest when the voltage conversion gain of DAB or semi-DAB converter is 1. And a wide output voltage range (270–400 V DC) is needed in many applications such as an on-board-charger. So a suitable turns ratio should be designed so that the voltage conversion gain is distributed on both sides of 1 (boost mode and buck mode) in the entire ac line period.

Firstly, the effect of turns ratio on the transformer RMS current is analyzed. In order to study the effect of turns ratio on switch conduction loss and magnetic components copper loss, the sum of the square of RMS value of the primary and the secondary side currents are taken as the optimization target,

$$i_{opt} = i_{rms}^2 = i_{prms}^2 + i_{srms}^2 = \frac{\int_0^{1000T_s} i_p^2(t)dt}{1000T_s} + \frac{\int_0^{1000T_s} i_s^2(t)dt}{1000T_s} \tag{9.7}$$

where, i_{prms} and i_{srms} represent the RMS value of the transformer primary and secondary side current within half the ac line period respectively. $i_p(t)$ and $i_s(t)$ refer to the primary and secondary side current respectively. In addition, in order to optimize the turns ratio under different output voltages and different loads condition, ten operating points are selected for analysis, they are 20% rated power, 40% rated power, 60% rated power, 80% rated power, 100% rated power under $V_o = 270$ V and same five load power points under $V_o = 400$ V. The following expression is calculated by averaging the square of $i_{rms}(k)$ for ten operating points.

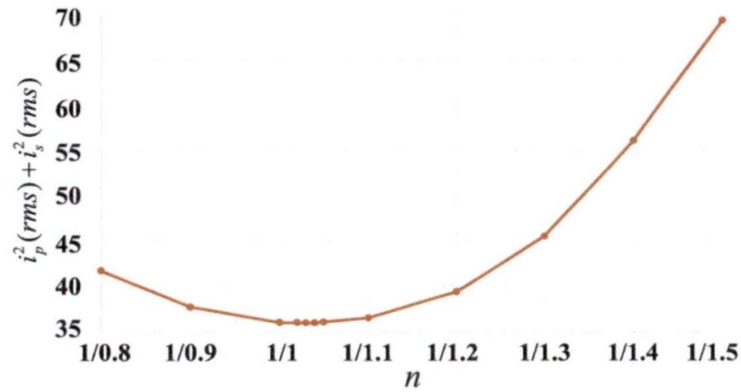

Fig. 9.10 $i_{\text{opt}}(n)$ versus the transformer turns ratio n

$$i_{\text{opt}}(n) = \frac{\sum_{i=1}^{10} i_{\text{rms}}^2(k)}{10} = \frac{\sum_{i=1}^{10} (i_{\text{prms}}^2(k) + i_{\text{srms}}^2(k))}{10} \tag{9.8}$$

where, n indicates the turns ratio, and k refers to the serial number of the operating point. Figure 9.10 shows the $i_{\text{opt}}(n)$ under different transformer turns ratio n. As seen, $i_{\text{opt}}(n)$ is the minimum as the turns ratio is 1:1.

Secondly, the influence of turns ratio on the envelope of the primary side peak current within the half ac line period is needed to be analyzed. Figure 9.11 shows the envelope of primary side peak current within the half ac line period under different n. As seen, the envelope of the primary side peak current is minimum as n is 1:1. The lower envelope of the primary side peak current could reduce the current stress and turn off loss of switches, causing the core loss of transformer.

So, the transformer turns ratio is selected as 1:1 in this paper through the above analysis of two aspects.

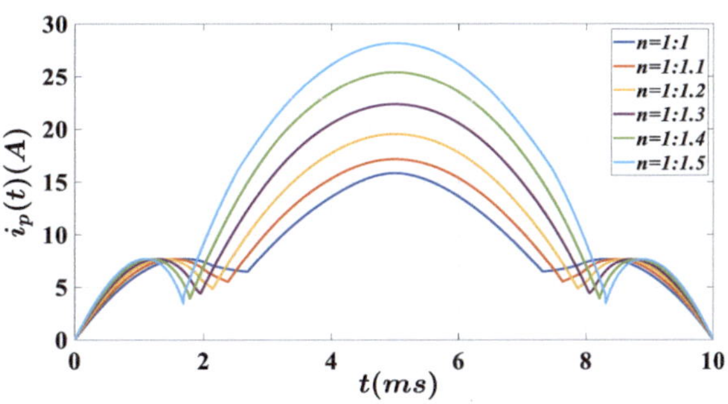

Fig. 9.11 Transformer primary side peak current with different turns ratio n

9.2.3 ZVS Range for Q_1–Q_4 Analysis and Mode Operating Region

For primary side switches to achieve ZVS, there are various ZVS criteria in the existing literatures, some methods are too simple to accurately judge whether ZVS are achieved, while other methods to calculate ZVS conditions are too complex. In order to make the ZVS criteria relatively precise and not too complicated, the ZVS criteria based on energy is adopted.

Taking the process of Q_1 achieving ZVS in mode 1 as an example, which has been illustrated in detail as Fig. 9.12. $I_{p.zvs}$ is the minimum current to achieve ZVS, $C_1 = C_2 = (1/2)C_{eq.ac}$. $C_{eq.ac}$ is the equivalent capacitance of a bridge leg on the AC side. The initial energy in the circuit is the sum of the leakage inductor energy and the energy of C_1 as (9.9), then the leakage inductor and C_1, C_2 start to be resonant as Q_2 is turned off. The power source $|v_{ac}|$ absorbs energy in this process as shown in (9.10). At the end of the resonant process, the leakage inductor current drops to zero, the final energy in the circuit is the energy of C_2 as shown in (9.11). The minimum ZVS current $I_{p.zvs}$ (shown in (9.13)) can be obtained according to the energy conservation in (9.12). The analytical process of other switches under different modes are similar.

$$E_{\text{initial}} = \frac{1}{2}(n^2L_r)I_{p.\text{ZVS}}^2 + \frac{1}{2}C_1|v_{\text{ac}}|^2 \tag{9.9}$$

$$\begin{aligned} E_{\text{delivered}} &= \int_0^t i_{C1}|v_{\text{ac}}|dt = \int_0^t \left(-C_1\frac{dv}{dt}\right)|v_{\text{ac}}|dt \\ &= -C_1|v_{\text{ac}}|\int_{|v_{\text{ac}}|}^0 dv = C_1|v_{\text{ac}}|^2 = \frac{1}{2}C_{\text{eq.ac}}|v_{\text{ac}}|^2 \end{aligned} \tag{9.10}$$

$$E_{\text{final}} = \frac{1}{2}C_2|v_{\text{ac}}|^2 \tag{9.11}$$

$$E_{\text{initial}} = E_{\text{final}} + E_{\text{delivered}} \tag{9.12}$$

$$I_{p.\text{ZVS}} = \frac{|v_{\text{ac}}|}{\sqrt{(n^2L_r)/C_{\text{eq.ac}}}} \tag{9.13}$$

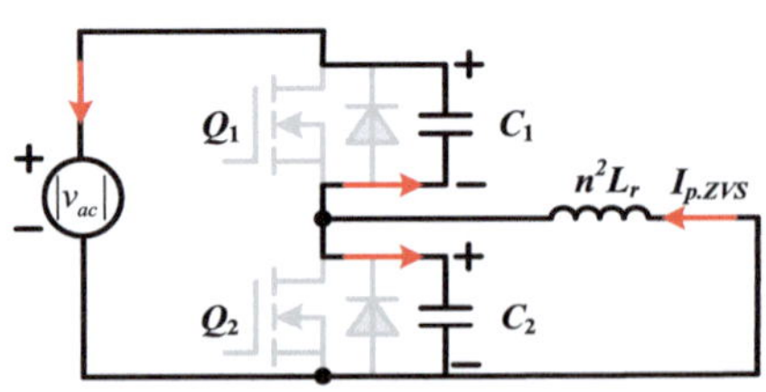

Fig. 9.12 The process of primary side switch Q_1 achieving ZVS in mode 1

Table 9.7 The current expression determining the soft switching of Q_1–Q_2

Mode 1	$\|i_p(T)\| = \frac{1}{n}\cdot\left[\frac{\|v_{ac}\|}{2nL_m}\cdot(D_1+D_2)T - v_O\sqrt{\frac{C_{OSS}+C_D}{L_r}}\right]$
Mode 2	$\|i_p(T)\| = \frac{1}{n}\cdot\left[\frac{\|v_{ac}\|}{2nL_m}\cdot(D_1+D_2)T\right]$
Mode 3	$\|i_p(T)\| = \frac{1}{n}\cdot\left[\frac{\|v_{ac}\|}{2nL_m}\cdot D_2T - v_O\sqrt{\frac{C_{OSS}+C_D}{L_r}}\right]$
Mode 4	$\|i_p(T)\| = \frac{1}{n}\cdot\left[\frac{\|v_{ac}\|}{2nL_m}\cdot(D_1+D_2)T\right]$

The primary side current $|i_p(T)|$ of the transformer determines the ZVS situation of Q_1–Q_2, so $|i_p(T)|$ in different modes have been listed in Table 9.7. The soft switching situation of Q_3–Q_4 of different operating modes are determined by the primary side current of the transformer at the corresponding time, which has been listed in Table 9.8.

The ZVS range of primary side switches under different output voltages ($V_o =$ 270 V and $V_o = 400$ V) are given in Fig. 9.13. Figure 9.13a is the case under $V_o =$ 270 V, the vertical black line is the boundary ($M = 1$) of $M > 1$ and $M < 1$, the red curve indicates the boundary of the quasi-ZCS on/off region and ZVS region within mode 1, the blue line indicates the boundary of mode 1 and mode 2, the green line is the boundary of mode 3 and mode 4. As seen in Fig. 9.13a, Q_1–Q_4 can realize ZVS at the right side of the red curve in mode 1, mode 2, mode 3 and mode 4, and Q_1–Q_4 can realize quasi-ZCS on/off at the left side of the red curve in mode 1 due to the very low turn on/off current. Figure 9.13a also reflects the mode operation region. As seen, under $V_o = 270$ V, there are three modes (mode 1, mode 2 and mode 4) when the load power is more than 350 W, while there will be four modes (mode 1, mode 2, mode 3 and mode 4) when the load power is less than 350 W.

The mode operation region and ZVS range of primary side switches under $V_o =$ 400 V is shown in Fig. 9.13b. As seen, there are two modes (mode 1 and mode 2) when the load power is more than 790 W, while there will be only one mode (mode 1) when the load power is less than 790 W.

Table 9.8 The current expression determining the soft switching of Q_3–Q_4

Mode 1	$\|i_p(t_2)\| = \frac{1}{n}\cdot\left[\frac{\|v_{ac}\|}{2nL_m}\cdot(D_1+D_2)T - v_O\sqrt{\frac{C_{OSS}+C_D}{L_r}}\right]$
Mode 2	$\|i_p(T)\| = \frac{1}{n}\cdot\left[\frac{v_oT}{L_r}\cdot(D_1M - M + 1) + \frac{\|v_{ac}\|}{2nL_m}\cdot(D_1+D_2)T\right]$
Mode 3	$\|i_p(t_1)\| = \frac{1}{n}\cdot\left[v_O\sqrt{\frac{C_{OSS}+C_D}{L_r}} + \frac{v_oT}{L_r}\cdot\frac{(1-M)D_2}{M} + \frac{\|v_{ac}\|}{2nL_m}\cdot D_2T\right]$
Mode 4	$\|i_p(T)\| = \frac{1}{n}\cdot\left[\frac{v_oT}{L_r}\cdot(D_1M - M + 1) + \frac{\|v_{ac}\|}{2nL_m}\cdot(D_1+D_2)T\right]$

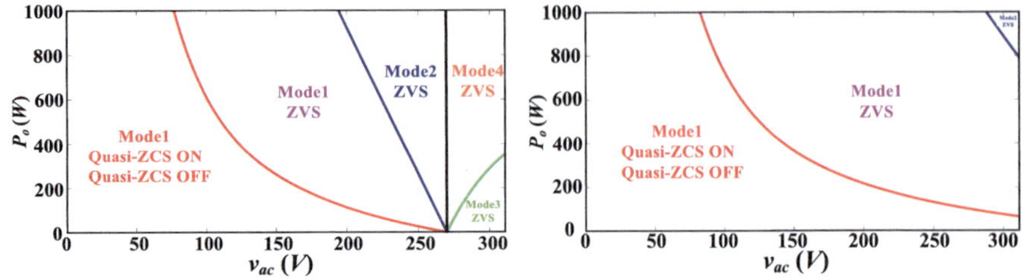

(a) V_o=270V (b) V_o=400V

Fig. 9.13 Soft switching range of Q_1–Q_4 and mode operating region

Fig. 9.14 The process of secondary side switch Q_5 achieving ZVS in mode 1

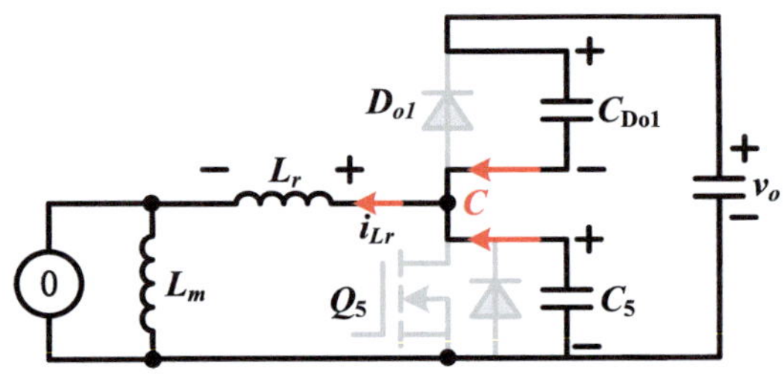

9.2.4 ZVS Range for Q_5–Q_6 Analysis and Mode Operating Region

For secondary side switches, the mechanism to achieve ZVS is different from primary side switches for the semi-DAB converter. The process of Q_5 achieving ZVS in the mode 1 is taken as an example, this is illustrated in detail as the following,

The differential equation to describe the resonant process is shown as (9.14) according to Kirchhoff's law, combining the initial condition (9.15) yields the analytical solution as (9.16). The resonant process is completed when v_{ds5} drops to zero. From (9.16), the resonant time $T_r/4$ could be solved in (9.17), and the leakage inductance current at the end moment of the resonant process I_{bias} is shown in (9.18). ZVS can be realized as long as Q_5 is turned on after the resonant process is completed. The analytical process of other switches under different modes are similar, which are not shown again here.

$$\begin{cases} L_r \frac{di_{Lr}}{dt} = v_{ds5} \\ i_{Lr} = C_{Do1} \frac{d(v_o - v_{ds5})}{dt} - C_5 \frac{dv_{ds5}}{dt} = -(C_{\text{oss}} + C_D) \frac{dv_{ds5}}{dt} \end{cases} \tag{9.14}$$

$$\begin{cases} i_{Lr}(0) = 0 \\ v_{ds5}(0) = v_o \end{cases} \tag{9.15}$$

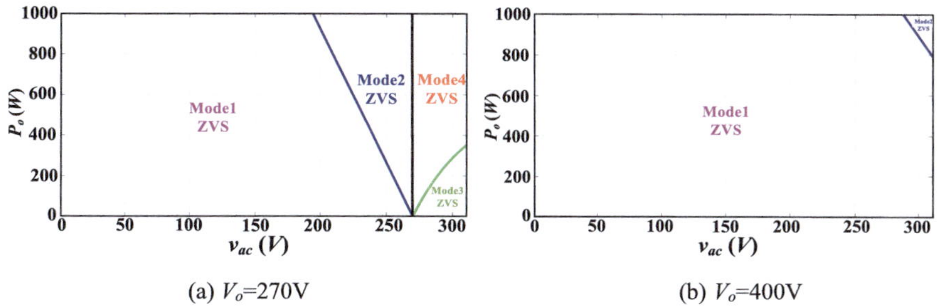

Fig. 9.15 Soft switching range of Q_5–Q_6 and mode operating region

$$\begin{cases} i_{Lr}(t) = v_o\sqrt{\frac{C_{\text{oss}}+C_D}{L_r}}\sin\left(\frac{t}{\sqrt{L_r(C_{\text{oss}}+C_D)}}\right) \\ v_{ds5}(t) = v_o\cos\left(\frac{t}{\sqrt{L_r(C_{\text{oss}}+C_D)}}\right) \end{cases} \tag{9.16}$$

$$\frac{T_r}{4} = \frac{\pi}{2}\cdot\sqrt{L_r\cdot(C_{\text{oss}}+C_D)} \tag{9.17}$$

$$i_{Lr}(\frac{T_r}{4}) = v_o\sqrt{\frac{C_{\text{oss}}+C_D}{L_r}} = I_{\text{bias}} \tag{9.18}$$

The ZVS range of secondary side switches under different output voltages are given in Fig. 9.15. It can be seen that Q_5–Q_6 can realize ZVS in the entire ac line period within the full load range. ZVS range is extremely wide that it is helpful to improve the conversion efficiency.

9.3 Comparison Between the Decoupled Leakage Inductor Current Modulation and the Optimal Peak Current Modulation Strategy

In previous work [9], the output voltage is fixed 400 V (DC), and the turns ratio is 1:1.2, which makes the converter only have two operating modes within the entire ac line period. While in order to improve the converter conversion efficiency and meet the requirement of wide output voltage (270–400 V) in many applications such as on-board-charger, two modes adopting the same optimal peak current strategy under $M < 1$ need to be added as the following Fig. 9.16. It is founded that it has the following shortcoming through calculation, simulation and experiment by using the optimal peak current modulation compared with using the decoupled inductor current modulation strategy,

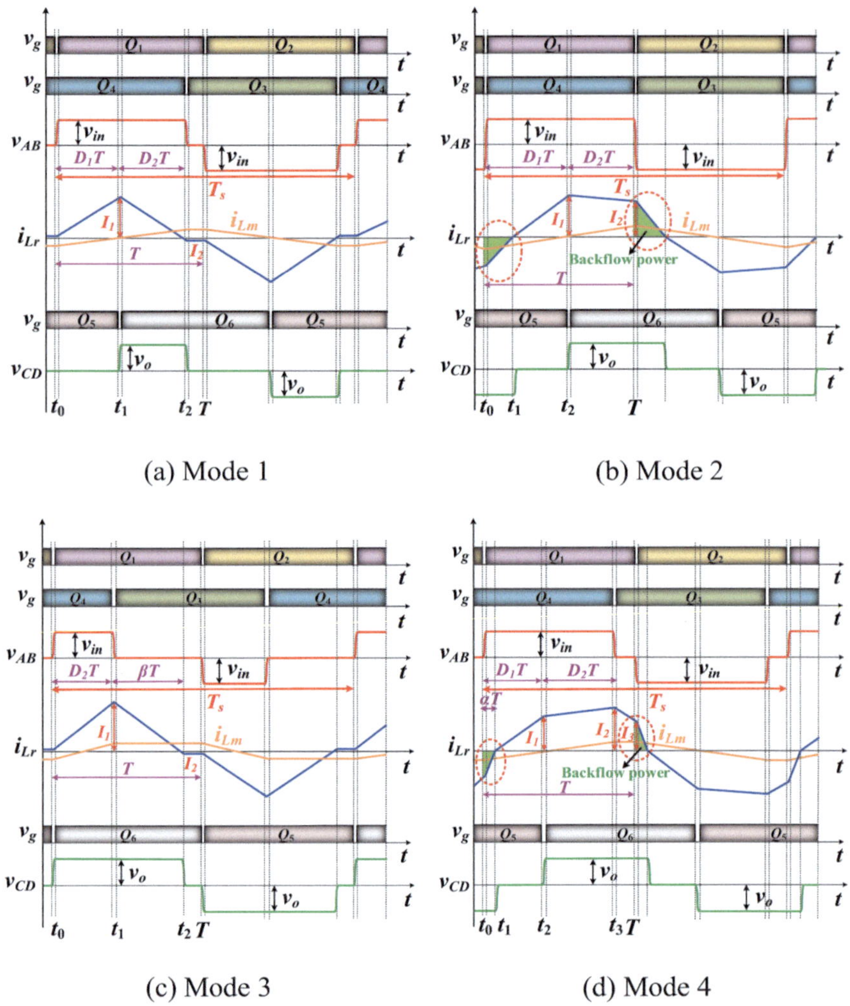

Fig. 9.16 The optimal peak current modulation strategy under $M > 1$ and $M < 1$

(1) The turn off current and switching loss of Q_1, Q_2 are larger.
(2) The pulse width of v_{ab} is wider, which results in the increase of the transformer core loss.
(3) The leakage inductor current of the transformer with two adjacent switching periods are coupled, which leads to a certain deviation between the actual average current within every switching period transferred by the converter and the required instantaneous current from the grid side, causing more distortion for the grid side current.
(4) Larger backflow power.

While the decoupled inductor current modulation strategy in this paper could solve the aforementioned disadvantage. The performance of the two modulation strategies will be compared and analyzed from various losses and the waveform quality of i_{ac}. For fair comparison, the same hardware condition for two control methods are listed in Table 9.6, the data were obtained from MATLAB calculation under the same working condition.

9.3.1 Comparison of Various Losses

Figure 9.17 shows the difference of Q_1, Q_2 switching loss and transformer core loss between two modulation strategies within the half ac line period under $V_o = 270$ V and $P_o = 1000$ W respectively. It can be seen that there is a large difference between two kinds of modulation strategies in the interval of 2–8 ms. According to the law of electromagnetic induction, the maximum flux density of a transformer can be estimated by:

$$B_m = \frac{|v_{ac}| \cdot (D_1 + D_2)T}{2N_2 S_T} \tag{9.19}$$

where S_T is the effective area of core. The transformer core loss can be estimated by Steinmetz empirical formula as following:

$$P_{fe\_trans} = C_m \cdot f_s^{\alpha} \cdot B_m^{\beta} \cdot V_{e\_trans} \tag{9.20}$$

As seen, the transformer core loss is related to $(D_1 + D_2)$, so the control variable of two modulation strategies under $V_o = 270$ V, $P_o = 1000$ W and $P_o = 200$ W are shown in Fig. 9.18, where solid lines and dotted lines indicate control variables of the decoupled inductor current modulation strategy and the optimal peak current

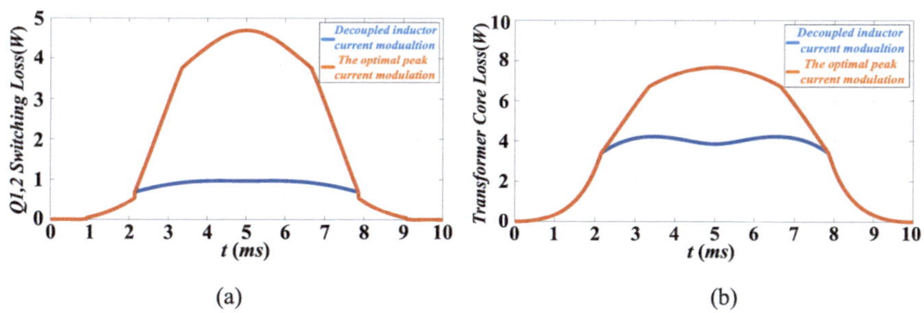

Fig. 9.17 **a** $Q_{1,2}$ switching loss. **b** Transformer core loss

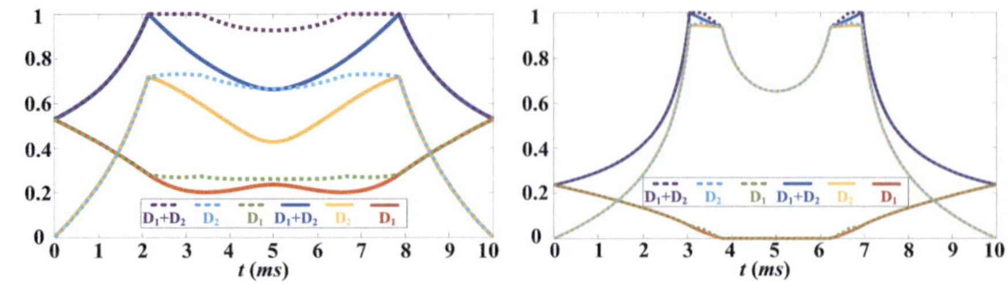

(a) 1000W (b) 200W

Fig. 9.18 The difference of control variables between two strategies

modulation strategy respectively. The difference of $D_1 + D_2$ is the reason of the existing difference in the transformer core loss between two modulation strategies.

Figure 9.19 shows the difference of the inductor core loss of the two modulation strategies within the half ac line period under $V_o = 270$ V, $P_o = 1000$ W and $P_o = 600$ W respectively. It can be observed that the difference is large at the heavy load, while is small under the light load. Similarly, the maximum flux density of the inductor can be calculated by:

$$B_m = \frac{L_r i_{Lr\max}}{2N_L S_L} \tag{9.21}$$

where N_L is the turns of the leakage inductor winding, and S_L is the effective area of the inductor core. The inductor core loss can be calculated by the Steinmetz empirical formula as shown in (9.20). The difference of inductor core loss is related to the peak current flowing through the leakage inductor.

Next, to further estimate various average power loss within half a line frequency period over the full range of load power, some calculations and comparisons are given as follows. Figure 9.20 shows the difference of average conduction loss of all switches and diodes, average switching loss of all switches, transformer average

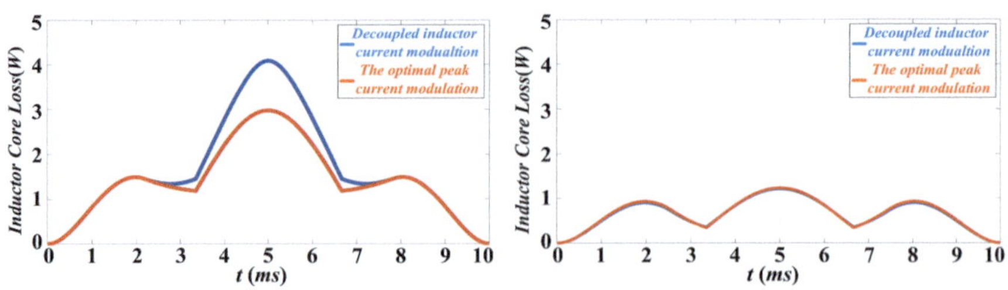

(a) 1000W (b) 600W

Fig. 9.19 Inductor core loss comparison

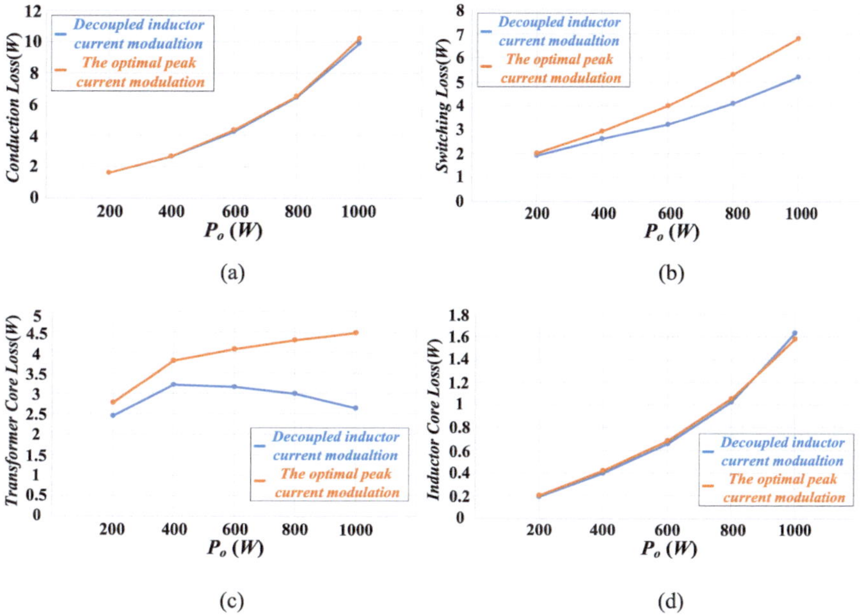

Fig. 9.20 Various loss comparisons. **a** The conduction loss of all switches and diodes of two strategies. **b** The switching loss of all switches between two strategies. **c** The transformer core loss of two strategies. **d** The inductor core loss of two strategies

core loss and inductor average core loss between two modulation strategies within the half a line frequency period under different loads respectively. It can be seen that there is only a slight difference between two strategies for average conduction loss and inductor average core loss. The difference of the average switching loss and transformer average core loss between the two strategies becomes larger as load power increases. In the rated load of 1 kW, the average switching loss is decreased by 28.6%, and the transformer average core loss is decreased by 45%, which are helpful to improve the conversion efficiency.

9.3.2 Comparison of the AC Input Current Waveform Quality

In order to achieve power factor correction, the reference value of AC input current varies continuously according to the sinusoidal law, the converter is in a continuous transition state as the input voltage value, the required instantaneous power and control variables. If the operation case of every switching period is regarded as independent, then the average value of the leakage inductor current is zero in every switching period, this causes i_{Lr} of two adjacent switching periods to be discontinuous, $i_{Lr.\text{end}}(i) \neq i_{Lr.\text{start}}(i+1)$. While the leakage inductor current must be continuous in actual operation, $i_{Lr.\text{end}}(i) = i_{Lr.\text{start}}(i+1)$, which will cause the average value of i_{Lr} in every switching period as non-zero. This is reason why inductor currents of two adjacent switching periods is coupled. In fact, the nature of coupled is that the actual starting current value of every switching period is affected by the current value at the end moment of the previous switching period, while the nature of decoupled is opposite, which means the starting current of each switching period is a fixed value (I_{bias} or zero), not affected by the previous switching period.

The control variable of the optimal peak current strategy is given in Table 9.9. As seen, the calculation is more complex compared with the decoupled inductor current modulation, this costs much longer calculation time in DSP.

Figure 9.21a is taken as an example to show the transition process of two adjacent switching periods and the coupled current mechanism. The red dotted line represents the ideal leakage inductor current waveform for each switching period, while the blue solid line indicates the actual current waveform. ΔI represents the DC bias current at the transition moment of two adjacent switching periods, which decays exponentially as the time constant of the equivalent RL circuit formed by the leakage inductor and total circuit ESR as the following,

$$\Delta i_{Lr}(t) = \Delta I \cdot e^{-\frac{R}{L}\cdot t} = (i_{\text{Lr.end}}(i) - i_{\text{Lr.start}}(i+1)) \cdot e^{-\frac{R}{L}\cdot t} \tag{9.22}$$

The actual average current transmitted by the converter is larger than the ideal average current in the first half of the switching period, the difference is the yellow area; while it is opposite in the second half of the period, the deviation is the green region. The analysis for Fig. 9.21b–d is similar to Fig. 9.21a. While for the four proposed modes, the starting current of every switching period is a fixed value (I_{bias}

Table 9.9 Control variables calculation of different operating modes

Mode 1	$\begin{cases} D_1 = \sqrt{i^*_{\text{ref}} \cdot (M-1)} = \sqrt{A^* \cdot \sin(\omega t) \cdot (M-1)} \\ D_2 = \frac{D_1}{M-1} \end{cases}$
Mode 2	$\begin{cases} D_1 = \frac{(M+2)^2}{2(M^2+2M+2)} \cdot \left[\frac{2(M^2+M+1)}{(M+2)^2} - \frac{\sqrt{2} \cdot \sqrt{M - 2 \cdot A^* \sin(\omega t) - M^2 \cdot A^* \sin(\omega t) - 2M \cdot A^* \sin(\omega t) + 1}}{(M+2)} \right] \\ D_2 = 1 - D_1 \end{cases}$
Mode 3	$\begin{cases} D_1 = 0 \\ D_2 = \sqrt{\frac{i^*_{\text{ref}} \cdot M}{1-M}} = \sqrt{\frac{A^* \cdot \sin(\omega t) \cdot M}{1-M}} \end{cases}$
Mode 4	$\begin{cases} D_1 = \frac{(M^2+M+1)^2}{M^6+3M^5+4M^4+M^3-M^2+2} \cdot \left[\frac{2M^6+4M^5+4M^4+2}{2(M^2+M+1)^2} - \frac{\sqrt{(M-M^4-M^3-1)(2 \cdot A^* \sin(\omega t) - M + M^2 \cdot A^* \sin(\omega t) + 2M \cdot A^* \sin(\omega t) - 1)}}{M^2+M+1} \right] \\ D_2 = M - D_1 \cdot \frac{[(M^2+3M+3)(M^2+M-1)+(M+1)]}{[(M^2+M-1)(M+1)+(M^2+3M+3)]} \end{cases}$

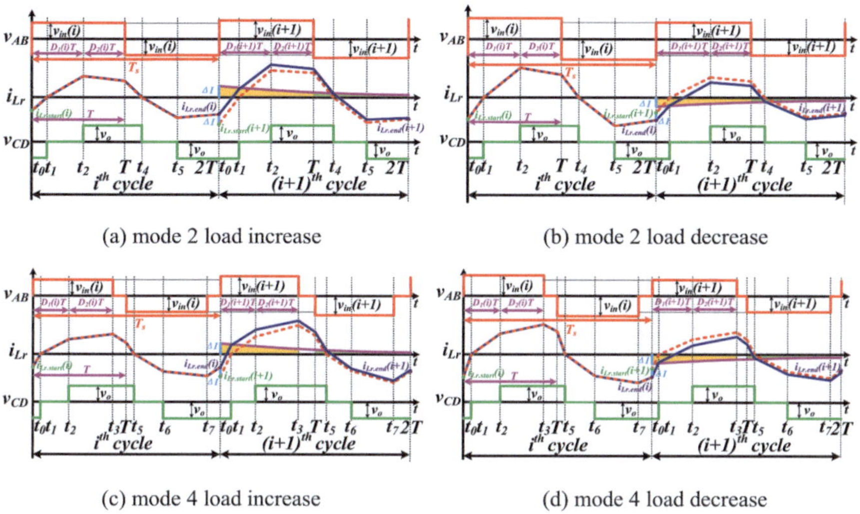

(a) mode 2 load increase (b) mode 2 load decrease

(c) mode 4 load increase (d) mode 4 load decrease

Fig. 9.21 Transient state transition waveform of different operating modes

is for mode 1/3 and zero is for mode 2/4), so i_{Lr} of two adjacent switching periods is decoupled.

Figure 9.22 shows the input current i_{ac} within the half ac line period for two modulation strategies. It can be observed that there are some deviations between them, that is the reason why the THD is different for two modulation strategies. In order to verify the above analysis, Fig. 9.23 shows the simulation result of two strategies under $V_o = 270$ V and $P_o = 1000$ W, the THD of two methods are 3.25% and 2.1% respectively. Furthermore, the experimental results of two modulation strategies under the same condition are shown in Fig. 9.24, the THD are 3.9% and 2.6% respectively. As can be seen, the decoupled inductor current modulation strategy can minimize the THD of the input current i_{ac}.

9.4 Experimental Results

9.4.1 Experimental Parameters and Control

A 1 kW experimental prototype was made to verify the effectiveness of the control strategy in this paper, and system parameters have been listed in Table 9.6. The control strategy is implemented by TI DSP TMS320F28335.

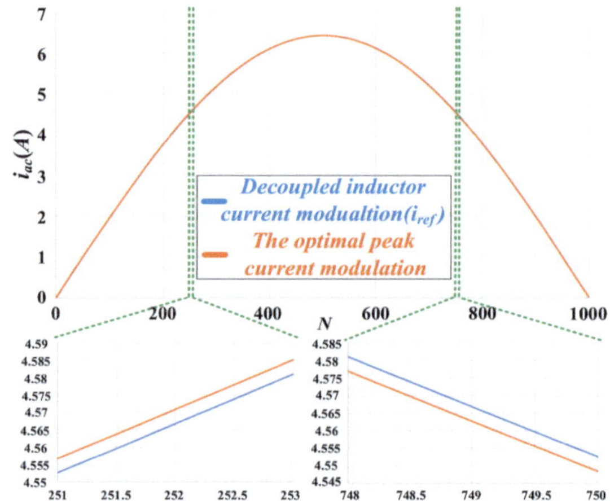

Fig. 9.22 Grid side current under two different modulation strategies

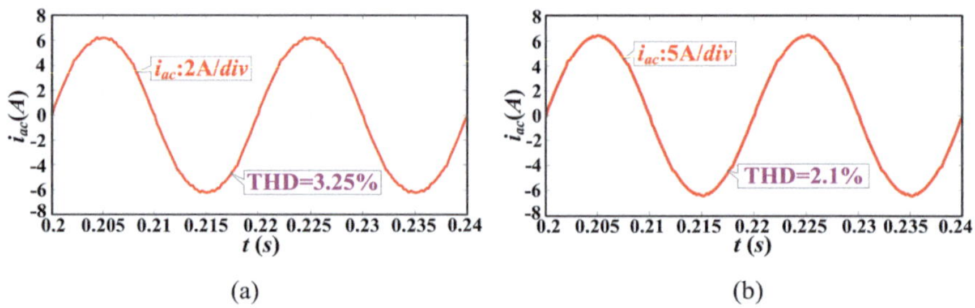

Fig. 9.23 The comparison of ac input current. **a** i_{ac} with optimal peak current modulation. **b** i_{ac} with decoupled inductor current modulation

9.4.2 Steady-State Operating Waveforms

Figure 9.25a–d show the steady state waveform of the grid voltage and current under the rated power and different output voltages respectively. As seen, high power quality can be obtained by using the proposed control.

9.4.3 Mode Operating Situations

Figure 9.26 shows the waveform of v_{ab}, v_{cd}, i_{Lr}, i_p within the entire a line frequency period and different operation modes under the rated power and different output voltage respectively. Figure 9.26a is the result for the rated power and $V_o = 270$ V. The amplitude of the input voltage varies from 0 to V_m cyclically, while the output voltage

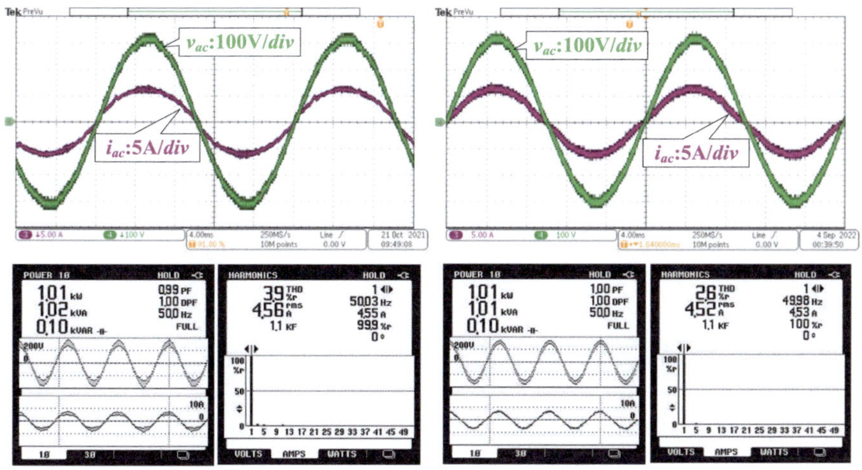

(a) (b)

Fig. 9.24 The comparison of ac input current. **a** i_{ac} with optimal peak current modulation. **b** i_{ac} with decoupled inductor current modulation

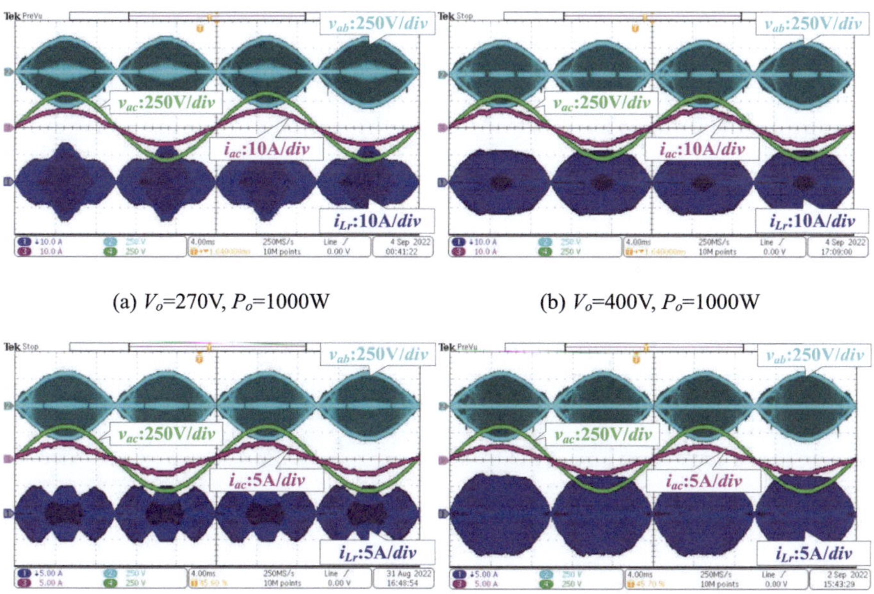

(a) V_o=270V, P_o=1000W (b) V_o=400V, P_o=1000W

(c) V_o=270V, P_o=400W (d) V_o=400V, P_o=360W

Fig. 9.25 Grid side current and voltage at steady state under different situations is constant as 270 V, which means the output voltage is more than the amplitude of input AC voltage ($M > 1$, Boost mode) and less than the amplitude of the input AC voltage ($M <= 1$, Buck mode) cyclically, so there will be three modes (mode 1, 2, 4).

Figure 9.26b is the result for the rated power and $V_o = 400$ V. The output voltage is always more than the amplitude of input AC voltage ($M > 1$, Boost mode), so there will be two operating modes. Experimental results are consistent with the previous theoretical analysis for the mode operating region shown in Figs. 9.13 and 9.15.

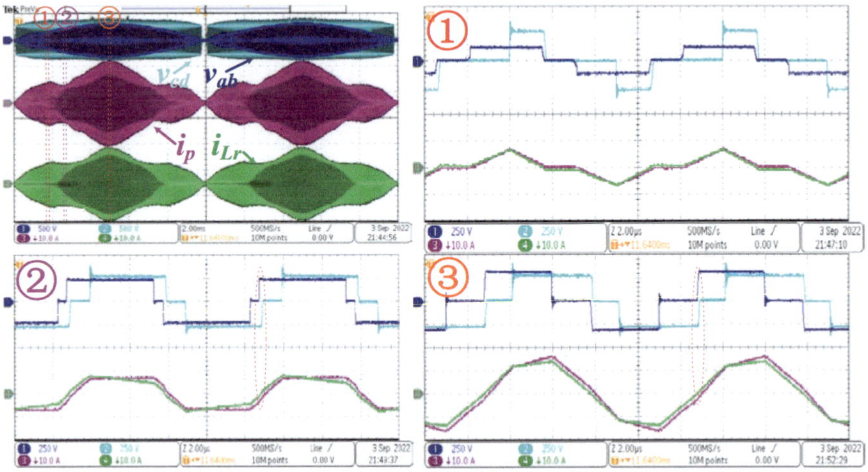

(a)V_o=270V, P_o=1000W

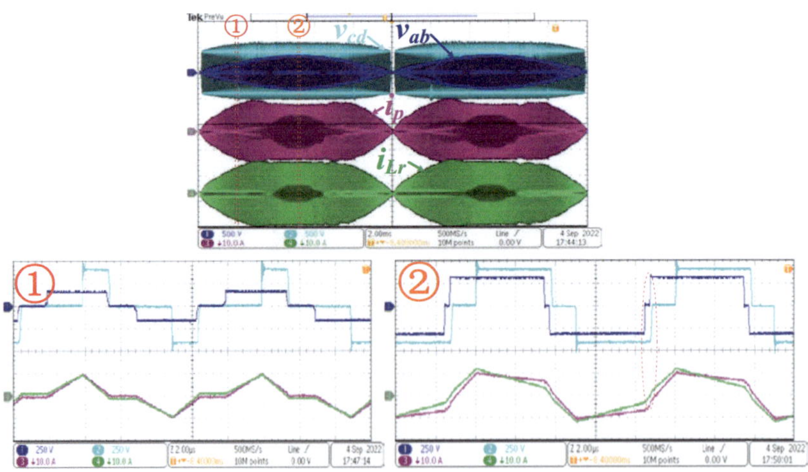

(b)V_o=400V, P_o=1000W

Fig. 9.26 Mode operation situations under the different operating conditions

9.4.4 Mode ZVS/ZCS Soft Switching Situations

Figure 9.27 shows soft switching situations of different switches at different voltage conversion gains under the rated load power and different output voltages respectively. Only the situation of Q_2, Q_6 and D_{o2} are given due to the symmetry between the upper switch and the bottom switch in a half bridge. As illustrated in Fig. 9.27, experimental results of the ZVS off switches and ZCS off diodes are consistent with the previous theoretical analysis as shown in Figs. 9.13 and 9.15.

9.4.5 Dynamic Response and Performance Comparisons

Dynamic response waveforms when the load power switches from 400 to 800 W are shown in Fig. 9.28. Figure 9.29 shows the conversion efficiency under different modulation strategies and output voltages respectively, it can be seen that the peak value is 96.38% under the decoupled inductor current modulation. Figure 9.30a, b show the THD and power factor (PF) of input current i_{ac} under two different control strategies, as can be seen, the power quality of i_{ac} is better under the decoupled inductor current modulation strategy of this chapter.

9.5 Conclusion

A novel decoupled inductor current modulation strategy is proposed for the single-stage AC-DC converter based on semi-DAB, which could reduce total harmonic distortion of i_{ac} and reduce losses, and improve the conversion efficiency. The optimal turns ratio could further improve the efficiency, and thus it works under boost mode and buck mode in the entire line frequency period in facing wide output voltages. Four operating modes are given and analyzed according to its operation. Only two control variables including the primary phase shift D_1 and the phase shift between two bridges D_2 are used in every operating mode. The mode transition is seamless. The control strategy can achieve PFC without mandatory inner current tracking loop under a wide output voltage range (270–400 V). Comparisons of various losses, THD and the backflow power by using two modulation strategies are shown in detail, the backflow power of the proposed modulation is minimized, which helps to improve the efficiency. The power transfer capability, mode transition mechanism, soft switching range and mode operation region, parameter design are given. The effectiveness of the control strategy is verified by a 1 kW experimental prototype.

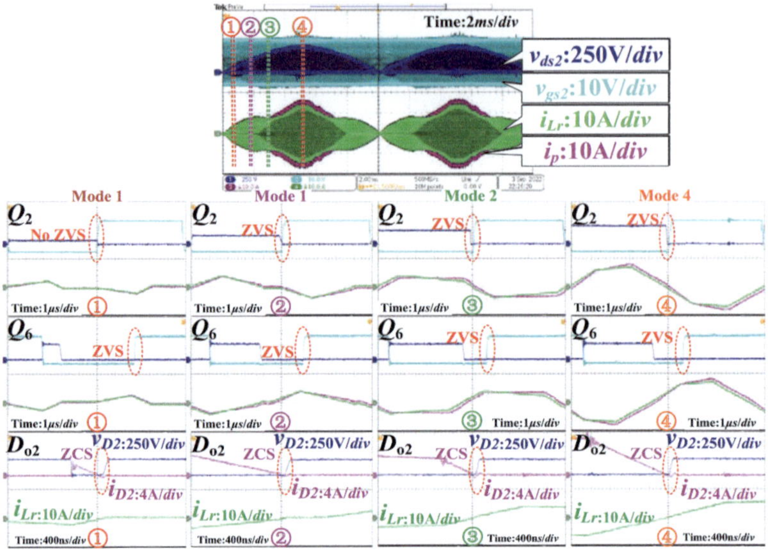

(a)V_o=270V, P_o=1000W

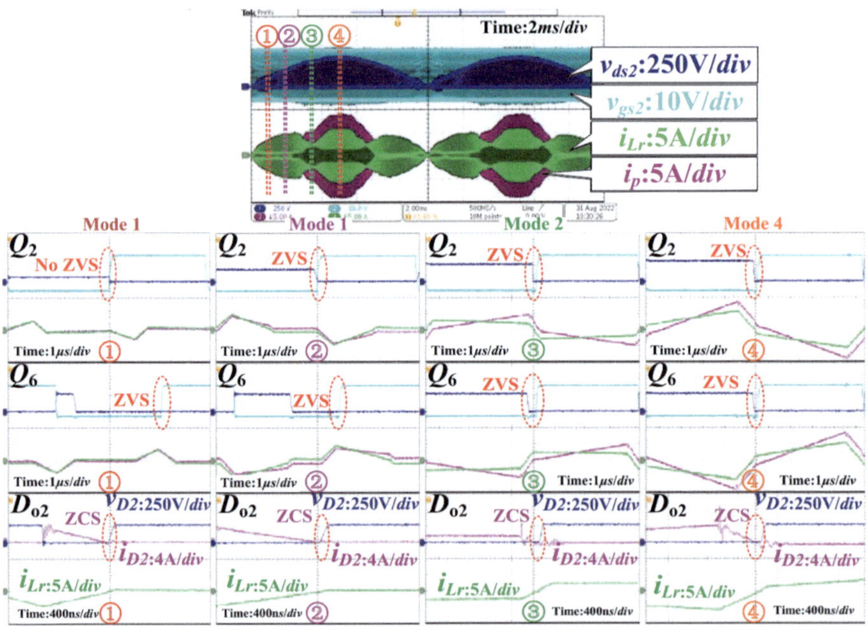

(b)V_o=270V, P_o=400W

Fig. 9.27 ZVS/ZCS situations under the different operating conditions

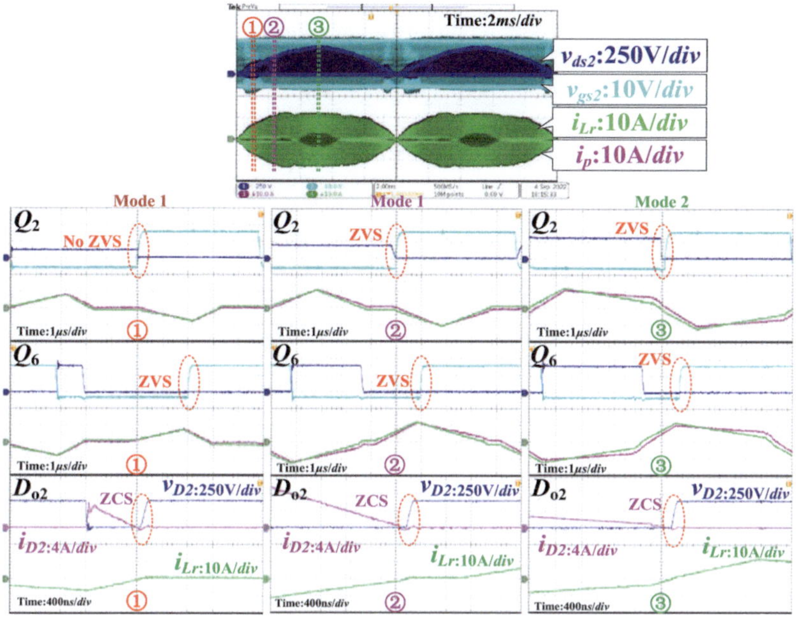

(c)V_o=400V, P_o=1000W

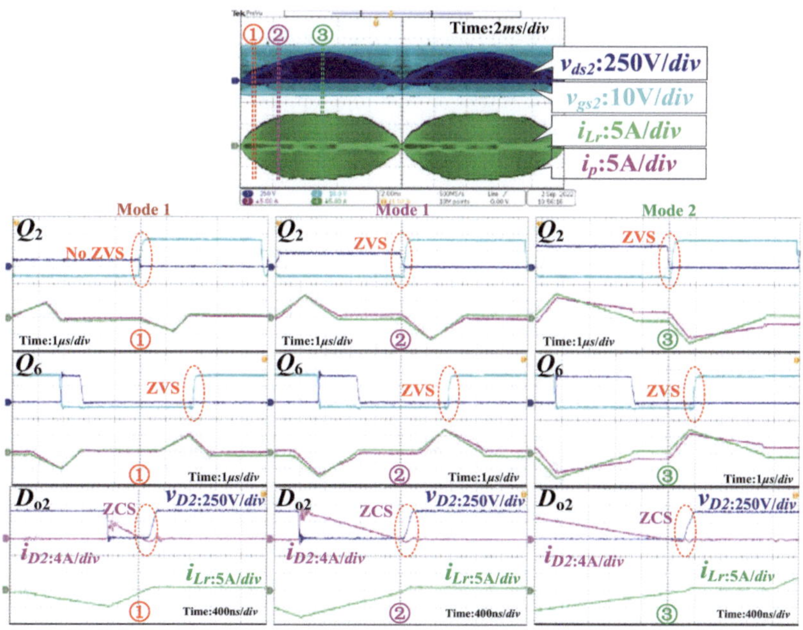

(d)V_o=400V, P_o=360W

Fig. 9.27 (continued)

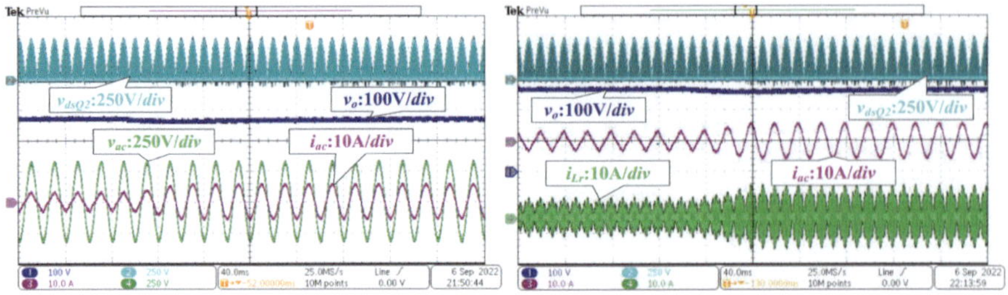

(a)V_o=270V, P_o=400W-800W (b)V_o=270V, P_o=400W-800W

Fig. 9.28 Dynamic response waveforms

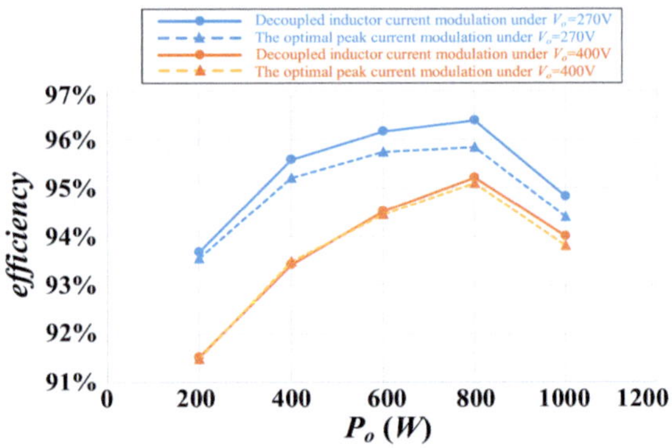

Fig. 9.29 Efficiency curve comparison

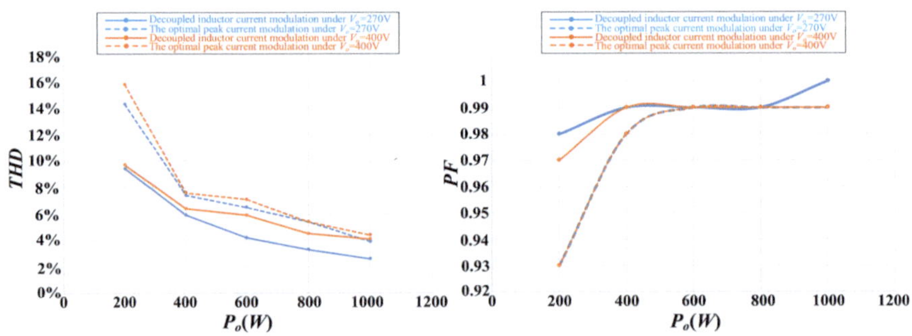

(a) THD comparison (b) PF comparison

Fig. 9.30 The THD and PF comparison of i_{ac}

References

1. Everts J, Krismer F, Van den Keybus J, Driesen J, Kolar JW (2014) Optimal ZVS modulation of single-phase single-stage bidirectional DAB AC–DC converters. IEEE Trans Power Electron 29(8):3954–3970
2. Everts J, Krismer F, Van den Keybus J, Driesen J, Kolar JW (2013) Charge-based ZVS soft switching analysis of a single-stage dual active bridge AC-DC converter. In: 2013 IEEE energy conversion congress and exposition, Denver, pp 4820–4829
3. Zengin S, Boztepe M (2020) A novel current modulation method to eliminate low-frequency harmonics in single-stage dual active bridge AC–DC converter. IEEE Trans Ind Electron 67(2):1048–1058
4. Sha D, Wang S (2021) A single-stage natural power factor corrector based on dual active bridge DC–DC converter without inner current tracking loop. IEEE Trans Power Electron 36(1):342–352
5. Zhang J, Sha D, Ma P (2022) A dual active bridge DC-DC based single stage AC-DC converter with seamless mode transition and high power factor. IEEE Trans Ind Electron 69(2):1411–1421
6. Everts J (2017) Closed-form solution for efficient ZVS modulation of DAB converters. IEEE Trans Power Electron 32(10):7561–7576
7. Singh A, Yadav AK, Khaligh A (2021) Steady-state modeling of a dual-active-bridge AC-DC converter considering circuit nonidealities and intracycle transient effects. IEEE Trans Power Electron 36(10):11276–11287
8. Ma P, Sha D, Song K (2023) A single-stage semi dual-active-bridge AC-DC converter with seamless mode transition and wide soft- switching range. IEEE Trans Ind Electron 70(2):9447–9458
9. Li F, Li Y, You X (2018) Optimal dual-phase-shift control strategy of an isolated buck-boost converter with a clamped inductor. IEEE Trans Power Electron 33(6):5374–5385

Chapter 10
Single-Stage Semi-DAB AC-DC Converter for Full Operating Range ZVS On

Abstract In Chaps. 8 and 9, the optimal peak current and decoupled inductor current modulation strategy are presented respectively, and the gird side current power quality improvement, reactive power optimization, transformer turns-ratio optimization and conversion efficiency are studied. But the primary side switches cannot achieve ZVS on within some operating conditions. So in order to achieve full operating range ZVS on, in this chapter, an improved modulation and variable frequency control strategy for a single-stage high frequency isolated semi dual active bridge (semi-DAB) based AC-DC converter is proposed, which can achieve a full operating range (the entire ac line cycle and full load range) zero-voltage-switching (ZVS) and high efficiency. Compared with the basic variable frequency control, improved variable frequency strategy can reduce the envelope of the leakage inductor peak current and reduce the dependence and sensitivity to the load power. More accurate ZVS criteria in view of the magnetizing inductor L_m is proposed. Seven operating modes are adopted and the transition among them is seamless. Three control variables are used in every mode. Wide voltage conversion gain and high quality grid side AC current can be obtained without mandatory current tracking loops. In addition, detailed comparisons among different control strategies and different mode operating regions are analyzed. The design of the transformer and inductor is not affected at all despite the varying switching frequency. A 250 W experimental prototype was built to verify the effectiveness of the proposed improved variable frequency control strategy.

Keywords Single stage AC-DC converter · Full operating range zero-voltage-switching (ZVS) · Improved variable switching frequency control · More accurate ZVS criteria

D. Sha et al., *High Frequency Isolated Semi Dual Active Bridge DC-DC and AC-DC Converter*, CPSS Power Electronics Series,
https://doi.org/10.1007/978-981-95-1641-4_10

10.1 Circuit Topology and ZVS Limitations of Constant Frequency Control Strategy

A single-stage isolated half-bridge matrix structure AC-DC converter with minimum peak current is proposed [1], but a full load operating range ZVS can not be obtained. In order to extend ZVS range, the switching frequency is introduced [2], but there is only one mode and lookup tables are needed. A simple open-loop control is proposed [3], but there is only one triangular mode and it causes a large peak current, high turn-off loss, inductor core loss and low efficiency. For the single-stage AC-DC converter based DAB, extra inductor is needed to assist ZVS achievement, but this results in additional loss, complex control and need of lookup tables [4]. A DPS modulation is proposed for the three-phase single-stage AC-DC converter, which can minimize the DAB root mean square (RMS) current, achieve high conversion efficiency and power density. However, only one mode exists and the switching frequency variation range is very wide [5].

For unidirectional power flow applications, an optimal peak current control strategy and decoupled inductor current modulation strategy are presented respectively for the Semi-DAB-based single-stage AC-DC converter [6, 7]. But there are some issues for the constant switching frequency control:

(1) The primary side switches cannot achieve ZVS within full operating range as shown in Fig. 10.1a, ZVS achievement around the zero-crossing-point and light load conditions is so difficult due to the positive platform current and the very low magnetizing current as shown in Fig. 10.1b.
(2) For secondary side switches Q_5–Q_6, the leakage inductor current i_{Lr} flows through the switch body diodes instead of the switches for some time during each switching period. Thus the conduction loss is very large.
(3) Magnetizing inductor L_m is not considered and thus the ZVS analysis is not accurate.

In order to achieve a full operating range ZVS for semi-DAB AC-DC converter, with reduced loss and improved efficiency, an improved modulation and variable frequency control strategy is proposed. The major novelties and contributions of the improved modulation and variable switching frequency control strategy can be summarized as the following,

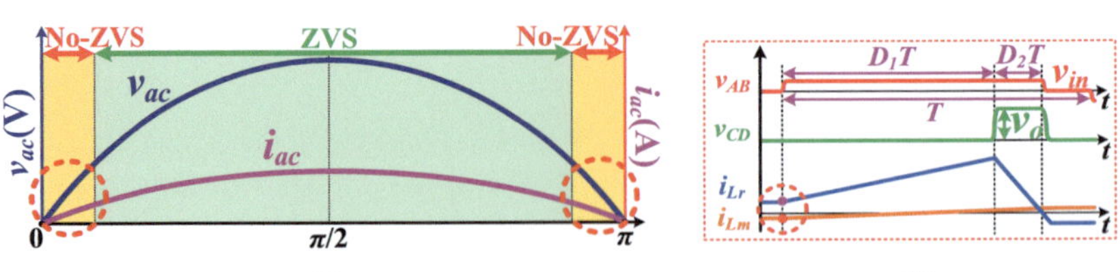

Fig. 10.1 **a** The ZVS situations within half AC line cycle. **b** The mode around the zero-crossing-point of v_{ac} in [7]

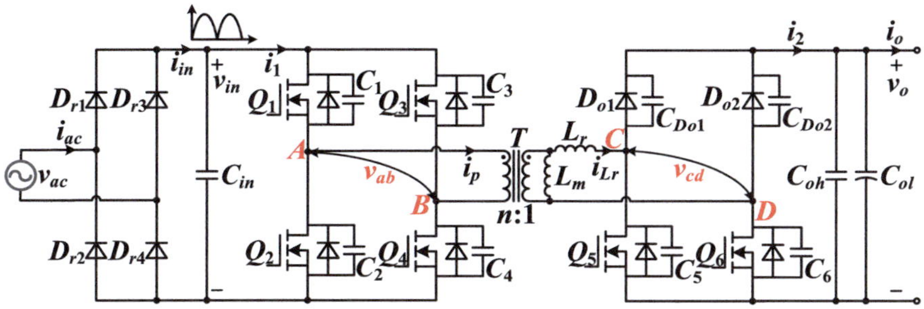

Fig. 10.2 Single-stage semi-DAB AC-DC converter topology

(1) Full operating range (the entire ac line cycle and full load range) ZVS and higher efficiency can be achieved.
(2) The envelope of the leakage inductor peak current can be reduced, as well as the dependence and sensitivity to load power compared with the basic variable frequency control.
(3) The design procedure of the transformer and inductor are not affected despite the switching frequency variation.
(4) The improved modulation for Q_5 and Q_6 helps to reduce the conduction loss of their body diodes to almost zero.
(5) The magnetizing inductor L_m is considered in ZVS analysis and ZVS analysis becomes more accurate.

10.1.1 Power Circuit Topology

The single-stage isolated semi-DAB AC-DC converter is shown in Fig. 10.2. The grid side is connected with an uncontrolled rectifier. The turns ratio of the transformer is n:1. L_r is the leakage inductor and L_m is the magnetizing inductor. M is the voltage conversion gain and is defined as $M = nv_o/|v_{ac}|$.

10.1.2 Proposed Working Modes and Circuit Operating Principle

The decoupled inductor current modulation modes for this converter have been proposed in [7], which could reduce the THD of ac input current and improve the power quality. While for secondary side switches Q_5 and Q_6, the leakage inductor current i_{Lr} flows through body diodes for a long period of time (t_2–$T + t_1$ for mode 1, t_2–T for mode 3, T–$T + t_1$ for mode 2 and mode 4) during each switching cycle, this causes a larger conduction loss. So in order to reduce this loss, an improved modulation strategy is proposed in Fig. 10.3. Figure 10.3a, b are two modes under $M > 1$, while Fig. 10.3c, d are two modes under $M <= 1$. As seen, the secondary side switches Q_5 and Q_6 are turned on immediately after the corresponding switches and diodes complete the resonant process, where t_r represents the resonant process time duration length of the secondary side diode junction capacitor C_{Do1} (C_{Do2}) and switch output junction capacitor C_5 (C_6) after the leakage inductor current turns to be zero. t_r is fixed in mode 1 and mode 3, while t_r is related to the instantaneous value of input voltage and output voltage in mode 2 and mode 4. Under constant frequency control, D_1 and D_2 indicate two control variables. $D_1T + D_2T$ represents the time length of Q_1 (Q_2) leading Q_3 (Q_4), the time length of Q_1 (Q_2) leading Q_6 (Q_5) is defined as Δt, which is different in different working modes as shown in (10.1). T represents half switching period and $T_S = 2T$.

$$\Delta t = \begin{cases} -(T - D_1T - D_2T - t_r) & \text{(Mode 1)} \\ t_r & \text{(Mode 2)} \\ -(T - D_2T - \beta T - t_r) & \text{(Mode 3)} \\ t_r & \text{(Mode 4)} \end{cases} \tag{10.1}$$

where, t_r has different expressions in different operating modes as expressed in (10.2), which is determined by the equivalent circuit and the resonant process of secondary side diodes and switches after the leakage inductor current turns to be zero. And the leakage inductor current at the end moment of resonant process I_{bias} in different operating modes is given in (10.3), which is useful for ZVS analysis.

$$t_r = \begin{cases} \frac{\pi}{2} \cdot \sqrt{L_r \cdot (C_{\text{oss}} + C_D)} & \text{(Mode 1/3)} \\ \sqrt{L_r \cdot (C_{\text{oss}} + C_D)} \cdot \arccos(\frac{v_{\text{in}}}{v_{\text{in}} + v_o}) & \text{(Mode 2/4)} \end{cases} \tag{10.2}$$

where, $C_5 = C_6 = C_{\text{oss}}$, $C_{Do1} = C_{Do2} = C_D$.

$$I_{\text{bias}} = \begin{cases} v_o\sqrt{\frac{C_{\text{oss}} + C_D}{L_r}} & \text{(Mode 1/3)} \\ (v_o + v_{\text{in}}) \cdot \sqrt{\frac{C_{\text{oss}} + C_D}{L_r}} \cdot \sin\left(\arccos\left(\frac{v_{\text{in}}}{v_{\text{in}} + v_o}\right)\right) & \text{(Mode 2/4)} \end{cases} \tag{10.3}$$

The control constraints and intersection points between D_1, D_2, and the power transfer ability of every mode have been given in [7]. The constraints of control variables are also listed in the following Table 10.1 for the convenience of analysis.

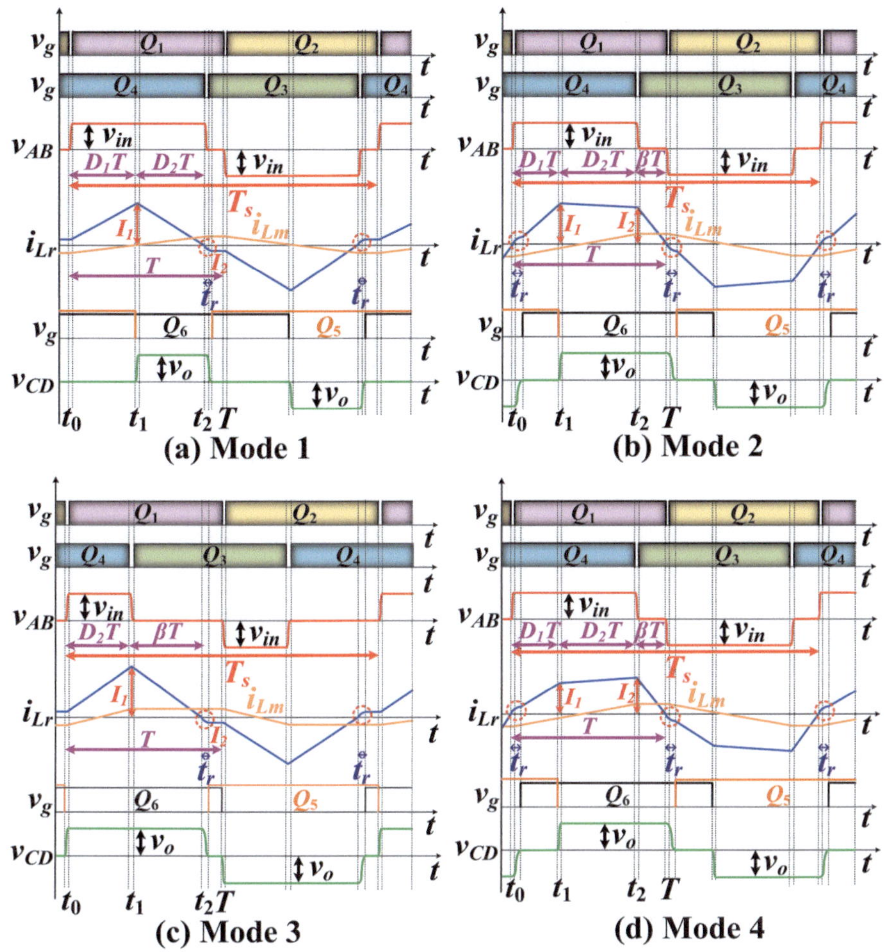

Fig. 10.3 Improved modulation strategy

Table 10.1 Control constraints and intersection point of different modes

Gain	Constraints of control variables	Intersection points
$M > 1$	$\begin{cases} D_2 = \frac{D_1}{M-1} & \text{(Mode 1)} \\ D_2 = M - (1+M)D_1 & \text{(Mode 2)} \end{cases}$	$\begin{cases} D_1 = 1 - \frac{1}{M} \\ D_2 = \frac{1}{M} \end{cases}$
$M < 1$	$\begin{cases} D_1 = 0 & \text{(Mode 3)} \\ D_2 = M - (1+M)D_1 & \text{(Mode 4)} \end{cases}$	$\begin{cases} D_1 = 0 \\ D_2 = M \end{cases}$

10.1.3 Power Factor Correction Control Strategy and Mode Transition Mechanism

The fundamental control strategy is to make the average input current $i_{in(av)}$ equal to the grid side reference i_{ref}. So, the expressions of $i_{in(av)}$ and i_{ref} in different modes are need to be given. In order to introduce the switching frequency as a new control degree of freedom, the corresponding expressions are no longer given in terms of P.U. (Per Unit) values. Their values are given in Table 10.2, two control variables D_1 and D_2 of four modes are obtained in the same form as Table 10.3. So, high power factor can be achieved naturally without the current tracking loop as long as the control is implemented according to the calculated control variables.

$$i_{ref} = A \cdot \sin(\omega t) = \left(\frac{2P_o}{V_m}\right) \cdot \sin(\omega t) \tag{10.4}$$

where, A is the amplitude of the required sinusoidal AC current.

The mode transition criterion can be derived through the transition point ($D_1 = 1\text{–}1/M$) between mode 1 and mode 2 when $M > 1$, the transition point ($D_1 = M$) between mode 3 and mode 4 when $M < 1$ as shown in the following Table 10.4 in the same form with $i_{in(av)}$ and i_{ref}.

Table 10.2 Average input currents of different operating modes

Mode 1	$i_{in(av)} = \frac{v_o T}{2nL_r} \cdot \frac{D_1^2}{M-1}$
Mode 2	$i_{in(av)} = -\frac{v_o T}{2nL_r} \cdot (D_1^2 M^2 + D_1^2 M + D_1^2 - 2D_1 M^2 + M^2 - M)$
Mode 3	$i_{in(av)} = \frac{v_o T}{2nL_r} \cdot \frac{D_2^2(1-M)}{M}$
Mode 4	$i_{in(av)} = -\frac{v_o T}{2nL_r} \cdot (D_1^2 M^2 + D_1^2 M + D_1^2 - 2D_1 M^2 + M^2 - M)$

Table 10.3 Control variable calculation of different operating modes

Mode 1	$\begin{cases} D_1 = \sqrt{\frac{2nL_r i_{ref}(M-1)}{v_o T}} = \sqrt{\frac{2nL_r A\sin(\omega t)(M-1)}{v_o T}} \\ D_2 = \frac{D_1}{M-1} \end{cases}$
Mode 2	$\begin{cases} D_1 = \frac{v_o TM^2 - \sqrt{Mv_o^2 T^2 - 2nL_r v_o T(M^2+M+1)A\sin(\omega t)}}{v_o T(M^2+M+1)} \\ D_2 = M - (1+M)D_1 \end{cases}$
Mode 3	$\begin{cases} D_1 = 0 \\ D_2 = \sqrt{\frac{2nL_r i_{ref} M}{v_o T(1-M)}} = \sqrt{\frac{2nL_r A\sin(\omega t)M}{v_o T(1-M)}} \end{cases}$
Mode 4	$\begin{cases} D_1 = \frac{v_o TM^2 - \sqrt{Mv_o^2 T^2 - 2nL_r v_o T(M^2+M+1)A\sin(\omega t)}}{v_o T(M^2+M+1)} \\ D_2 = M - (1+M)D_1 \end{cases}$

Table 10.4 Mode transition mechanism

$M > 1$	$\begin{cases} 0 < i_{\text{ref}} = A\sin(\omega t) < \frac{v_o T(M-1)}{2nL_r M^2} & \text{(Mode 1)} \\ \frac{v_o T(M-1)}{2nL_r M^2} < i_{\text{ref}} = A\sin(\omega t) < \frac{2P_{o\max}}{V_m} & \text{(Mode 2)} \end{cases}$
$M < 1$	$\begin{cases} 0 < i_{\text{ref}} = A\sin(\omega t) < \frac{v_o T(1-M)M}{2nL_r} & \text{(Mode 3)} \\ \frac{v_o T(1-M)M}{2nL_r} < i_{\text{ref}} = A\sin(\omega t) < \frac{2P_{o\max}}{V_m} & \text{(Mode 4)} \end{cases}$

10.1.4 ZVS Range of Constant Frequency Control

To simplify the analysis, the ZVS criteria based on energy is adopted for primary side switches Q_1–Q_4, and the magnetizing inductor L_m is not considered in the previous work [7], which is not precise. So in order to describe the ZVS condition of primary side switches more accurately, the ZVS process of Q_1 in mode 1 is taken as an example, the resonant equivalent circuit in view of magnetizing inductor L_m can be established as shown in Fig. 10.4, then the corresponding differential equations are expressed according to Kirchhoff's law as shown in (10.5). To solve it yields the minimum primary side ZVS current. But different switches in different operating modes have different minimum ZVS current expressions because each equivalent resonant circuit model is different, this is given in the Table 10.5. The derivation process of resonant equivalent circuits and calculation are similar for other switches under different operating modes, so it is not given here for brevity.

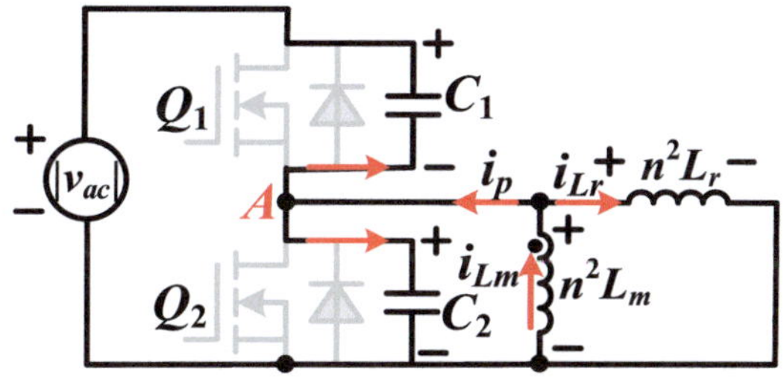

Fig. 10.4 Resonant equivalent circuit of discharging Q_1 in mode 1

Table 10.5 Minimum primary side ZVS current expressions

Q_1 in Mode 1, 3	$I_{p\text{.ZVS}} = \sqrt{\frac{(L_m+L_r)C_{\text{eq.ac}}\lvert v_{\text{ac}}\rvert^2}{n^2L_mL_r}}$
Q_1 in Mode 2, 4	$I_{p\text{.ZVS}} = \sqrt{\frac{C_{\text{eq.ac}}\left(2L_m\lvert v_{\text{ac}}\rvert v_o+L_m\lvert v_{\text{ac}}\rvert^2+L_r\lvert v_{\text{ac}}\rvert^2\right)}{n^2L_mL_r}}$
Q_4 in Mode 1	$I_{p\text{.ZVS}} = \sqrt{\frac{C_{\text{eq.ac}}\left(2L_m\lvert v_{\text{ac}}\rvert v_o-L_m\lvert v_{\text{ac}}\rvert^2-L_r\lvert v_{\text{ac}}\rvert^2\right)}{n^2L_mL_r}}$

$$\begin{cases} \frac{dv_{ds1}}{dt} = -\frac{1}{2C_{\text{oss}}} i_p = -\frac{1}{C_{\text{eq.ac}}} i_p \\ \frac{di_p}{dt} = \left(\frac{1}{n^2 L_m} + \frac{1}{n^2 L_r}\right) v_{ds1} - \left(\frac{1}{n^2 L_m} + \frac{1}{n^2 L_r}\right) |v_{\text{ac}}| \\ v_{ds1}(0) = |v_{\text{ac}}| \\ i_p(0) = I_{p.\text{ZVS}} \end{cases} \tag{10.5}$$

where $I_{p.\text{ZVS}}$ is the minimum current to achieve ZVS for primary side switches, and $C_{\text{eq.ac}}$ is the equivalent capacitance in the resonant process of a bridge leg on the AC side, $C_{\text{eq.ac}} = 2C_{\text{oss}}$.Only the minimum ZVS current of Q_1 in mode 1–mode 4 are given due to the symmetry. For Q_4, ZVS in mode 1 is the most difficult to achieve, so only the minimum ZVS current value of Q_4 in mode 1 is given. The specification is listed as follows: $v_{\text{ac}} = 110$ V (RMS), $V_o = 135$–200 V, $L_m = 210$ μH, $n = 1{:}1$, $fs =$ 100 kHz and $P_{\text{rated}} = 250$ W, $L_r = 34$ μH. The ZVS range of primary side switches is shown in Fig. 10.5a, b. Since the secondary side switches can achieve ZVS easily, so their ZVS range is not demonstrated.

10.2 Proposed Basic Variable Frequency Control

As seen in Fig. 10.5, based on constant frequency (100 kHz), the switches cannot achieve ZVS at the left side of the red line region especially at light load conditions. The primary side current $|i_p(T)|$ determines the ZVS achievement, which could realize ZVS once $|i_p(T)| > I_{p.\text{ZVS}}$. In order to extend ZVS range, the switching frequency needs to be introduced as a new degree of freedom. D_{c1}, D_{c2} and T_{c1} are two control variables and the half switching period under the basic variable frequency control, which can be obtained by making the grid side required power equal to the transferred power in mode 1 is shown in (10.6) and (10.7). Making $|i_p(T)| = I_{p.\text{ZVS}}$ yields the suitable switching frequency f_{s1} to achieve ZVS as shown in (10.8) and (10.9) based on the premise of achieving a high power factor and controllable output voltage.

$$\begin{cases} p_{\text{require}} = 2P_o\left(\frac{|v_{\text{ac}}|}{V_m}\right)^2 \\ p_{\text{mode 1}} = \frac{v_o^2 T_{c1}}{2L_r} \cdot \frac{D_{c1}^2}{M(M-1)} \end{cases} \tag{10.6}$$

$$\begin{cases} D_{c1} = \frac{2\sqrt{nL_r P_o (nv_o - |v_{\text{ac}}|)}}{V_m \sqrt{v_o T_{c1}}} \\ D_{c2} = \frac{D_{c1}}{M-1} = \frac{2|v_{\text{ac}}|\sqrt{nL_r P_o}}{V_m \sqrt{v_o (nv_o - |v_{\text{ac}}|) T_{c1}}} \end{cases} \tag{10.7}$$

$$\begin{cases} I_{p.\text{ZVS}} = \frac{|v_{\text{ac}}|}{\sqrt{\frac{n^2 L_m L_r}{(L_m + L_r) C_{\text{eq.ac}}}}} \\ |i_p(T)| = \frac{1}{n} \cdot [\frac{|v_{\text{ac}}|}{2nL_m} \cdot (D_{c1} + D_{c2}) T_{c1} - I_{\text{bias}}] \end{cases} \tag{10.8}$$

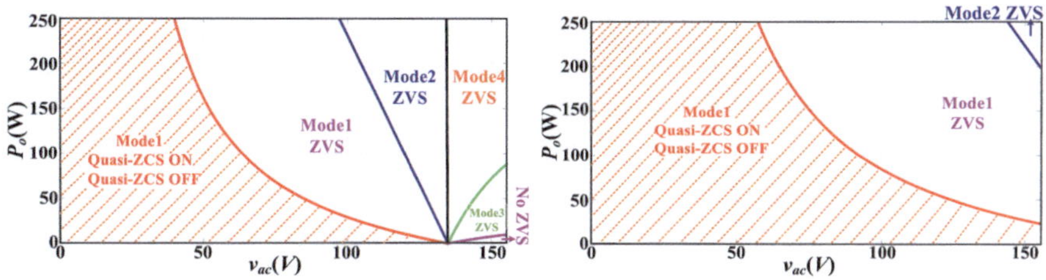

(a) V_o=135V (b) V_o=200V

Fig. 10.5 ZVS range of primary side switches with constant frequency control

$$f_{s1} = \frac{n^3 L_r^2 |v_{\mathrm{ac}}|^2 v_o P_o}{2L_m V_m^2 (nv_o - |v_{\mathrm{ac}}|) C_{\mathrm{eq.ac}} (L_m + L_r) \left[n|v_{\mathrm{ac}}| + I_{\mathrm{bias}} \sqrt{\frac{n^2 L_m L_r}{C_{\mathrm{eq.ac}}(L_m + L_r)}} \right]^2} \quad (10.9)$$

In order to demonstrate the relationship among f_{s1}, P_o and v_{ac} more clearly, three-dimensional variable frequency surfaces under $V_o = 135$ V and $V_o = 200$ V are given respectively in Fig. 10.6. As seen, in order to achieve ZVS in the entire ac line cycle and full load range, the switching frequency needs to be reduced to a very low value or even zero. However, it is not practical. The extreme low switching frequency causes noises to human ears, and the magnetic components owns the risk of becoming saturated. So the lower limit of the switching frequency f_{s1} is set to be 20 kHz which is immune to human ears in this paper, and the switching frequency variation range is set to be: 20–100 kHz. Two-dimensional variable frequency curves under different output voltages and loads are shown in Fig. 10.7. Although the full range ZVS (the entire ac line cycle and full load range) is still not achieved due to the lower limit of f_{s1}, but ZVS range could be extended greatly as long as the switching frequency varies according to (10.9) compared with the constant frequency control strategy.

After adopting the basic variable frequency control, the ZVS range of primary side switches under different output voltages ($V_o = 135$ V and $V_o = 200$ V) are

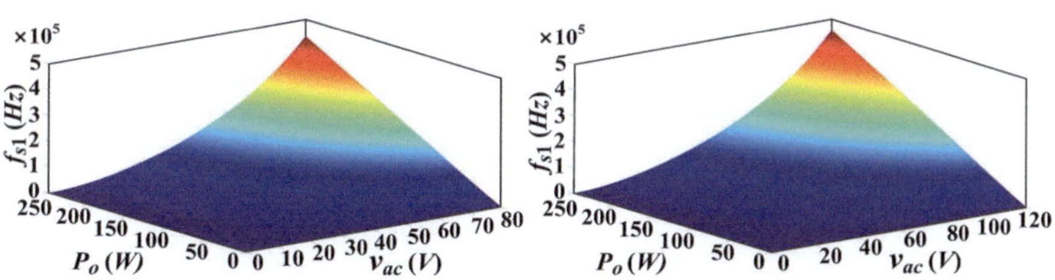

(a) V_o=135V (b) V_o=200V

Fig. 10.6 The switching frequency f_{s1} versus P_o and v_{ac} under different V_o

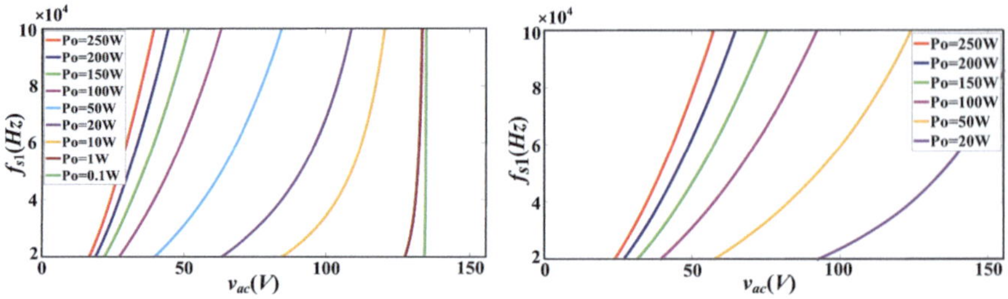

(a) V_o=135V (b) V_o=200V

Fig. 10.7 Two-dimensional variable frequency curves

given respectively in Fig. 10.8. Figure 10.8a is the case under $V_o = 135$ V. As seen, five curves at the left side are the boundary lines between ZVS and quasi-ZCS on/off in mode 1 by using different a switching frequency (20, 40, 60, 80 and 100 kHz). The ZVS range can be extended from the right side red curve to the left side purple curve. As illustrated, the switches can't achieve ZVS as $v_{\mathrm{ac}} < 39$ V under the rated load power by using constant frequency control of 100 kHz, but the switches can achieve ZVS when $18 < v_{\mathrm{ac}} < 39$ V by using the basic variable frequency strategy. The switches can also achieve ZVS when $25 < v_{\mathrm{ac}} < 64$ V under $P_o = 100$ W similarly. The case under $V_o = 200$ V is the same as shown in Fig. 10.8b.

Figure 10.8a also reflects the mode operation region, the vertical black line indicates the boundary line of $M > 1$(Boost mode) and $M < 1$(Buck mode), the blue line refers to the boundary of mode 1 and mode 2, the green line is the boundary of mode 3 and mode 4. As seen, Q_1–Q_4 can realize ZVS at the right side of the purple curve in mode 1, mode 2, mode 3 and mode 4. There are three operating modes (mode 1, mode 2 and mode 4) as the load power is beyond 81 W, but there are four modes (mode 1, mode 2, mode 4 and mode 3) as the load power is less than 81 W.

The case under $V_o = 200$ V is similar as shown in Fig. 10.8b. As seen, there are two modes (mode 1 and mode 2) as the load power is more than 198 W. However, there is only one mode (mode 1) as the load power is less than 198 W.

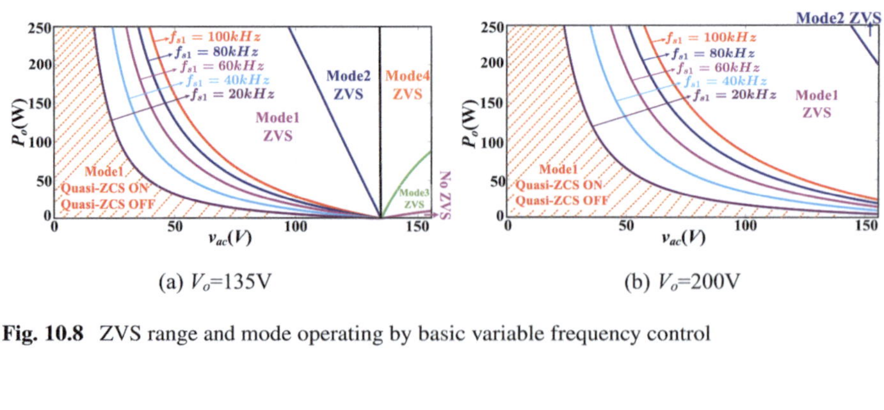

(a) V_o=135V (b) V_o=200V

Fig. 10.8 ZVS range and mode operating by basic variable frequency control

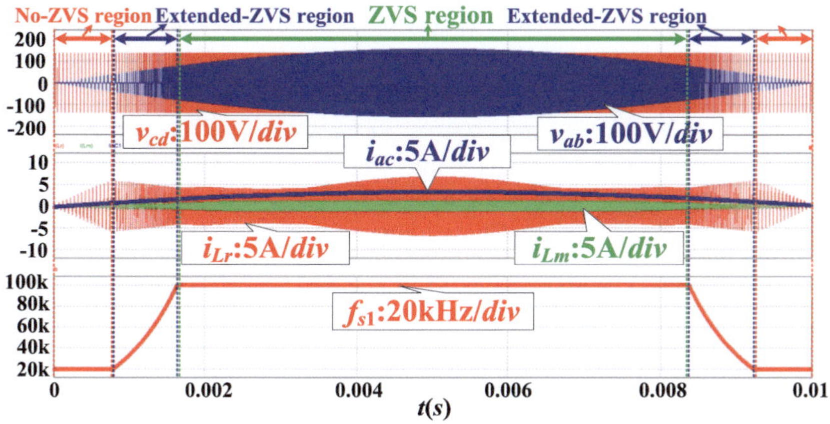

(a)V_o=135V, P_o=250W

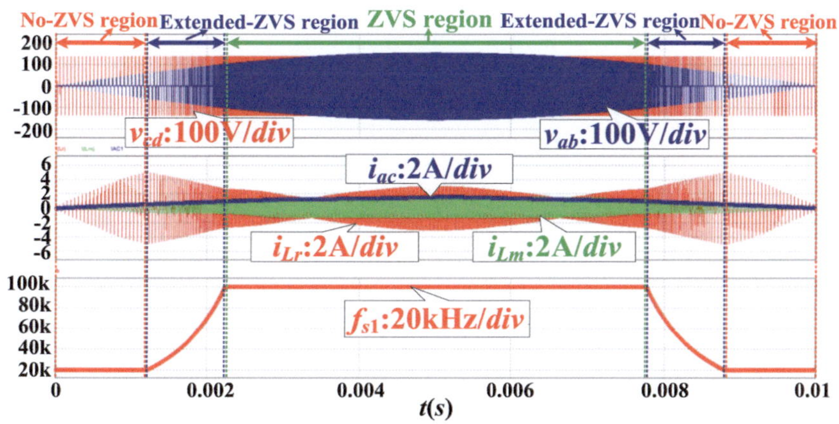

(b)V_o=135V, P_o=100W

Fig. 10.9 Steady state simulation waveforms under different situations

The steady state simulation waveforms under $V_o = 135$ V and different loads are given in Fig. 10.9a, b. As seen, the simulation results agree well with the previous theoretical analysis. It can be found that the basic variable frequency control strategy has the following shortcomings:

(1) Although ZVS range can be extended, a full operating range ZVS cannot be achieved.
(2) With the decrease of the switching frequency, the envelope of the leakage inductor current in mode 1 increases as well. This increases the current stress and turn-off current values especially under light load conditions.

(3) The switching frequency curve is very sensitive to the load, and it will change sharply under extreme light load power, which is not good for the system control stability.

In order to overcome these shortcomings of the basic variable frequency control strategy, an improved variable frequency control strategy is proposed.

10.3 Improved Variable Frequency Control Strategy

10.3.1 Improved Variable Frequency Control Strategy of M > *1*

The primary side switches cannot achieve ZVS as $|v_{ac}|$ is relatively low by using constant frequency or the basic variable frequency strategy. The reason is that the pulse width time of v_{ab} is too narrow, which makes the magnetizing inductor current i_{Lm} and primary side current i_p become very low, thus the energy is not enough to charge/discharge the switches of the output junction capacitors. The gradual change process of mode 1 with the basic variable frequency control is shown in the above five red dotted boxes of Fig. 10.10a, where, D_{c1}, D_{c2} and T_{c1} indicate the two control variables and half switching period. The switching frequency varies as the red curve in the Fig. 10.10b. As seen, ZVS of mode 111 and mode 112 are not achieved due to the lower limit of f_{s1}. In order to achieve ZVS of the two modes, the switching frequency must vary along green dotted curve (defined as the virtual variable switching frequency curve f_{sv}), then the pulse width of v_{ab} to achieve the ZVS at any voltage gain needs to be solved. First, the virtual half switching period T_v at any electrical angle can be obtained from (10.10) according to (10.6)–(10.8), the pulse width during length of v_{ab} is defined as $t_{12} = (D_{v1} + D_{v2})T_v$, and the detailed expression of t_{12} is shown in (10.12). Where, D_{v1}, D_{v2} and T_v indicate two virtual control variables and virtual half switching period. Five figures in the middle green dotted box of Fig. 10.10a are mode 1 adopting the virtual variable frequency curve. As seen, although the pulse width of v_{ab} has been enlarged to achieve ZVS, the transferred instantaneous power is fixed at a certain $|v_{ac}|$ to achieve PFC, so the leakage inductor peak current increases sharply.

$$T_v = \frac{nL_m^2 V_m^2 (nv_o - |v_{ac}|)\left[\frac{I_{bias}}{n} + \frac{|v_{ac}|}{\sqrt{\frac{n^2 L_m L_r}{C_{eq.ac}(L_m+L_r)}}}\right]}{L_r |v_{ac}|^2 v_o P_o} \tag{10.10}$$

$$t_{12} = (D_{v1} + D_{v2})T_v \tag{10.11}$$

$$
\begin{aligned}
t_{12} &= (D_{v1} + D_{v2})T_v \\
&= \frac{2C_{\mathrm{eq.ac}}L_m V_m \sqrt{nv_o} - |v_{\mathrm{ac}}|\left[n|v_{\mathrm{ac}}| + I_{\mathrm{bias}}\sqrt{\frac{n^2 L_m L_r}{C_{\mathrm{eq.ac}}(L_m+L_r)}}\right]^2 (L_m + L_r)}{|v_{\mathrm{ac}}|^2\sqrt{n^3 L_r^3 P_o v_o}\sqrt{\frac{C_{\mathrm{eq.ac}}L_m V_m^2(nv_o - |v_{\mathrm{ac}}|)(L_m+L_r)\left[n|v_{\mathrm{ac}}| + I_{\mathrm{bias}}\sqrt{\frac{n^2 L_m L_r}{C_{\mathrm{eq.ac}}(L_m+L_r)}}\right]^2}{n^3 L_r^2 |v_{\mathrm{ac}}|^2 P_o v_o}}}
\end{aligned}
\quad (10.12)
$$

In order to minimize the envelope of the leakage inductor peak current under the premise of achieving ZVS for primary side switches in mode 1, the virtual half switching period T_v needs to be reduced as much as possible. Improved variable frequency strategy is proposed as the five figures in the bottom blue dotted box of Fig. 10.10a, which shows the derivation and gradual change of mode 1*a*–mode 1*d*. Where, d_1, d_2 and T_{c2} indicate the two control variables and half switching period of the improved variable frequency strategy.

(1) When $t_{12} = (D_{v1} + D_{v2})T_v > 25$ µs, the new half switching period T_{c2} is set to be 25 µs, the new switching frequency $f_{s2} = 1/(2T_{c2})$ is set to be the lower limit 20 kHz. The new half switching period T_{c2} has been minimized, the transferred power is fixed at a certain electrical angle in order to achieve PFC, this makes $d_1 T_{c2} < D_{v1} T_v$, and the leakage inductor peak current I_1 will decrease greatly. The modulation mode is mode 1*a* in this case as shown in Fig. 10.10, the new control variables d_1 and d_2 are calculated as the following,

$$
\begin{cases}
d_1 = \sqrt{\frac{2nL_r i_{\mathrm{ref}}(M-1)}{v_o T_{c2}}} = \sqrt{\frac{2nL_r A \sin(\omega t)(M-1)}{v_o T_{c2}}} \\
d_2 = 1 - d_1
\end{cases}
\quad (10.13)
$$

The leakage inductor current at the rising edge of v_{ab} in mode 1*a* is zero instead of platform current I_{bias}. Figure 10.11 gives the ZVS boundary and primary side current $|i_p(T_{c2})|$ around the zero-crossing-point. As seen, $|i_p(T_{c2})|$ or $|i_p(t_0)|$ is larger than the minimum ZVS current in all conditions. So the ZVS of the primary side switches can be achieved although $(d_1 + d_2)T_{c2} < (D_{v1} + D_{v2})T_v$.

(2) When 5 µs $< t_{12} = (D_{v1} + D_{v2})T_v <$ 25 µs, the new half switching period $T_{c2} = t_{12} = (D_{v1} + D_{v2})T_v$, the new switching frequency $f_{s2} = 1/(2T_{c2})$ is between 20 and 100 kHz. The principle of reducing the leakage inductor peak current I_1 is the same. The modulation modes are mode 1*b*1 and mode 1*b*2 in this case as shown in Fig. 10.10. The new control variables d_1 and d_2 are calculated as follows,

$$
\begin{cases}
d_1 = \sqrt{\frac{2nL_r i_{\mathrm{ref}}(M-1)}{v_o T_{c2}}} = \sqrt{\frac{2nL_r A \sin(\omega t)(M-1)}{v_o T_{c2}}} \\
d_2 = 1 - d_1
\end{cases}
\quad (10.14)
$$

(3) When $t_{12} = (D_{v1} + D_{v2})T_v < 5$ µs, the new half switching period T_{c2} is set to be 5 µs, the new switching frequency $f_{s2} = 1/(2T_{c2})$ is set to be in the upper limit

Fig. 10.10 The improved variable frequency principle

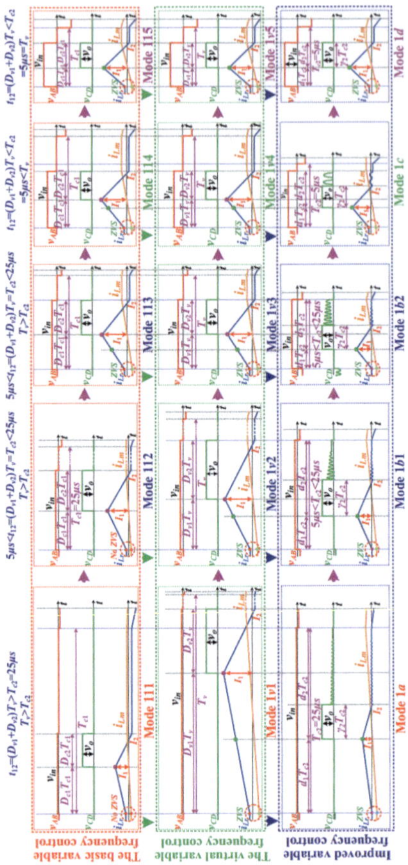

(a) Improved modulations of reducing the leakage inductor peak current for mode1 and the mechanism to achieve full operating range ZVS

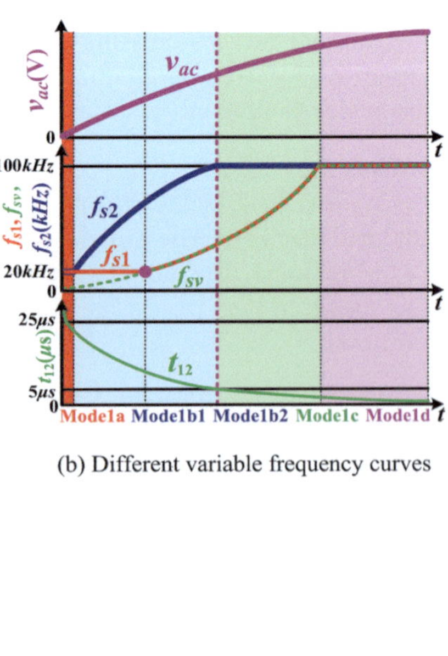

(b) Different variable frequency curves

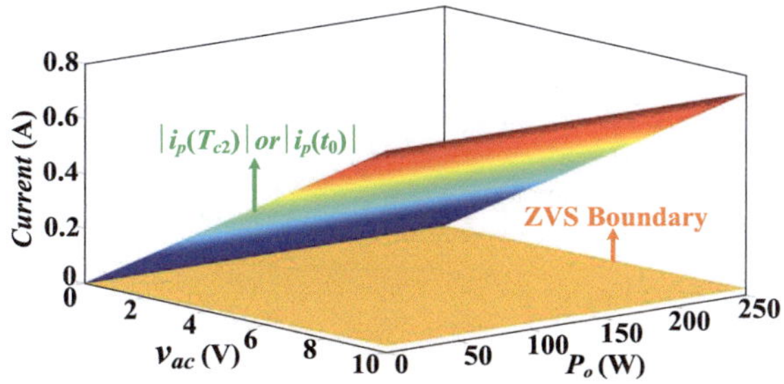

Fig. 10.11 The ZVS achievement proof of mode 1*a* around the zero-crossing-point

100 kHz. If $T_v > T_{c2}$, the modulation mode is mode 1*c*. The derivation process of control variables d_1, d_2 is given by,

$$\begin{cases} p = \frac{v_o^2 T_v}{2L_r} \cdot \frac{D_{v1}^2}{M(M-1)} = \frac{v_o^2 T_{c2}}{2L_r} \cdot \frac{d_1^2}{M(M-1)} \\ d_2 T_{c2} = \left(D_{v1} + \frac{D_{v1}}{M-1}\right) T_v - d_1 T_{c2} \end{cases} \quad (10.15)$$

$$\begin{cases} d_1 = \sqrt{\frac{2nL_r i_{\text{ref}}(M-1)}{v_o T_{c2}}} = \sqrt{\frac{2nL_r A \sin(\omega t)(M-1)}{v_o T_{c2}}} \\ D_{v1} = \sqrt{\frac{2nL_r i_{\text{ref}}(M-1)}{v_o T_v}} = \sqrt{\frac{2nL_r A \sin(\omega t)(M-1)}{v_o T_v}} \\ d_2 = \left(D_{v1} + \frac{D_{v1}}{M-1}\right) \cdot \frac{d_1^2}{D_{v1}^2} - d_1 \end{cases} \quad (10.16)$$

(4) When $t_{12} = (D_{v1} + D_{v2})T_v < 5$ μs, the new half switching period T_{c2} is set to be 5 μs, the new switching frequency $f_{s2} = 1/(2T_{c2})$ is set to be the upper limit 100 kHz. If $T_v = T_{c2}$, the modulation mode is mode 1*d* as shown in Fig. 10.10. In this case, $D_{v1} = d_1$, $D_{v2} = d_2$, and the new control variables d_1, d_2 are expressed as,

$$\begin{cases} p = \frac{v_o^2 T_v}{2L_r} \cdot \frac{D_{v1}^2}{M(M-1)} = \frac{v_o^2 T_{c2}}{2L_r} \cdot \frac{d_1^2}{M(M-1)} \\ d_2 T_{c2} = \left(D_{v1} + \frac{D_{v1}}{M-1}\right) T_v - d_1 T_{c2} \end{cases} \quad (10.17)$$

$$\begin{cases} d_1 = \sqrt{\frac{2nL_r i_{\text{ref}}(M-1)}{v_o T_{c2}}} = \sqrt{\frac{2nL_r A \sin(\omega t)(M-1)}{v_o T_{c2}}} \\ D_1 = \sqrt{\frac{2nL_r i_{\text{ref}}(M-1)}{v_o T_v}} = \sqrt{\frac{2nL_r A \sin(\omega t)(M-1)}{v_o T_v}} \\ d_2 = \left(D_{v1} + \frac{D_{v1}}{M-1}\right) \cdot \frac{d_1^2}{D_{v1}^2} - d_1 = \frac{d_1}{M-1} \end{cases} \quad (10.18)$$

Based on the analysis of improved variable frequency control strategy, the relationship between the new switching frequency f_{s2} and t_{12} is shown in (10.19), and the detailed expression of f_{s2} is given in (10.20). The switching frequency variation range is 20–100 kHz. The primary side switches could achieve ZVS in mode 1 within all operating conditions even at the zero-crossing point of v_{ac} under no load condition. Besides, the leakage inductor peak current is reduced greatly simultaneously by taking the proposed improved variable frequency control.

$$f_{s2} = \frac{1}{2t_{12}} \tag{10.19}$$

$$f_{s2} = \frac{|v_{ac}|^2\sqrt{n^3L_r^3P_ov_o}\sqrt{\frac{C_{eq.ac}L_mV_m^2(nv_o-|v_{ac}|)(L_m+L_r)\left[n|v_{ac}|+I_{bias}\sqrt{\frac{n^2L_mL_r}{C_{eq.ac}(L_m+L_r)}}\right]^2}{n^3L_r^2|v_{ac}|^2P_ov_o}}}{4C_{eq.ac}L_mV_m\sqrt{nv_o-|v_{ac}|}\left[n|v_{ac}|+I_{bias}\sqrt{\frac{n^2L_mL_r}{C_{eq.ac}(L_m+L_r)}}\right]^2(L_m+L_r)} \tag{10.20}$$

10.3.2 Improved Variable Frequency Control Strategy of M < 1

As seen in Fig. 10.7a, ZVS of primary side switches cannot be achieved in mode 3 under extremely light load power ($P_o < 9$ W) as $V_o = 135$ V. The reason is similar to mode 1. So the suitable switching frequency f_{s2} to achieve the ZVS of primary side switches in mode 3 under the premise of achieving power factor correction and controllable output voltage needs to be calculated as follows,

$$\begin{cases} p_{require} = 2P_o\left(\frac{|v_{ac}|}{V_m}\right)^2 \\ p_{mode\ 3} = \frac{v_o^2T_{c2}}{2L_r}\cdot\frac{d_2^2(1-M)}{M^2} \end{cases} \tag{10.21}$$

$$\begin{cases} d_1 = 0 \\ d_2 = \frac{2n\sqrt{L_rP_o|v_{ac}|}}{V_m\sqrt{(|v_{ac}|-v_o)T_{c2}}} \end{cases} \tag{10.22}$$

$$\begin{cases} I_{p.ZVS} = \frac{|v_{ac}|}{\sqrt{\frac{n^2L_mL_r}{(L_m+L_r)C_{eq.ac}}}} \\ |i_p(T)| = \frac{1}{n}\cdot\left(\frac{|v_{ac}|}{2nL_m}\cdot d_2T_{c2} - I_{bias}\right) \end{cases} \tag{10.23}$$

$$f_{s2} = \frac{n^2L_r^2|v_{ac}|^3P_o}{2C_{eq.ac}L_mV_m^2(|v_{ac}|-nv_o)\left[n|v_{ac}|+I_{bias}\sqrt{\frac{n^2L_mL_r}{C_{eq.ac}(L_m+L_r)}}\right]^2(L_m+L_r)} \tag{10.24}$$

Similarly, in order to achieve ZVS within the full operating range, the switching frequency f_{s2} needs to be reduced. The lower limit of the switching frequency f_{s2} is set to be 20 kHz in this paper, and the switching frequency variation range is 20–100 kHz. The ZVS of primary side switches in mode 3 could be achieved as long as the switching frequency varies according to (10.24).

Based on the comprehensive analysis under $M > 1$ and $M < 1$, using the improved variable frequency control, the primary and secondary side switch can achieve ZVS in the full operating range (the entire ac line cycle and full load range) even at the zero-crossing-point of v_{ac} and no load condition as shown in Fig. 10.12.

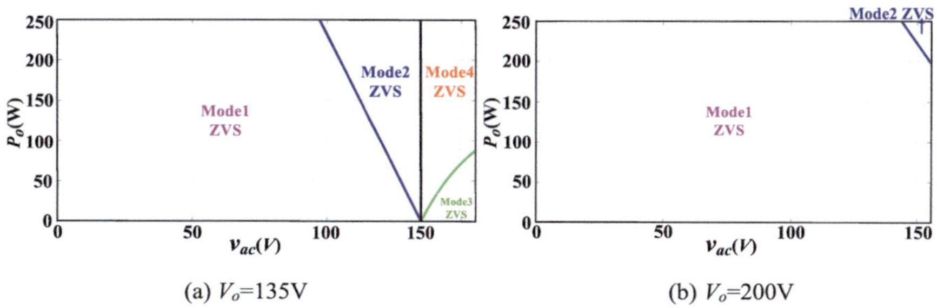

Fig. 10.12 ZVS range of switches under different output voltages with improved variable frequency control

The switching frequency under different voltage gains and power can be calculated according to the flow chart in Fig. 10.13, in which $f_{s\min} = 20$ kHz, $f_{s\max} = 100$ kHz, $T_{c2\min} = 1/(2f_{s\max}) = 5$ μs and $T_{c2\max} = 1/(2f_{s\min}) = 25$ μs. And the proposed closed loop diagram is shown in Fig. 10.14. As seen, there are seven modes in total, including mode 1*a*, mode 1*b*, mode 1*c*, mode 1*d* in Fig. 10.10 and mode 2, mode 3, mode 4 in Fig. 10.3.

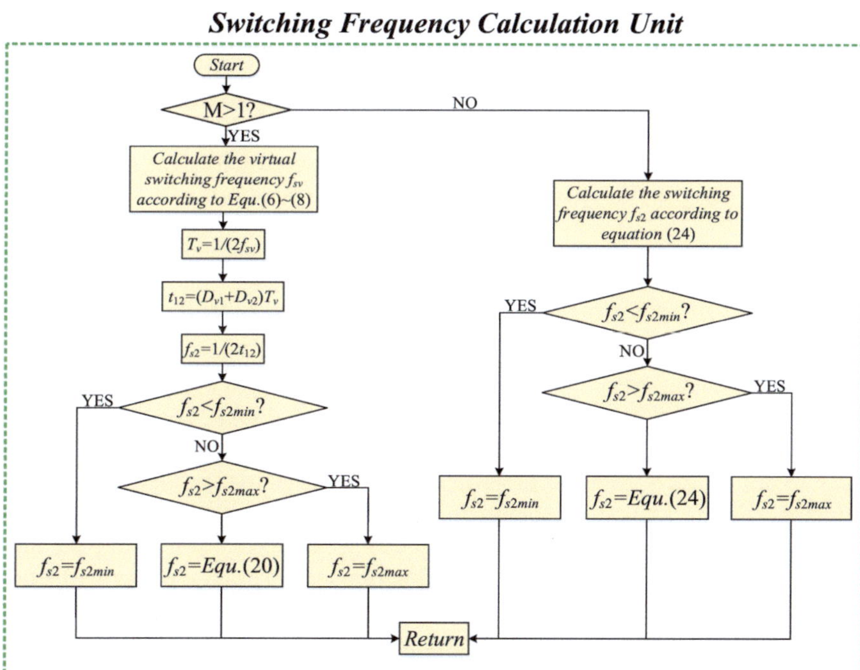

Fig. 10.13 Flow chart of the improved switching frequency calculation

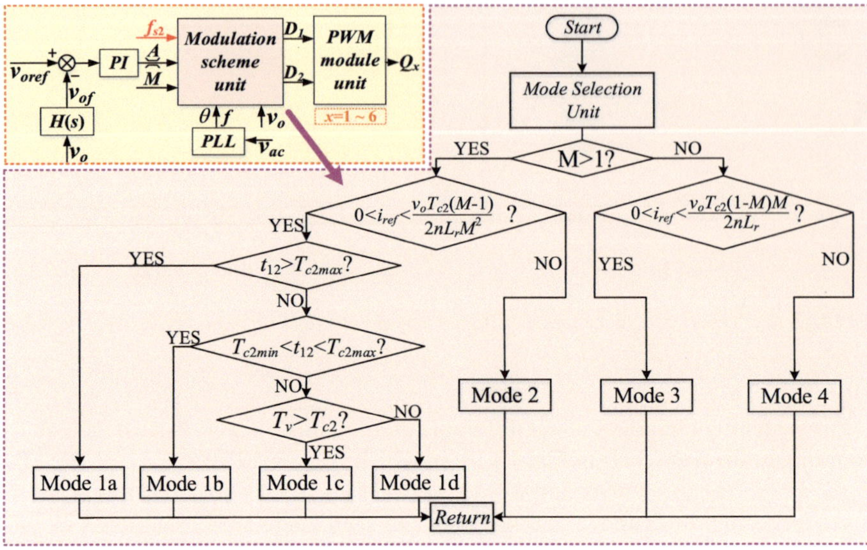

Fig. 10.14 Closed loop control diagram of improved variable frequency control

10.4 Comparison Between Different Control Strategies

A series of performance and design comparisons among the constant frequency control [7], the basic variable frequency and the proposed improved variable frequency control strategy have been done in this section to prove the advantages of the control strategy in this chapter.

10.4.1 Magnetic Components Design Consideration

The volt-second area of the transformer within the complete half ac line cycle under different control strategies are given in Fig. 10.15a, b, which determines the maximum flux density as (10.25). The volt-second area increases slightly at lower electrical angles to ensure ZVS with the proposed control, but the design of transformer is determined by the maximum value of volt-second area in the entire half ac line cycle. The maximum values of the volt-second area are the same by using the three different controls. Thus the improved variable frequency strategy will not affect the design of the transformer although the frequency variation range is 20–100 kHz. Its design is the same as that the switching frequency is fixed as 100 kHz.

$$B_m = \frac{|v_{\mathrm{ac}}| \cdot (D_1 + D_2)T_{c2}}{2N_2S_T} \tag{10.25}$$

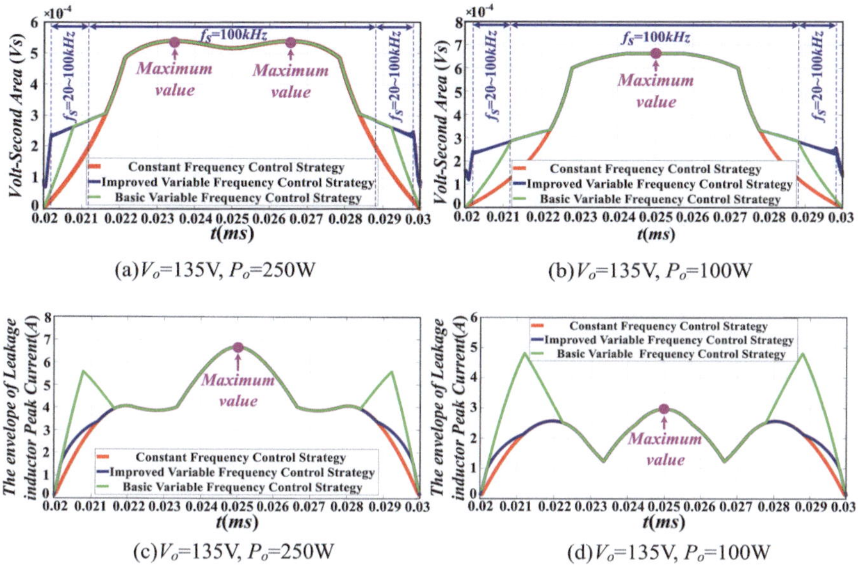

Fig. 10.15 Volt-second area and leakage inductor peak current comparisons

The design principle of the leakage inductor L_r is similar, the envelope of the leakage inductor peak current within the half ac line cycle under different control strategies are given in Fig. 10.15c, d, which is determined by the maximum flux density as shown in (10.26). The leakage inductor peak current increase slightly at low $|v_{ac}|$ in order to achieve ZVS with the proposed control. But the design of L_r is determined by the maximum value of the leakage inductor peak current in the entire half ac line cycle. As seen in Fig. 10.15, the maximum peak current with the proposed improved variable frequency control is the same as that obtained by the constant frequency control in spite of the load power. Thus, the proposed improved variable frequency control strategy will not affect the design of the leakage inductor.

$$B_m = \frac{L_r i_{Lr\max}}{2N_L S_L} \tag{10.26}$$

10.4.2 The Comparison for the Envelope of i_{Lr}, i_p and ZVS Range

The transformer primary side current and leakage inductor current waveform using the basic and improved variable frequency control strategy under different operating conditions are given in Fig. 10.16. Figure 10.16a, c are obtained with the basic variable frequency strategy, while Fig. 10.16b, d are got by an improved variable

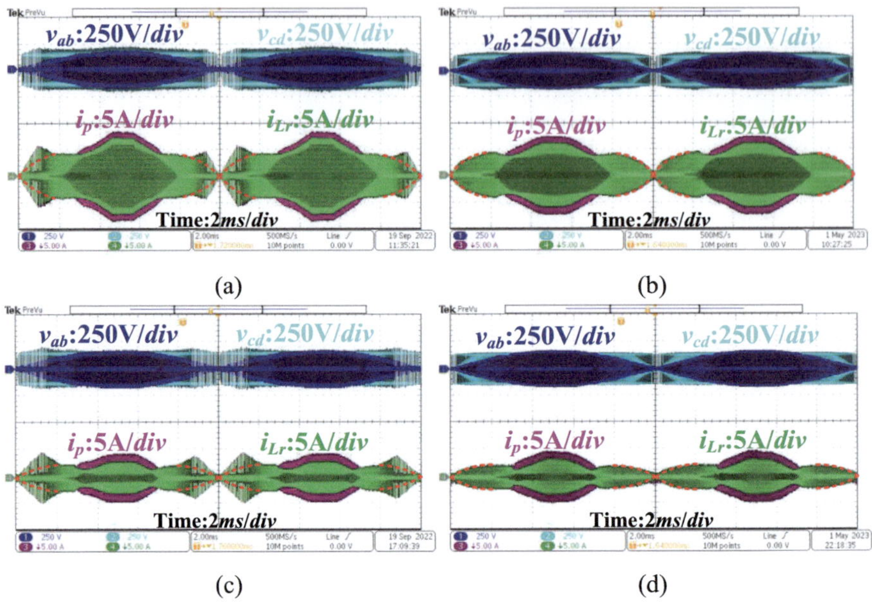

Fig. 10.16 The comparison of i_p and i_{Lr} with two variable frequency control. **a** $V_o = 135$ V, $P_o = 250$ W with the basic variable frequency control. **b** $V_o = 135$ V, $P_o = 250$ W with the improved variable frequency control. **c** $V_o = 135$ V, $P_o = 100$ W with the basic variable frequency control. **d** $V_o = 135$ V, $P_o = 100$ W with the improved variable frequency control

frequency control strategy. As seen, the envelope of i_{Lr} and i_p are minimized significantly as shown in red dotted lines after taking the improved variable frequency strategy compared with the basic variable frequency strategy.

In addition, as seen from Fig. 10.8, although ZVS range can be extended greatly using the basic variable frequency strategy, ZVS is still lost in partial regions. In comparison, the full operating range ZVS can be achieved by the improved variable frequency control strategy, which is shown in Fig. 10.12.

10.4.3 Comparison of Loss Breakdown

Various loss comparisons among the three control strategies are given in Fig. 10.17. As seen, only the transformer core loss with the improved variable frequency control increases slightly due to achieving ZVS in the full operating range, while other losses can be reduced. So the total loss is reduced by the proposed improved variable frequency control strategy compared with using the other two control strategies.

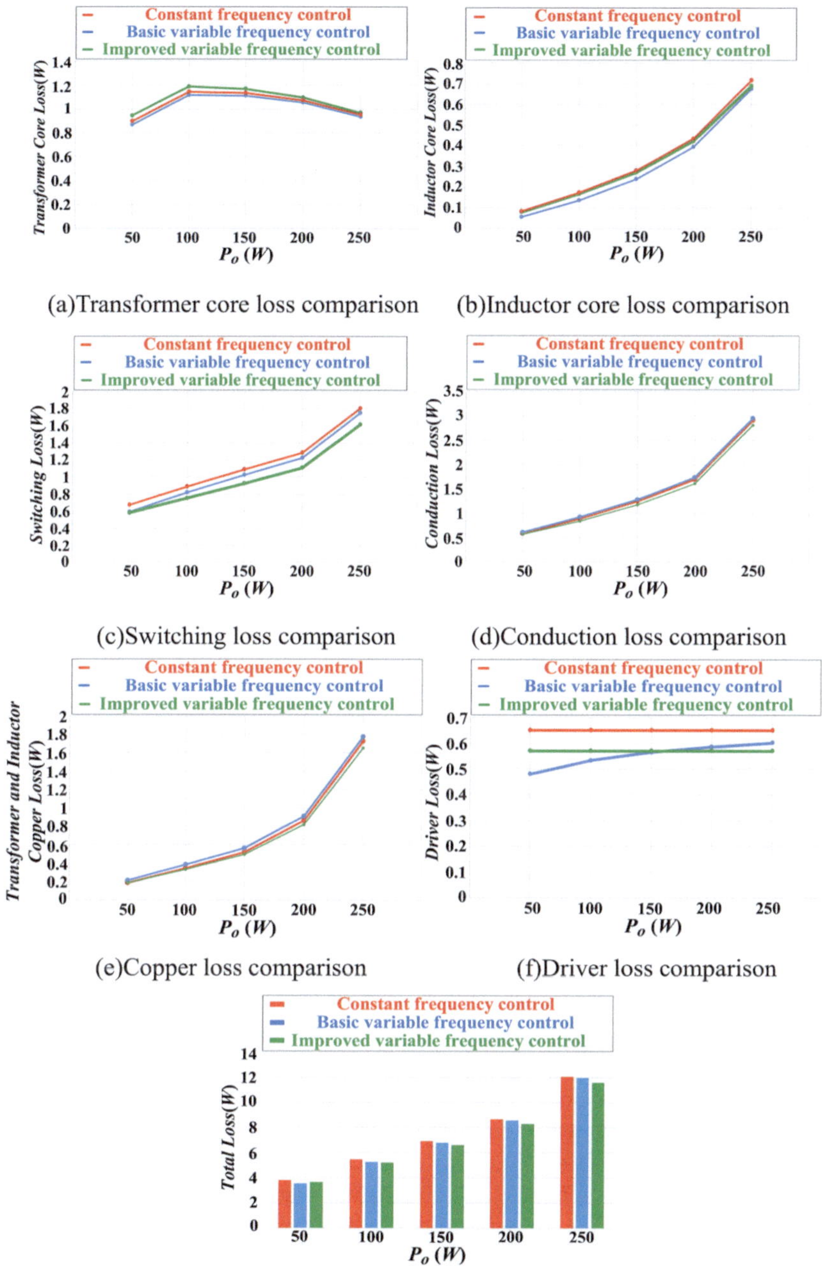

(a)Transformer core loss comparison (b)Inductor core loss comparison

(c)Switching loss comparison (d)Conduction loss comparison

(e)Copper loss comparison (f)Driver loss comparison

(g)Total loss comparison

Fig. 10.17 Comparison of various loss

10.5 The Design of Converter Various Components

10.5.1 The Design of Leakage Inductance L_r

The leakage inductance determines the power transmission capability. The maximum instantaneous power $p_{o,\max}$ with the proposed control strategy lies in mode 4 at the maximum absolute value of the grid voltage. So the leakage inductance value can be calculated as follows,

$$\begin{cases} p_{o,\max} > 2P_{\text{rated}} \\ p = -\frac{v_o^2 T}{2L_r} \cdot \left[\frac{D_1^2 M^2 + D_1^2 M + D_1^2 - 2D_1 M^2 + M^2 - M}{M} \right] \\ D_{1,\max}(M) = \frac{M^2}{M^2+M+1} \\ M_{\min} = n \frac{V_o}{|v_{\text{ac}}|_{\max}} \end{cases} \tag{10.27}$$

Substituting system parameters into the above equations yields the maximum value of L_r,

$$L_r < \frac{v_o^2 T}{4P_{\text{rated}}\left(M_{\min}^2 + M_{\min} + 1\right)} = 34.76\,\mu\text{H} \tag{10.28}$$

So the leakage inductance value is selected as $L_r = 34\ \mu\text{H}$.

10.5.2 The Design of Magnetizing Inductance L_m

An important objective of this paper is that all switches can achieve ZVS within the full operating range. Based on previous analysis, Q_1, Q_2 are the most difficult to achieve ZVS under no load and the $M < 1$(especially around the $|v_{\text{ac}}|_{\max}$), so the variable switching frequency control strategy is introduced to achieve ZVS within the full operating range, which is affected by the magnetizing inductance L_m.

The variable switching frequency curves for $M < 1$ under no load and different magnetizing inductance values can be obtained as shown in the following Fig. 10.18 according to (10.24). As seen, the primary side switches Q_1 and Q_2 can achieve ZVS within the full operating range even around $|v_{\text{ac}}|_{\max}$ with the variable switching frequency (20–100 kHz) when $L_m < 210\ \mu\text{H}$, but when $L_m > 210\ \mu\text{H}$, Q_1 and Q_2 lose ZVS around $|v_{ac}|_{\max}$ as shown in the dotted circle of Fig. 10.18 even though the switching frequency reduces to 20 kHz. In addition, lower L_m causes higher magnetizing current and primary side current i_p, which causes a higher current stress and turn off loss. So L_m cannot be designed too small. Thus L_m is selected as 210 μH.

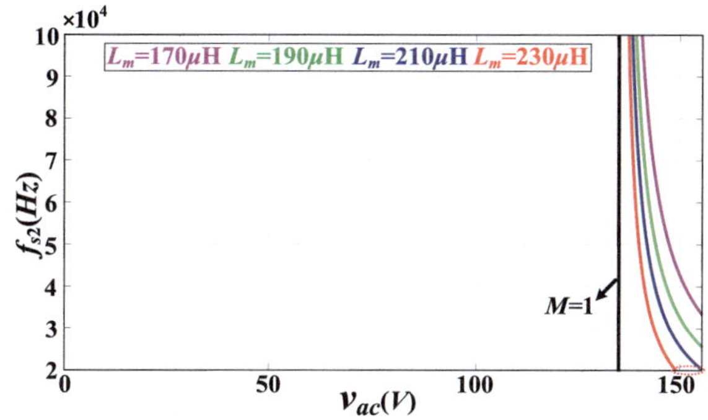

Fig. 10.18 The variable frequency curves for $M < 1$ under no load (consider the output voltage sampling power) and different L_m

10.5.3 The Design of Output Capacitor C_o and Input Capacitor C_{in}

Assuming that the power factor (PF) is 1 and loss is zero, the output power is expressed as the following,

$$P_o = P_{\text{in}} = V_m \sin(\omega t) \cdot I_m \sin(\omega t) = \frac{V_m I_m}{2}[1 - \cos(2\omega t)] \tag{10.29}$$

Furthermore, the output current $i_o(t)$ is given by,

$$\begin{cases} i_o(t) = \frac{1}{2}\frac{V_m I_m}{V_o}[1 - \cos(2\omega t)] = I_o - i_{\text{rip}}(t) \\ I_o = \frac{1}{2}\frac{V_m I_m}{V_o} \\ i_{\text{rip}}(t) = \frac{1}{2}\frac{V_m I_m}{V_o}\cos(2\omega t) \end{cases} \tag{10.30}$$

where, I_o is the average value of the output current, and $i_{\text{rip}}(t)$ is the ripple current, so the voltage ripple ΔV_c on the output filter capacitor is expressed as the following,

$$\Delta V_c = \frac{\int_{-\frac{\pi}{4\omega}}^{\frac{\pi}{4\omega}} \frac{1}{2}\frac{V_m I_m}{V_o}\cos(2\omega t)dt}{C_o} = \frac{V_m I_m}{2V_o C_o \omega} \tag{10.31}$$

where, V_m, V_o, I_m and ω are constant, so the larger the output capacitor is, the smaller the output voltage ripple is. In general, $\Delta V_c / V_o < 2\%$, so the output filter capacitor can be calculated as the following expression according to the system parameters,

$$C_o \geq \frac{V_m I_m}{2V_o \omega \Delta V_c} = 9952.23\,\mu\text{F} \tag{10.32}$$

The output filter capacitor includes two parts, the first is a high frequency film capacitor C_{oh} to bypass the high frequency current, and the second is an electrolytic capacitor C_{ol} used to reduce the low frequency voltage ripple. So based on the above analysis and calculation, the high frequency capacitor C_{oh} is selected as 20 μF, and the electrolytic capacitor is designed as10,000 μF.

The input current of semi-DAB consists of high frequency ripple, so a film capacitor C_{in} is utilized to bypass it. A lager value of C_{in} is preferred to reduce the voltage ripple, but this affects the grid current i_{ac}. So C_{in} is designed as 2 μF based on the tradeoff between by passing the high frequency current ripple and the effect on quality of i_{ac}.

10.5.4 The Design of Power Devices

According to the system parameters and calculation, $|v_{ac}|_{max} = \sqrt{2}\ 110$ V, $i_{pmax} =$ 7.9 A, so the primary side switches are selected as SiC MOSFET C3M0065045K; $V_{omax} = 200$ V, $i_{Lrmax} = 6.7$ A, so the secondary side switches are also selected as C3M0065045K, and the secondary side diodes are selected as MBR20300FCT. Since $|i_{ac}|_{max} = 3.2146$ A, so the rectifier is selected as KBPC2510.

10.6 Experimental Results

10.6.1 Steady-State Operating Waveforms

A 250 W prototype as shown in Fig. 10.19 was built to verify the effectiveness of the proposed improved variable frequency control strategy, the system parameters are listed in Table 10.6. The experiments are implemented by DSP TMS320F28335.

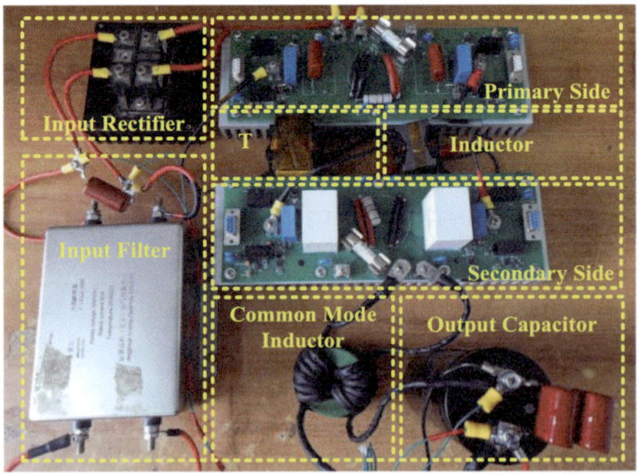

Fig. 10.19 The experimental prototype

Table 10.6 System parameters

Symbol	Value	Symbol	Value
v_{ac}	110 $\mathrm{V_{rms}}$	C_{ol}	10,000 μF
f_{ac}	50 Hz	Q_1, Q_2, Q_3, Q_4	C3M0045065K
v_o	135–200 V	Q_5, Q_6	C3M0045065K
C_{in}	2 μF	$D_{\mathrm{o1}}, D_{\mathrm{o2}}$	MBR20300FCT
L_r	34 μH	$D_{\mathrm{r1}}, D_{\mathrm{r2}}, D_{\mathrm{r3}}, D_{\mathrm{r4}}$	KBPC2510
L_m	210 μH	f_{s2}	20–100 kHz
n	1: 1	t_{dead}	200 ns
C_{oh}	20 μF	P_{rated}	250 W

Figure 10.20 shows the steady state experimental waveforms of v_{ac}, i_{ac}, i_{Lr} and v_{ab} at the different operating conditions respectively with the proposed improved variable frequency control strategy. As seen, the input power factor is high.

Figure 10.21 shows the experimental waveforms of v_{ab}, v_{cd}, i_{Lr} and i_p within the entire ac line cycle, the variable frequency process and different operation modes under the rated power and different output voltages respectively. As seen, the experimental results agree well with the theoretical analysis.

The ZVS situation of different switches at different electrical angles under the rated load and no load are shown in Fig. 10.22 respectively. Only ZVS states of Q_2, Q_6 and D_{o2} are given due to the symmetry between the upper and bottom switches in a half bridge of primary side, and the symmetry of secondary side switches. As seen, the switches can achieve ZVS within full operating range, even at the zero-crossing point of v_{at} under no load condition, and the secondary side diodes can obtain ZCS off under all operation conditions, which agrees well with the previous theoretical analysis as shown in Fig. 10.12.

10.6.2 Dynamic Response and Conversion Efficiency

Figure 10.23 shows the dynamic response waveform as the load steps from 40% load to the rated load. Figure 10.24 shows the efficiency curve at different output voltages. As seen, the peak efficiency value is 96.08% with the proposed improved variable frequency control and its conversion efficiency is higher than that with the other two control strategies.

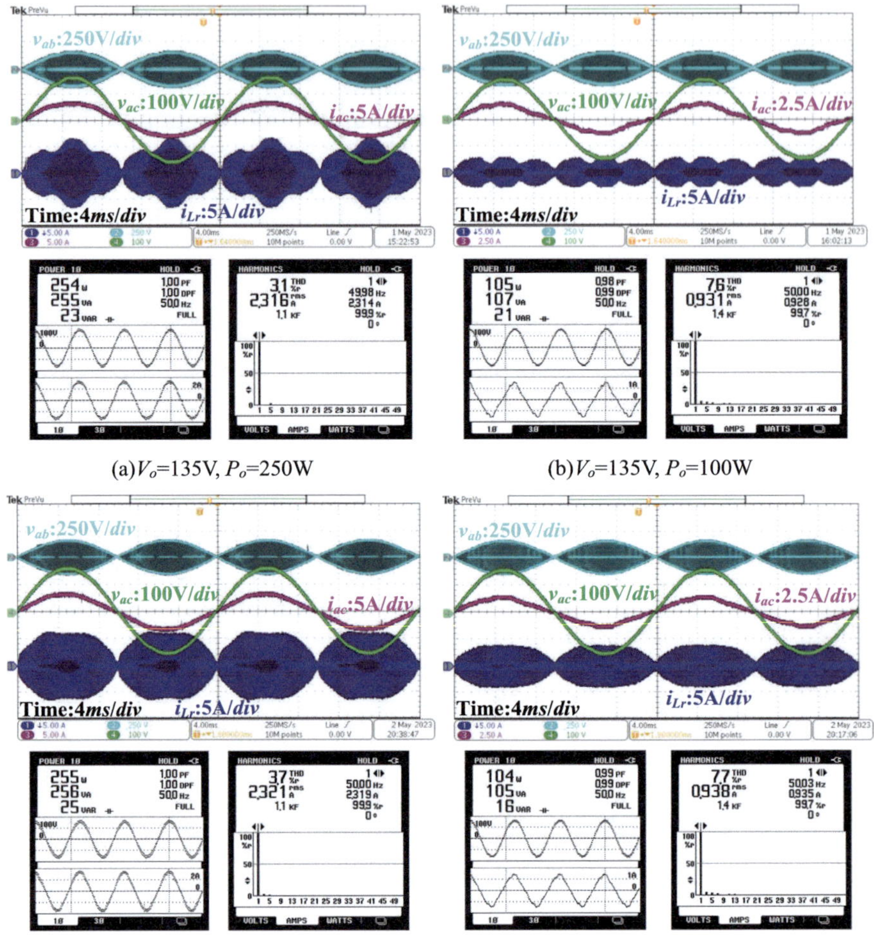

(a) V_o=135V, P_o=250W

(b) V_o=135V, P_o=100W

(c) V_o=200V, P_o=250W

(d) V_o=200V, P_o=100W

Fig. 10.20 Grid side voltage and current by improved variable frequency control

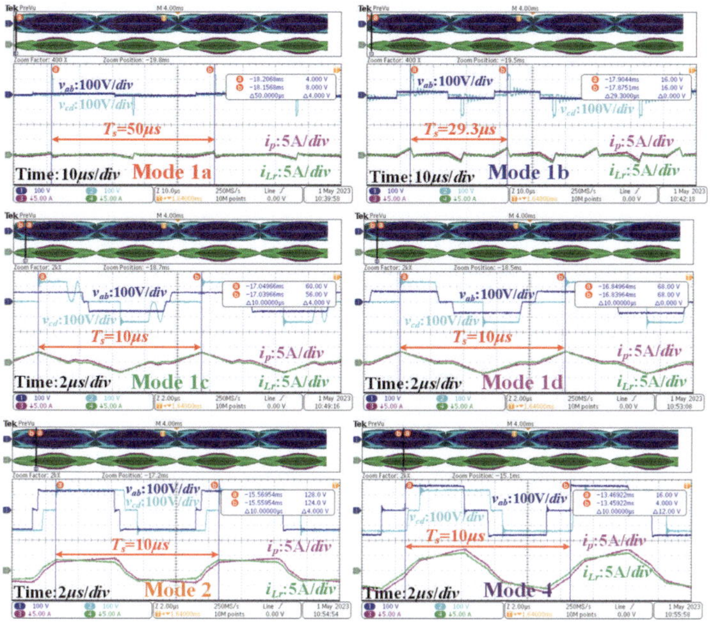

(a) V_o=135V, P_o=250W

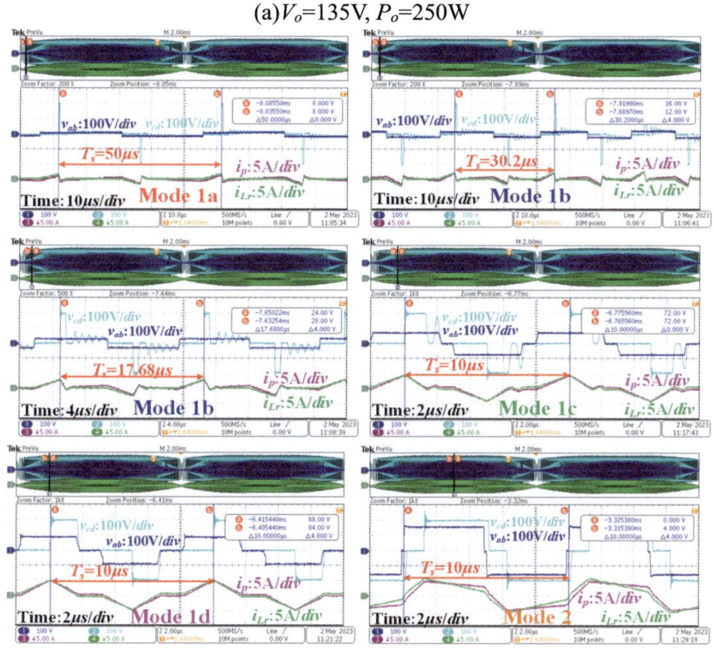

(b) V_o=200V, P_o=250W

Fig. 10.21 Mode operating situations under different operating conditions

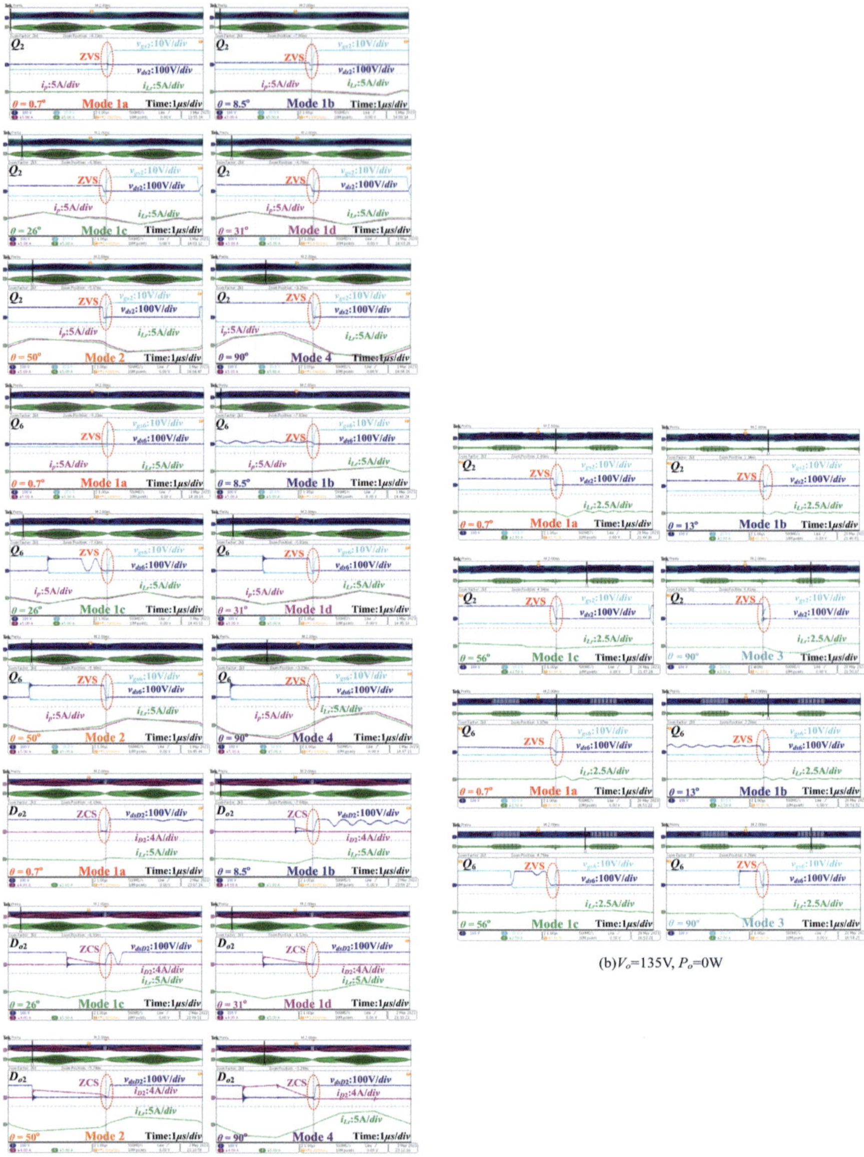

(a) V_o=135V, P_o=250W

(b) V_o=135V, P_o=0W

Fig. 10.22 Soft-switching situations under different operating conditions

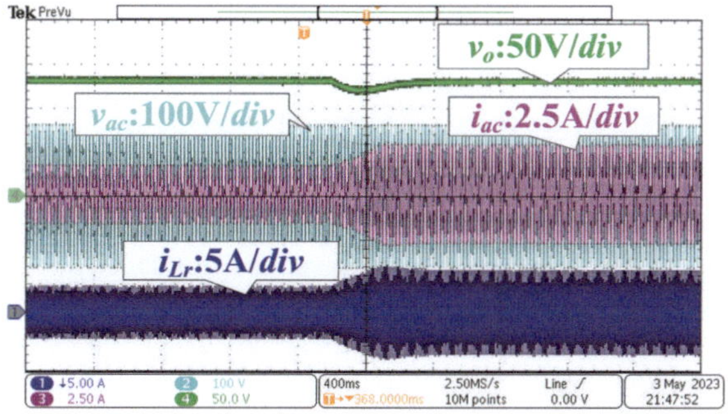

Fig. 10.23 Dynamic response waveform

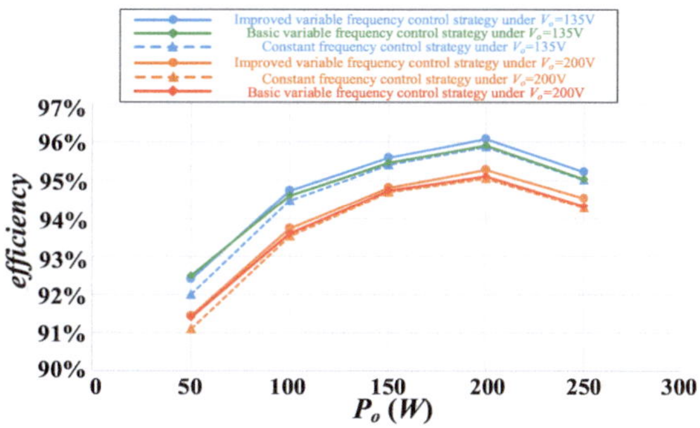

Fig. 10.24 Efficiency Curve

10.7 Conclusion

A high efficiency and full operating range ZVS control is proposed for single-stage semi-DAB AC-DC converter by using the improved modulation and variable frequency control. This could minimize the conduction loss, reduce the envelope value of l_{ira} and i_s significantly compared with using the basic variable frequency control. The ZVS around the zero-crossing-point of v_{at} under no load is guaranteed. It is worth noting that no special concern is needed for the transformer and inductor design although the switching frequency varies. Their design will follow the maximum switching frequency, thus the power density is the same as that using constant frequency control which is actually the maximum frequency. The transition among different operating modes is seamless. The control strategy could achieve PFC without a mandatory inner current loop in facing a wide output voltage range. The effectiveness of the improved variable frequency strategy is verified by a 250 W experimental prototype result.

Appendix

A. The Detail Derivation Process of t_{ry} in Different Modes

The equivalent circuit to describe the resonant process for Q_5 and D_{o1} in mode 1 and mode 3 are same as shown in Fig. 10.25 the differential equation to describe the resonant process can be set out as equation (10.33) according to Kirchhoff's law, combing the initial condition (10.34) yields the analytical solutions as (10.35). The resonant process is complete when v_{ds5} drops to zero. From (10.35), the resonant time t_{ry} can be solved as (10.36), and the leakage inductance current at the end moment of the resonant process I_{bis} is shown in (10.37).

$$\begin{cases} L_r \frac{di_{Lr}}{dt} = v_{ds5} \\ i_{Lr} = C_D \frac{d(v_o - v_{ds5})}{dt} - C_{oss} \frac{dv_{ds5}}{dt} = -(C_{oss} + C_D) \frac{dv_{ds5}}{dt} \end{cases} \tag{10.33}$$

$$\begin{cases} i_{Lr}(0) = 0 \\ v_{ds5}(0) = v_o \end{cases} \tag{10.34}$$

$$\begin{cases} i_{Lr}(t) = v_o \sqrt{\frac{C_{oss}+C_D}{L_r}} \sin\left(\frac{t}{\sqrt{L_r(C_{oss}+C_D)}}\right) \\ v_{ds5}(t) = v_o \cos\left(\frac{t}{\sqrt{L_r(C_{oss}+C_D)}}\right) \end{cases} \tag{10.35}$$

$$t_r = \frac{\pi}{2} \cdot \sqrt{L_r \cdot (C_{oss} + C_D)} \tag{10.36}$$

$$I_{bias} = i_{Lr}(t_r) = v_o \sqrt{\frac{C_{oss} + C_D}{L_r}} \tag{10.37}$$

The equivalent circuit to describe the resonant process for Q_5 and D_{o1} in mode 2 and mode 4 are same as shown in Fig. 10.26 a similar analysis and calculation process are shown in equations (10.38)–(10.42).

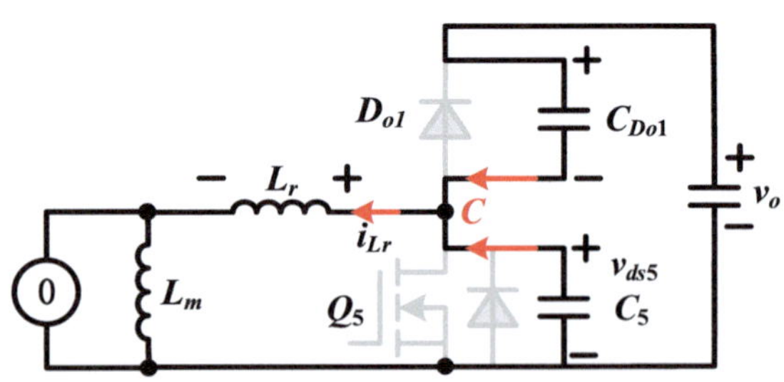

Fig. 10.25 The equivalent circuit of the resonant process for secondary side diode and switch in mode 1 and mode 3

Fig. 10.26 The equivalent circuit of the resonant process for secondary side diode and switch in mode 2 and mode 4

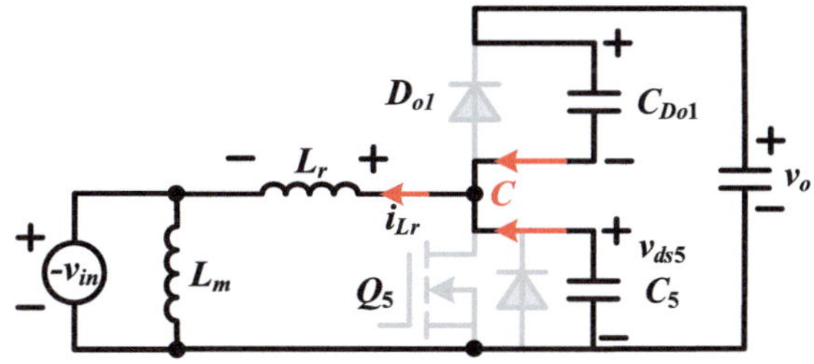

$$\begin{cases} L_r \frac{di_{Lr}}{dt} = v_{ds5} + v_{\text{in}} \\ i_{Lr} = C_D \frac{d(v_o - v_{ds5})}{dt} - C_{\text{oss}} \frac{dv_{ds5}}{dt} = -(C_{\text{oss}} + C_D) \frac{dv_{ds5}}{dt} \end{cases} \tag{10.38}$$

$$\begin{cases} i_{Lr}(0) = 0 \\ v_{ds5}(0) = v_o \end{cases} \tag{10.39}$$

$$\begin{cases} i_{Lr}(t) = (v_o + v_{\text{in}}) \cdot \sqrt{\frac{C_{\text{oss}} + C_D}{L_r}} \sin\left(\frac{t}{\sqrt{L_r(C_{\text{oss}} + C_D)}}\right) \\ v_{ds5}(t) = (v_o + v_{\text{in}}) \cdot \cos\left(\frac{t}{\sqrt{L_r(C_{\text{oss}} + C_D)}}\right) - v_{\text{in}} \end{cases} \tag{10.40}$$

$$t_r = \sqrt{L_r \cdot (C_{\text{oss}} + C_D)} \cdot \arccos\left(\frac{v_{\text{in}}}{v_{\text{in}} + v_o}\right) \tag{10.41}$$

$$I_{\text{bias}} = i_{Lr}(t_r) = (v_o + v_{\text{in}}) \cdot \sqrt{\frac{C_{\text{oss}} + C_D}{L_r}} \cdot \sin\left(\arccos\left(\frac{v_{\text{in}}}{v_{\text{in}} + v_o}\right)\right) \tag{10.42}$$

B. The Detail Derivation of the Minimum Primary Side Current to Achieve ZVS

Mode 1/3 Q_1

The equivalent circuit to describe the resonant process for Q_1 and Q_2 in mode 1 and mode 3 are same as shown in Fig. 10.27 the differential equation to describe the resonant process can be set out as (10.43)–(10.46) according to Kirchhoff's law, the systems of differential equations (10.47) can be obtained by derivation.

$$i_p = C_{\text{oss}} \frac{d(|v_{\text{ac}}| - v_{ds1})}{dt} - C_{\text{oss}} \frac{dv_{ds1}}{dt} = -2C_{\text{oss}} \frac{dv_{ds1}}{dt} = i_{Lm} - i_{Lr} \tag{10.43}$$

$$n^2 L_r \frac{di_{Lr}}{dt} = |v_{\text{ac}}| - v_{ds1} \tag{10.44}$$

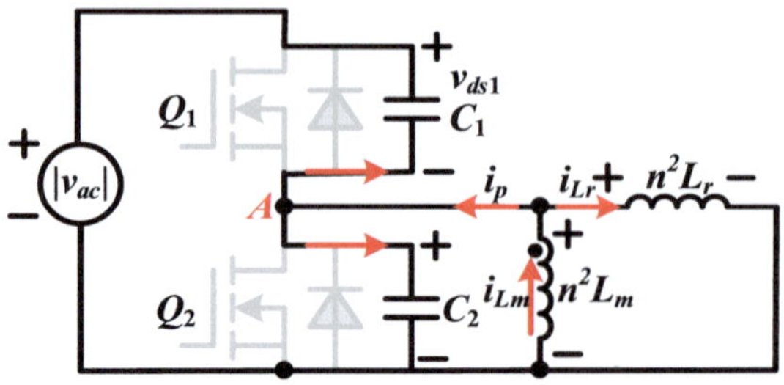

Fig. 10.27 The equivalent circuit of resonant process for primary side switches Q_1 and Q_2 in mode 1 and mode 3

$$n^2 L_m \frac{di_{Lm}}{dt} = -(|v_{\text{ac}}| - v_{ds1}) \tag{10.45}$$

$$n^2 L_m \frac{di_{Lm}}{dt} - n^2 L_r \frac{di_{Lr}}{dt} = -2(|v_{\text{ac}}| - v_{ds1}) = -2|v_{\text{ac}}| + 2v_{ds1} \tag{10.46}$$

$$\begin{cases} \frac{dv_{ds1}}{dt} = \frac{1}{-2C_{\text{oss}}} i_p = \frac{1}{-C_{\text{eq.ac}}} i_p \\ \frac{di_p}{dt} = \left(\frac{1}{n^2 L_m} + \frac{1}{n^2 L_r}\right) v_{ds1} - \left(\frac{1}{n^2 L_m} + \frac{1}{n^2 L_r}\right) |v_{\text{ac}}| \end{cases} \tag{10.47}$$

Initial conditions of resonant process are as following (10.48), $I_{p.\text{ZVS}}$ is the minimum primary side current to achieve ZVS of Q_1.

$$\begin{cases} v_{ds1}(0) = |v_{\text{ac}}| \\ i_p(0) = I = I_{p.\text{ZVS}} \end{cases} \tag{10.48}$$

Combining the initial condition (10.48) yields the analytical solutions as following (10.49). The equation of the circle (10.50) can be got after further derivation.

$$\begin{cases} v_{ds1}(t) = |v_{\text{ac}}| - I\sqrt{\frac{n^2 L_m L_r}{(L_m + L_r) C_{\text{eq.ac}}}} \sin\left(\sqrt{\frac{L_m + L_r}{n^2 L_m L_r C_{\text{eq.ac}}}} \cdot t\right) \\ i_p(t) = I \cos\left(\sqrt{\frac{L_m + L_r}{n^2 L_m L_r C_{\text{eq.ac}}}} \cdot t\right) \end{cases} \tag{10.49}$$

$$(v_{ds1}(t) - |v_{\text{ac}}|)^2 + \left(i_p(t)\sqrt{\frac{n^2 L_m L_r}{(L_m + L_r) C_{\text{eq.ac}}}}\right)^2 = \left(I\sqrt{\frac{n^2 L_m L_r}{(L_m + L_r) C_{\text{eq.ac}}}}\right)^2 \tag{10.50}$$

Figure 10.28 is the geometry of a circle drawn according to (10.50), the coordinates of the center of this circle are ($|v_{\text{ac}}|$, 0). The resonant process for Q_1 and Q_2 in mode 1 and mode 3 are actually a part of the circle as the red curve. The initial point of the resonant process is A, the ideal resonant process is that the primary side current i_p decrease to zero when the voltage of Q_1 v_{ds1} decrease to zero, so the ideal end point of the resonant process is B, the coordinates are (0, 0).

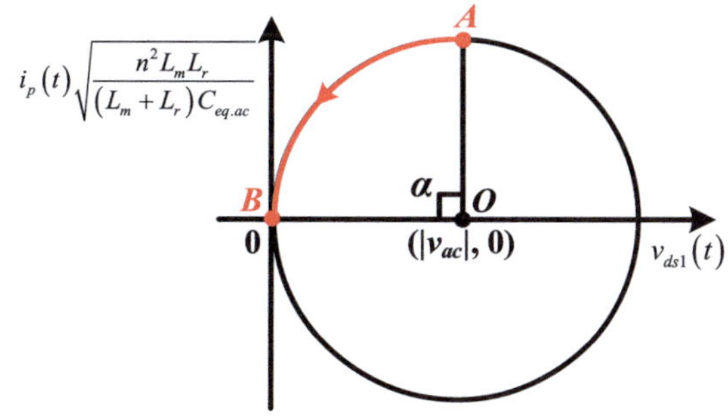

Fig. 10.28 The steady-state trajectory of single-stage high frequency isolated AC-DC converter based on semi-DAB

Since point B is on the circle, so the minimum primary side current to achieve ZVS $I_{p.\text{ZVS}}$ can be obtained by substituting point B into the equation of the circle (10.50), and the corresponding resonant time can be derived as (10.52).

$$|v_{\text{ac}}|^2 = \left(I\sqrt{\frac{n^2L_mL_r}{2(L_m+L_r)C_{\text{eq.ac}}}}\right)^2 \Rightarrow I = \sqrt{\frac{(L_m+L_r)C_{\text{eq.ac}}|v_{\text{ac}}|^2}{n^2L_mL_r}} = I_{p.\text{ZVS}} \tag{10.51}$$

$$t_r = \frac{\alpha}{\omega} = \frac{\frac{\pi}{2}}{\omega} = \frac{\pi}{2}\sqrt{\frac{n^2L_mL_rC_{\text{eq.ac}}}{L_m+L_r}} \tag{10.52}$$

Mode 2/4 Q*1*

Similarly, the equivalent circuit to describe the resonant process for Q_1 and Q_2 in mode 2 and mode 4 are same as shown in Fig. 10.29. The difference is that the voltage from the secondary side to primary side is no longer zero. The differential equation to describe the resonant process are shown as (10.53)–(10.56), the systems of differential equations (10.57) can be got by derivation.

$$i_p = C_{\text{oss}}\frac{d(|v_{\text{ac}}| - v_{ds1})}{dt} - C_{\text{oss}}\frac{dv_{ds1}}{dt} = -2C_{\text{oss}}\frac{dv_{ds1}}{dt} = i_{Lm} - i_{Lr} \tag{10.53}$$

$$n^2L_r\frac{di_{Lr}}{dt} = (|v_{\text{ac}}| - v_{ds1}) + v_o \tag{10.54}$$

$$n^2L_m\frac{di_{Lm}}{dt} = -(|v_{\text{ac}}| - v_{ds1}) \tag{10.55}$$

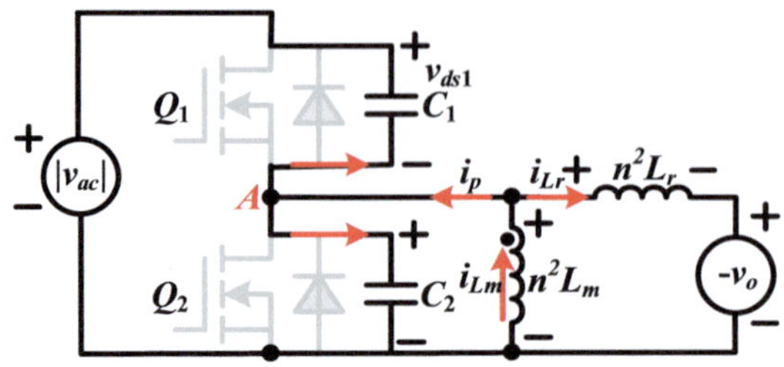

Fig. 10.29 The equivalent circuit of resonant process for primary side switches Q_1 and Q_2 in mode 2 and mode 4

$$n^2L_m\frac{di_{Lm}}{dt} - n^2L_r\frac{di_{Lr}}{dt} = -2|v_{ac}| + 2v_{ds1} - v_o \tag{10.56}$$

$$\begin{cases} \frac{dv_{ds1}}{dt} = \frac{1}{-2C_{oss}}i_p = \frac{1}{-C_{eq.ac}}i_p \\ \frac{di_p}{dt} = \left(\frac{1}{n^2L_m} + \frac{1}{n^2L_r}\right)v_{ds1} - \left(\frac{1}{n^2L_m} + \frac{1}{n^2L_r}\right)|v_{ac}| - \frac{1}{n^2L_r}v_o \end{cases} \tag{10.57}$$

Initial conditions of resonant process are shown as (10.58), $I_{p.ZVS}$ is the minimum primary side current to achieve ZVS of Q_1.

$$\begin{cases} v_{ds1}(0) = |v_{ac}| \\ i_p(0) = I = I_{p.ZVS} \end{cases} \tag{10.58}$$

Combining the initial condition (10.58) yield the analytical solutions as the following (10.59). The equation of the circle (10.60) can be got after the further derivation.

$$\begin{cases} v_{ds1}(t) = \left(|v_{ac}| + \frac{L_m}{L_m + L_r}v_o\right) - \left(\frac{L_m}{L_m + L_r}v_o\right)\cos\left(\sqrt{\frac{L_m + L_r}{n^2L_mL_rC_{eq.ac}}}\cdot t\right) \\ \quad - I\sqrt{\frac{n^2L_mL_r}{(L_m + L_r)C_{eq.ac}}}\sin\left(\sqrt{\frac{L_m + L_r}{n^2L_mL_rC_{eq.ac}}}\cdot t\right) \\ i_p(t)\sqrt{\frac{n^2L_mL_r}{(L_m + L_r)C_{eq.ac}}} = -\left(\frac{L_m}{L_m + L_r}v_o\right)\sin\left(\sqrt{\frac{L_m + L_r}{n^2L_mL_rC_{eq.ac}}}\cdot t\right) \\ \quad + I\sqrt{\frac{n^2L_mL_r}{(L_m + L_r)C_{eq.ac}}}\cos\left(\sqrt{\frac{L_m + L_r}{n^2L_mL_rC_{eq.ac}}}\cdot t\right) \end{cases} \tag{10.59}$$

$$\left(v_{ds1}(t) - \left(|v_{ac}| + \frac{L_m}{L_m + L_r}v_o\right)\right)^2 + \left(i_p(t)\sqrt{\frac{n^2L_mL_r}{(L_m + L_r)C_{eq.ac}}}\right)^2$$

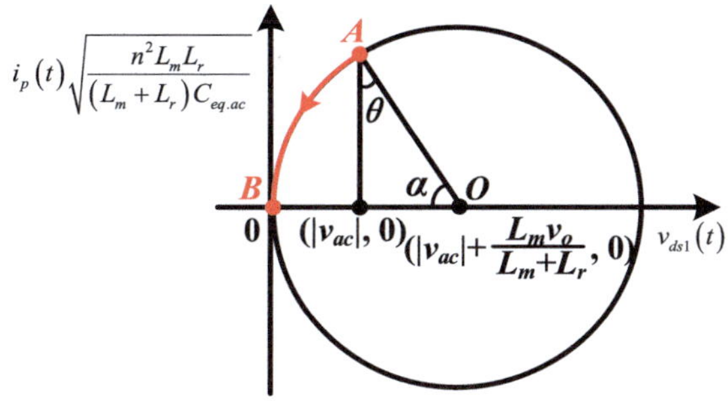

Fig. 10.30 The steady-state trajectory of single-stage high frequency isolated AC-DC converter based on semi-DAB

$$= \left(\frac{L_m}{L_m + L_r}v_o\right)^2 + \left(I\sqrt{\frac{n^2L_mL_r}{(L_m + L_r)C_{\text{eq.ac}}}}\right)^2 \tag{10.60}$$

Figure 10.30 is the steady-state trajectory drawn according to (10.60), the center of this circle are ($|v_{\text{ac}}| + L_m v_o/(L_m + L_r)$, 0). The resonant process for Q_1 and Q_2 in mode 2 and mode 4 are actually a part of the circle as the red curve. The initial point of the resonant process is A, the ideal resonant process is that the primary side current i_p decreases to zero when the voltage of Q_1 v_{ds1} decreases to zero, so the ideal end point of the resonant process is B, the coordinates are (0,0).

Since point B is on the circle, so the minimum primary side current to achieve ZVS $I_{p.ZVS}$ can be obtained by substituting point B into the equation of the circle, and the corresponding resonant time can be derived as (10.62).

$$\begin{aligned} &\left(|v_{\text{ac}}| + \frac{L_m}{L_m + L_r}v_o\right)^2 = \left(\frac{L_m}{L_m + L_r}v_o\right)^2 + \left(I\sqrt{\frac{n^2L_mL_r}{(L_m + L_r)C_{\text{eq.ac}}}}\right)^2 \\ &\Rightarrow I = \sqrt{\frac{C_{\text{eq.ac}}\left(2L_m|v_{\text{ac}}|v_o + L_m|v_{\text{ac}}|^2 + L_r|v_{\text{ac}}|^2\right)}{n^2L_mL_r}} = I_{p.\text{ZVS}} \end{aligned} \tag{10.61}$$

$$t_r = \frac{\alpha}{\omega} = \frac{\frac{\pi}{2} - \arctan\left(\frac{\frac{L_m}{L_m+L_r}v_o}{\sqrt{\frac{2L_m}{L_m+L_r}|v_{\text{ac}}|v_o+|v_{\text{ac}}|^2}}\right)}{\sqrt{\frac{L_m+L_r}{n^2L_mL_rC_{\text{eq.ac}}}}} \tag{10.62}$$

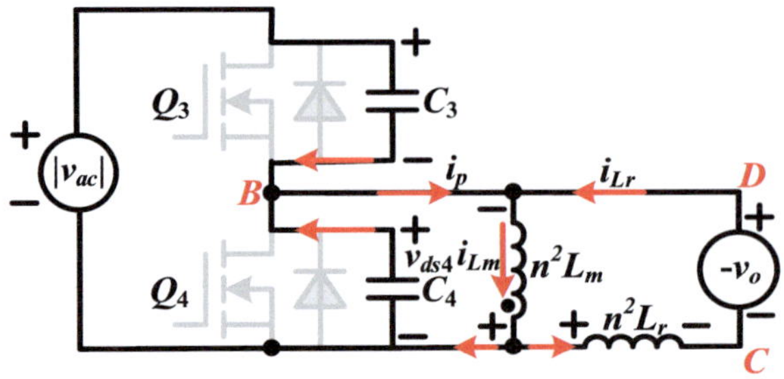

Fig. 10.31 The equivalent circuit of the resonant process for primary side switches Q_3 and Q_4 in mode 1

Mode 1 Q_4

The equivalent circuit to describe the resonant process for Q_3 and Q_4 in mode 1 is shown in Fig. 10.31. A similar analysis and calculation process are shown in equations (10.62)–(10.72).

$$i_p = C_{\text{oss}} \frac{d(|v_{\text{ac}}| - v_{ds4})}{dt} - C_{\text{oss}} \frac{dv_{ds4}}{dt} = -2C_{\text{oss}} \frac{dv_{ds4}}{dt} = i_{Lm} - i_{Lr} \tag{10.63}$$

$$n^2 L_r \frac{di_{Lr}}{dt} = v_o - v_{ds4} \tag{10.64}$$

$$n^2 L_m \frac{di_{Lm}}{dt} = v_{ds4} \tag{10.65}$$

$$n^2 L_m \frac{di_{Lm}}{dt} - n^2 L_r \frac{di_{Lr}}{dt} = 2v_{ds4} - v_o \tag{10.66}$$

$$\begin{cases} \frac{dv_{ds4}}{dt} = \frac{1}{-2C_{\text{oss}}} i_p = \frac{1}{-C_{\text{eq.ac}}} i_p \\ \frac{di_p}{dt} = \left(\frac{1}{n^2 L_m} + \frac{1}{n^2 L_r} \right) v_{ds4} - \frac{1}{n^2 L_r} v_o \end{cases} \tag{10.67}$$

Initial conditions of the resonant process are shown as followings:

$$\begin{cases} v_{ds1}(0) = |v_{\text{ac}}| \\ i_p(0) = I = I_{p.\text{ZVS}} \end{cases} \tag{10.68}$$

The analytical solutions are as following (10.69). And the equation of the circle (10.70) can be obtained after further derivation (Fig. 10.32).

Fig. 10.32 The steady-state trajectory of single-stage high frequency isolated AC-DC converter based on semi-DAB

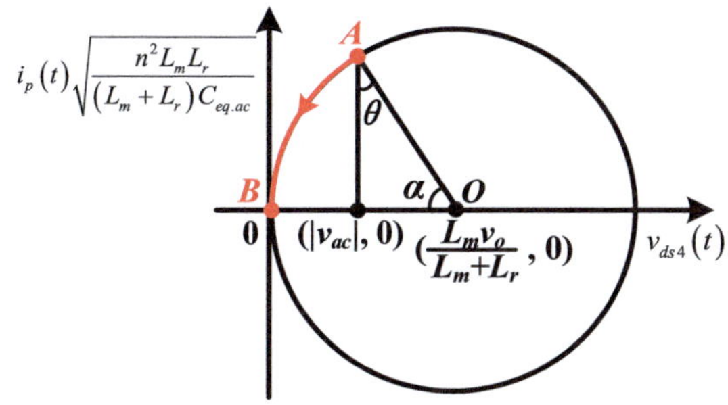

$$\begin{cases} v_{ds4}(t) = \dfrac{L_m}{L_m+L_r}v_o + \left(|v_{\text{ac}}| - \dfrac{L_m}{L_m+L_r}v_o\right)\cos\left(\sqrt{\dfrac{L_m+L_r}{n^2L_mL_rC_{\text{eq.ac}}}}\cdot t\right) \\ \quad - I\sqrt{\dfrac{n^2L_mL_r}{(L_m+L_r)C_{\text{eq.ac}}}}\sin\left(\sqrt{\dfrac{L_m+L_r}{n^2L_mL_rC_{\text{eq.ac}}}}\cdot t\right) \\ i_p(t)\sqrt{\dfrac{n^2L_mL_r}{(L_m+L_r)C_{\text{eq.ac}}}} = \left(|v_{\text{ac}}| - \dfrac{L_m}{L_m+L_r}v_o\right)\sin\left(\sqrt{\dfrac{L_m+L_r}{n^2L_mL_rC_{\text{eq.ac}}}}\cdot t\right) \\ \quad + I\sqrt{\dfrac{n^2L_mL_r}{(L_m+L_r)C_{\text{eq.ac}}}}\cos\left(\sqrt{\dfrac{L_m+L_r}{n^2L_mL_rC_{\text{eq.ac}}}}\cdot t\right) \end{cases} \tag{10.69}$$

$$\begin{aligned} &\left(v_{ds4}(t) - \frac{L_m}{L_m+L_r}v_o\right)^2 + \left(i_p(t)\sqrt{\frac{n^2L_mL_r}{(L_m+L_r)C_{\text{eq.ac}}}}\right)^2 \\ &= \left(|v_{\text{ac}}| - \frac{L_m}{L_m+L_r}v_o\right)^2 + \left(I\sqrt{\frac{n^2L_mL_r}{(L_m+L_r)C_{\text{eq.ac}}}}\right)^2 \end{aligned} \tag{10.70}$$

The minimum primary side current to achieve ZVS $I_{p.\text{ZVS}}$ can be obtained as (10.71), and the corresponding resonant time can be derived as (10.72).

$$\begin{aligned} &\left(\frac{L_m}{L_m+L_r}v_o\right)^2 = \left(|v_{\text{ac}}| - \frac{L_m}{L_m+L_r}v_o\right)^2 + \left(I\sqrt{\frac{n^2L_mL_r}{(L_m+L_r)C_{\text{eq.ac}}}}\right)^2 \\ &\Rightarrow I = \sqrt{\frac{C_{\text{eq.ac}}\left(2L_m|v_{\text{ac}}|v_o - L_m|v_{\text{ac}}|^2 - L_r|v_{\text{ac}}|^2\right)}{n^2L_mL_r}} = I_{p.\text{ZVS}} \end{aligned} \tag{10.71}$$

$$t_r = \frac{\alpha}{\omega} = \frac{\frac{\pi}{2} - \arctan\left(\frac{\frac{L_m}{L_m+L_r}v_o - |v_{\text{ac}}|}{\sqrt{2\frac{L_m}{L_m+L_r}|v_{\text{ac}}|v_o - |v_{\text{ac}}|^2}}\right)}{\sqrt{\frac{L_m+L_r}{n^2L_mL_rC_{\text{eq.ac}}}}} \tag{10.72}$$

References

1. Sha D, Zhang D, Zhang J (2021) A single stage dual-active-bridge AC–DC converter employing mode transition based on real-time calculation. IEEE Trans Power Electron 36(9):10081–10088
2. Jauch F, Biela J (2016) Combined phase-shift and frequency modulation of a dual-active-bridge AC-DC converter with PFC. IEEE Trans Power Electron 31(12):8387–8397
3. Weise ND, Castelino G, Basu K, Mohan N (2014) A single-stage dual-active-bridge-based soft switched AC-DC converter with open-loop power factor correction and other advanced features. IEEE Trans Power Electron 29(8):4007–4016
4. Everts J, Krismer F, Keybus Van den J, Driesen J, Kolar JW (2014) Optimal ZVS modulation of single-phase single-stage bidirectional DAB AC–DC converters. IEEE Trans Power Electron 29(8):3954–3970
5. Chen T, Yu R, Huang AQ (2022) Variable switching-frequency single-stage bidirectional GaN AC-DC converter for the grid-tied battery energy storage system. IEEE Trans Ind Electron 69(11):10776–10786
6. Ma P, Sha D, Song K (2023) A single-stage semi dual-active-bridge AC-DC converter with seamless mode transition and wide soft-switching range. IEEE Trans Ind Electron 70(2):1387–1397
7. Ma P, Sha D (2023) A single-stage AC-DC converter based on semi dual-active-bridge with decoupled inductor current modulation strategy. IEEE Trans Power Electron 38(8):10170–10182

Chapter 11
Single-Stage Resonant Semi-DAB AC-DC Converter with Multi-Mode Control

Abstract For conventional semi DAB DC-DC converters, the turn off current for some switches are pretty high. Besides, ZVS cannot be ensured within the full operating range with constant switching control. EMI is relatively high due to the sharp change of the current in transformers. To deal with the issues brought by conventional semi DAB converters, a resonant semi-DAB-based single-stage high frequency isolated AC-DC converter is proposed. Its multi-mode control strategy with smooth transition is presented. Four decoupled resonant inductor current modulation modes with fixed switching frequency control strategy are given. The resonant tank is designed in an optimized way for better performance and higher conversion efficiency. The converter could get a wider output voltage range with the proposed control method. The primary side switches can achieve ZVS within the full operating range with the proposed fixed switching frequency control. In addition, the RMS value of the resonant inductor current, the maximum current stress of switches and the turn-off current are compared with other modulation strategies under different operating conditions. The power transfer ability, mode transition mechanism and control implementation are also analyzed. A 1000 W experimental prototype was built to verify the effectiveness of the proposed control.

Keywords Single-stage AC-DC converter · Resonant semi-DAB · Fixed switching frequency · Multi-mode control strategy · Smooth transition

11.1 Circuit Topology and Control Strategy

Resonant single-stage AC-DC converters are researched extensively recently. A full bridge resonant AC-DC converter using T-type voltage doubler is proposed [1], which could minimize the switching loss, while only the DCM mode is used, and the voltage gain is always larger than 1, causing higher peak current of transformer. The similar topologies are given [2], and the same problems still exist due to only one operating mode. A series of quasi-single-stage bridgeless electrolytic capacitor-free AC-DC converters for supplying multiple light-emitting diode strings in parallel are

D. Sha et al., *High Frequency Isolated Semi Dual Active Bridge DC-DC and AC-DC Converter*, CPSS Power Electronics Series,
https://doi.org/10.1007/978-981-95-1641-4_11

proposed [3], which is the integration of boost PFC and half-bridge LLC, reducing device numbers and cost, while the grid side current quality is relatively poor. A simple synchronous rectification strategy is proposed [4], but the proposed SR driving strategy needs accurate time-domain analysis, thus lookup tables are needed and the control is complex. A power boundary control strategy is proposed for the single-stage isolated AC-DC converter based half-bridge LLC [5], the parameters are designed optimally. Although rms current of the resonant tank is be achieved, the grid side current quality is relatively poor. An interleaved totem-pole LLC single-stage AC-DC converter with a novel hybrid control strategy is proposed [6], utilizes PWM to shape the input current, and PFM to regulate the output voltage. LLC-based single stage AC-DC converters have less freedom of control.

So in order to increase the flexibility of control, resonant DAB-based single-stage AC-DC converters have drawn some attention. A variable switching frequency control strategy employing EPS modulation is proposed [7], which could reduce the turn-off current of switches and obtain high efficiency, while only one mode exists within the entire ac line period, which is hard to optimize the performance in facing varying grid side voltages, so the variable frequency range is relatively wide.

In order to further reduce the turn off current of switches, a resonant semi-DAB-based single-stage high frequency isolated AC-DC converter is proposed in this chapter. In order to optimize the performance for all voltage conversion gains, a multi-mode control with fixed switching frequency is proposed in this chapter.

11.1.1 Power Circuit Topology

The proposed single-stage high frequency isolated AC-DC converter based resonant semi-DAB is shown in Fig. 11.1, which consists of a diode rectifier followed by a resonant semi-DAB. As seen, the resonant semi-DAB is different from the traditional non-resonant one, which adds a resonant tank which consists of a resonant inductor L_r (which represents the sum of the leakage inductance of transformer and an external inductance) and a resonant capacitor C_r. It is worth noting that the magnetizing inductor L_m is not a part of the resonant tank, which is only used to assist ZVS achievement within the full operating range. The turns ratio of the transformer is n:1. M is the voltage conversion gain and it is defined as $M = nv_o/|v_{ac}|$.

11.1.2 Operating Modes and Circuit Operating Principle

There are a lot of modulation modes including SPS and DPS for the resonant semi-DAB DC-DC converter, six basic modulation modes under $M > 1$ and $M < 1$ are given in [8], which are also shown in the following Fig. 11.2. Where, mode 1 and mode 5 are suitable to the application of $M > 1$ and $M < 1$, mode 2, mode 3 and mode 4 are only suitable for $M < 1$, mode 6 is only suitable for $M > 1$. The control variable

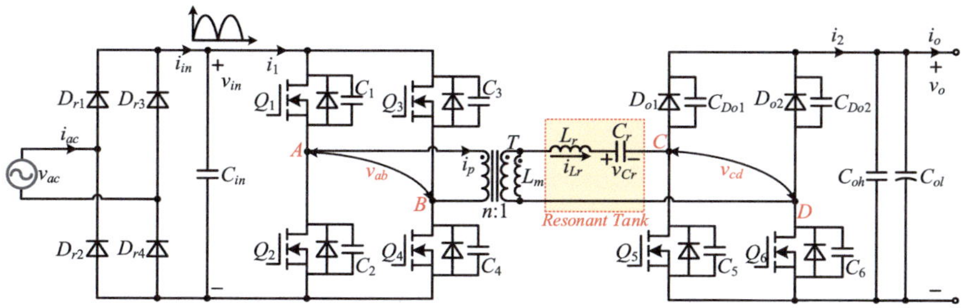

Fig. 11.1 The single-stage AC-DC converter topology based resonant semi-DAB

α represents the angle of Q_1 leading Q_4, Φ represents the angle of Q_1 leading Q_6, and β is the angle from the start of a switching period to the zero-crossing-point of the resonant inductor current.

Nine boundary modulation modes based on the above six basic modulation modes shown in Fig. 11.3 are derived, which is the foundation of selecting the mode combinations. The definitions of the control variables are the same as the above six basic modulation modes. The constraint relationship of control variables for the nine modulation modes are summarized in the following Table 11.1.

The above nine boundary modes have different characteristics. Mode 1b and Mode 1–5 are all suitable for high power, and they are all suitable as $M > 1$ and $M < 1$. As seen, Mode 1b has a large backflow power, which is not good to improve the

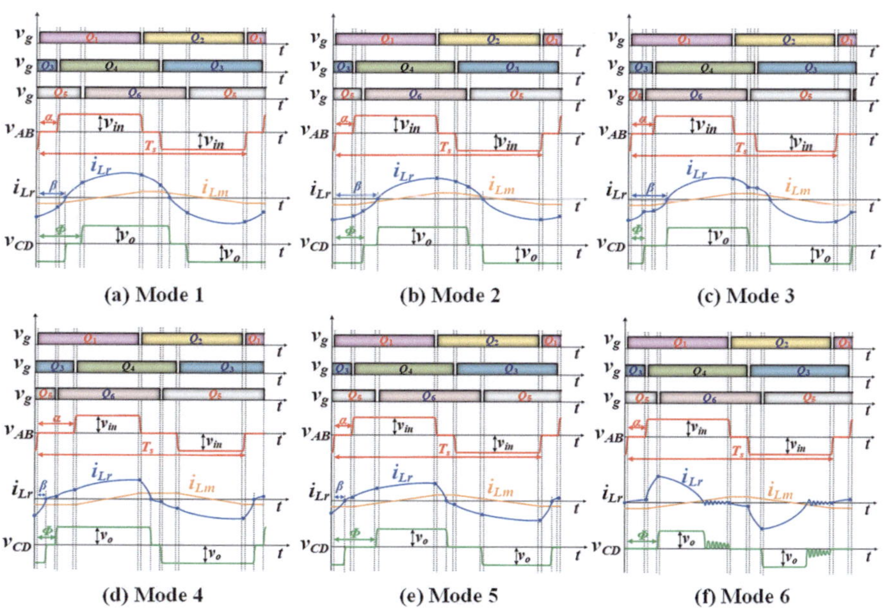

Fig. 11.2 The six basic modulation modes under $M > 1$ and $M < 1$ in [8]

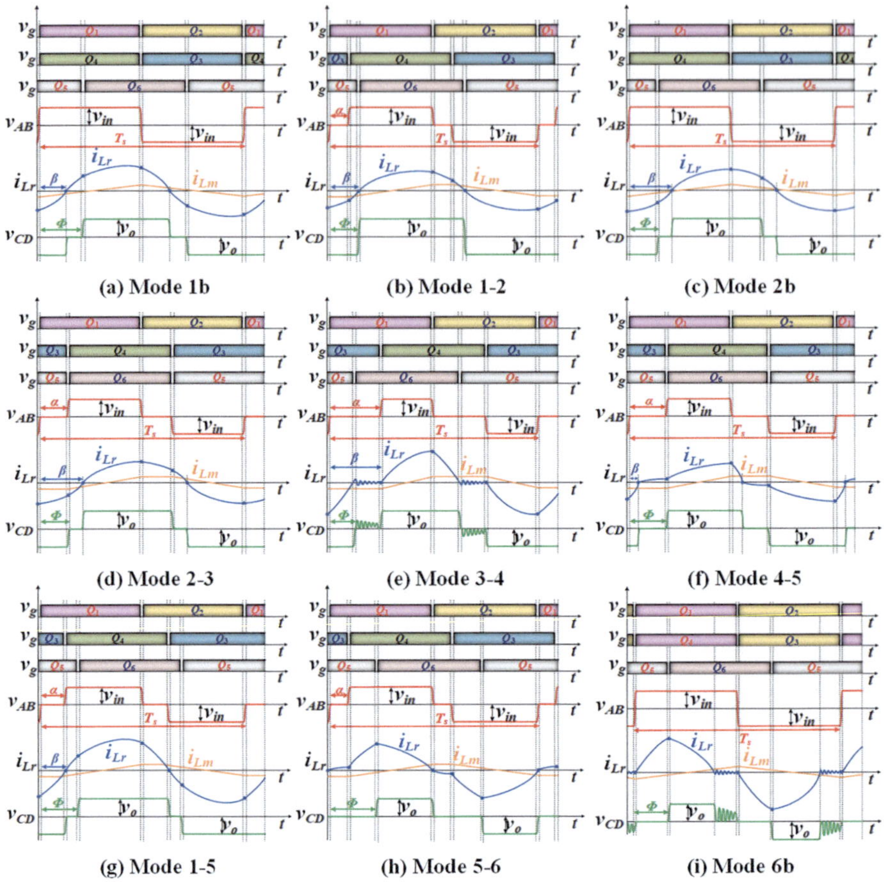

Fig. 11.3 The nine special boundary modulation modes

Table 11.1 The constraint relationship of control variables for nine boundary modes

Mode	Constraint relationship	Mode	Constraint relationship
Mode 1–2	Mode 1 and Mode 2 $\beta = \Phi$	Mode 1–5	Mode 1 and Mode 5 $\alpha = \beta$
Mode 2–3	Mode 2 and Mode 3 $\alpha = \Phi$	Mode 1b	Mode 1 $\alpha = 0$
Mode 3–4	Mode 3 and Mode 4 $\alpha = \beta$	Mode 2b	Mode 2 $\alpha = 0$
Mode 4–5	Mode 4 and Mode 5 $\alpha = \Phi$	Mode 6b	Mode 6 $\alpha = 0$
Mode 5–6	Mode 5 and Mode 6 $\beta = 0$		

conversion efficiency, while the backflow power of Mode 1–5 is relatively small. According to the analysis of the previous work [9], the resonant inductor current of Mode 1b is coupled, which is not helpful to reduce the THD of input current i_{ac}, but Mode 1–5 is decoupled.

The other seven boundary modes are all suitable for low power. Mode 1–2 and Mode 2–3 all have small backflow power, while Mode 2b has large backflow power. The resonant inductor currents of Mode 1–2, Mode 2–3 and Mode 2b are all coupled, but ZVS of primary side switches can be achieved easily due to the large ZVS current. Mode 3–4 and Mode 6b all belong to the decoupled modulation modes, and the backflow power are all relatively small. Mode 4–5 and Mode 5–6 all have a bias current, which are not helpful to achieve ZVS. Since the bias current of Mode 3–4 and Mode 6b is almost zero, so ZVS of the primary side switches can be easily achieved with help of the magnetizing current. The performance characteristics of the above nine boundary modes are summarized in the following Table 11.2.

According to the above analysis and introduction, there are a lot of modulation mode combinations within the entire ac line cycle. Figure 11.4 shows the decoupled inductor current modulation modes for the semi-DAB-based single-stage AC-DC converter in [9], which are helpful to obtain a better grid current quality. The similar resonant inductor current modulation mode combinations are shown in Fig. 11.5.

The common features of Figs. 11.4 and 11.5 are the bias current, which are denoted with blue dotted circle in Mode 1 and Mode 3. The bias current polarity is opposite to ZVS achievement of primary side switches. So in order to achieve ZVS, the magnetizing current is used. While at the small grid phase angle region even around the zero-crossing-point of v_{ac}, the input voltage is very small, the magnetizing current is also too small, so the ZVS of the primary side switches are hard to be achieved although with the help of the magnetizing current.

So in order to extend the ZVS range and minimize the turn-off currents of switches, novel decoupled resonant inductor current modulation modes are proposed as shown in Fig. 11.6. As seen, the bias current in Mode 1 and Mode 3 are approaching zero, so ZVS can be achieved easily with the help of the small magnetizing current.

Table 11.2 The performance characteristics of the above nine boundary modes

Mode	Gain	Backflow power	Power ability	Coupled/decoupled	ZVS achievement
Mode 1b	Any	Large	High	Coupled	Easy
Mode 1-5	Any	No	High	Decoupled	Easy
Mode 1-2	M < 1	Small	Low	Coupled	Easy
Mode 2-3	M < 1	Small	Low	Coupled	Easy
Mode 2b	M < 1	Large	Low	Coupled	Easy
Mode 3-4	M < 1	No	Low	Decoupled	Easy
Mode 4-5	M < 1	No	Low	Coupled	Difficult
Mode 5-6	M > 1	No	Low	Coupled	Difficult
Mode 6b	M > 1	No	Low	Decoupled	Easy

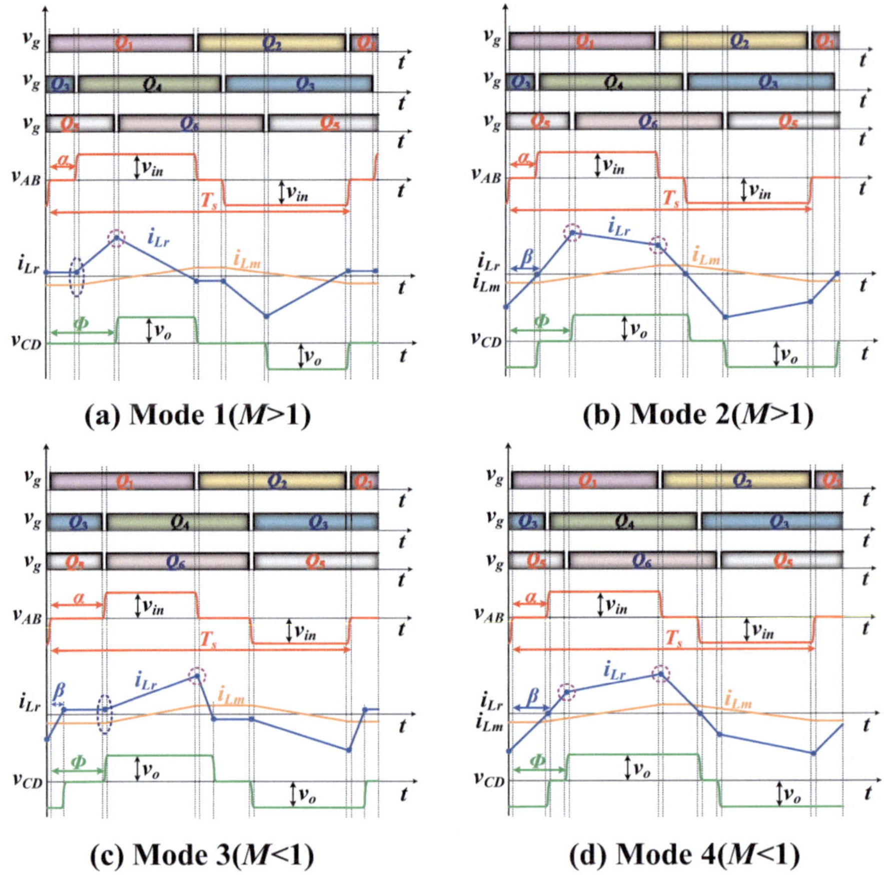

Fig. 11.4 The traditional decoupled inductor current modulation modes in [9]

The control trajectories or constraint relationships between α and Φ of four modulation modes are derived according to the analysis and introduction in [8], which are shown as the following four expressions, and the intersections between Mode 1 and Mode 2, Mode 3 and Mode 4 are listed in the following Table 11.3,

$$\begin{cases} \alpha = 0 & \text{(Mode 1)} \\ \alpha = \arcsin\left[\frac{1-M}{\sqrt{(M\sin\phi)^2+(M\cos\phi-1)^2}}\right] - \arctan\left[\frac{M\cos\phi-1}{M\sin\phi}\right] & \text{(Mode 2)} \end{cases} \tag{11.1}$$

$$\begin{cases} \alpha = \arcsin\left[\frac{1-M}{\sqrt{(M\sin\phi)^2+(M\cos\phi-1)^2}}\right] - \arctan\left[\frac{M\cos\phi-1}{M\sin\phi}\right] & \text{(Mode 3)} \\ \alpha = \arcsin\left[\frac{1-M}{\sqrt{(M\sin\phi)^2+(M\cos\phi-1)^2}}\right] - \arctan\left[\frac{M\cos\phi-1}{M\sin\phi}\right] & \text{(Mode 4)} \end{cases} \tag{11.2}$$

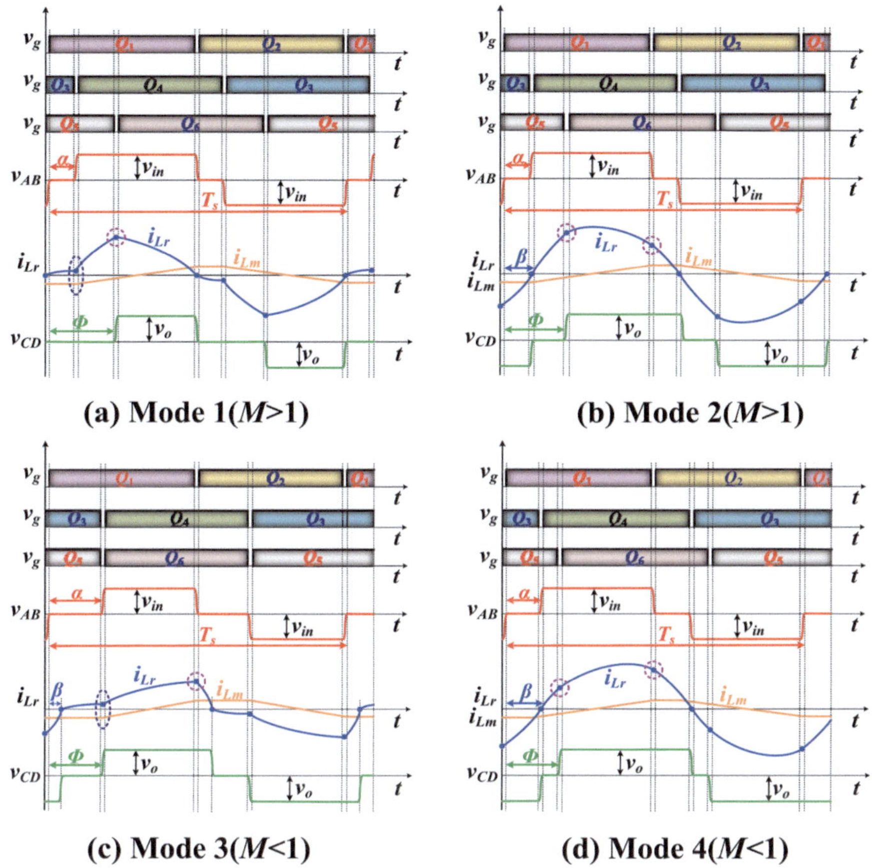

Fig. 11.5 The similar resonant inductor current modulation modes

In order to express the constraint relationships of the control variables α and Φ for different modes more intuitively, the two-dimensional (2-D) control trajectories under $M > 1$ and $M < 1$ are shown in Fig. 11.7a and b respectively. As seen, the different regions represent the six basic modulation modes in Fig. 11.2, and the different boundary lines indicate nine boundary modulation modes in Fig. 11.3. Where, the purple solid line and the green line in Fig. 11.7a refer to Mode 1 and Mode 2 of the proposed modulation modes respectively, and the green solid line in Fig. 11.7b indicates Mode 3 and Mode 4 of the proposed modulation modes respectively.

The AC input voltage is rectified to a steamed bun wave DC voltage through the diode rectifier bridge and it works as the input for the resonant semi-DAB DC-DC converter, so the semi-DAB converter needs to handle the extremely wide input voltage (from near zero to the amplitude of the ac input voltage V_m) and conversion gain range (from infinite to the minimum value M_{in}).

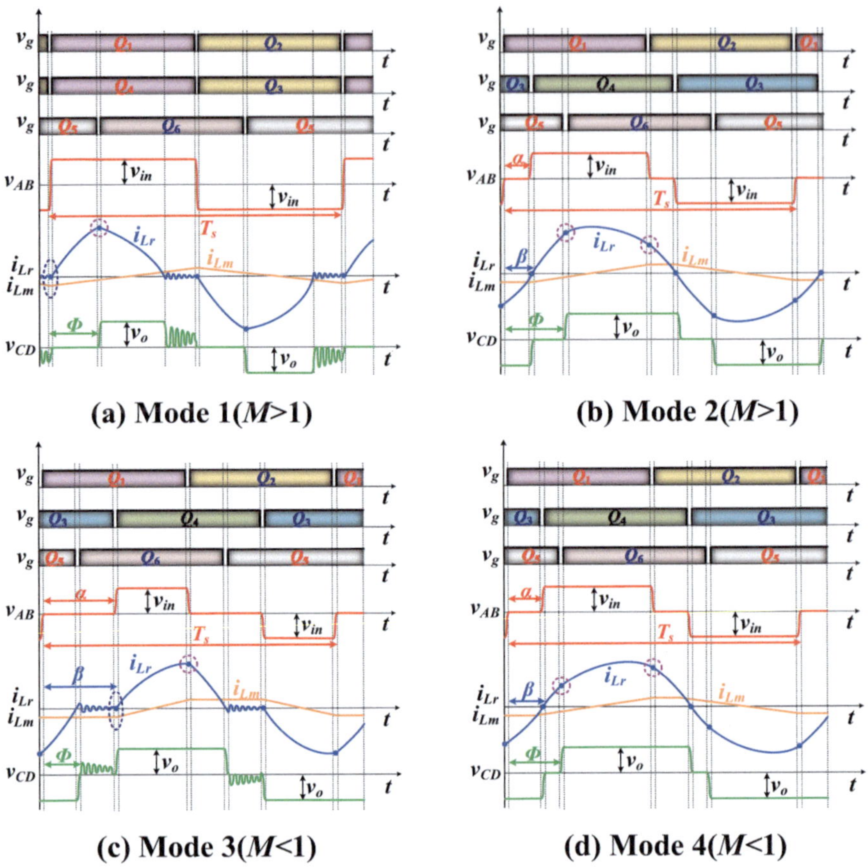

Fig. 11.6 The proposed decoupled resonant inductor current modulation modes

Table 11.3 The control trajectory intersection points of different modes

Gain	Intersection point
$M > 1$	$\begin{cases} \phi = \arccos\left(\frac{2-M}{M}\right) \\ \alpha = 0 \end{cases}$ (Mode 1–Mode 2)
$M < 1$	$\begin{cases} \phi = 2\arccos\left(\sqrt{M}\right) \\ \alpha = 2\arccos\left(\sqrt{M}\right) \end{cases}$ (Mode 3–Mode 4)

So the control trajectories of the proposed modulation modes are needed to be considered as the conversion gain M varies from the infinite to the minimum value. The three-dimensional (3-D) control trajectories under $M > 1$ and $M < 1$ are shown in Fig. 11.8a and b respectively. Figure 11.8a is taken as an example, the bottom plane ($\alpha = 0$) is a set of two-dimensional (2-D) control trajectories of Mode 1 under

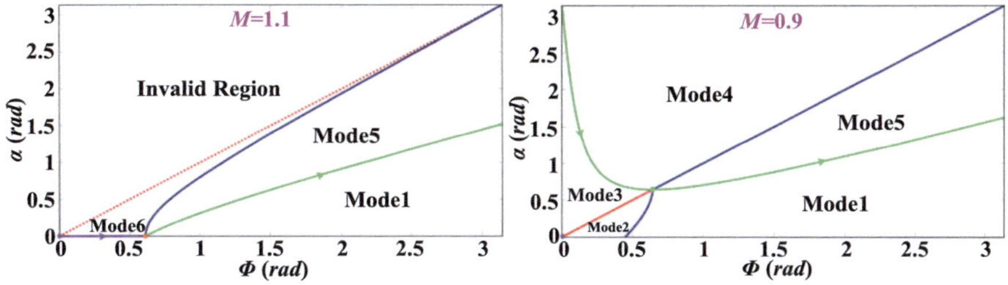

(a)M>1 (b)M<1

Fig. 11.7 The six basic mode distribution region, boundary line and the control trajectories of the decoupled resonant inductor current modulation modes

different gains, the surface 2 is a set of two-dimensional control trajectories of Mode 2 under different gains. The purple intersection line of the bottom plane 1 and the orange vertical plane 3 refers to Mode 1, the green intersection line indicates Mode 2. The mode transition point at different voltage conversion gains moves along the pink dotted line. The principle under $M < 1$ is similar and it is not elaborated for simplicity.

Table 11.4 gives the used system parameters. Figure 11.9 shows the power transfer capability of the proposed four modulation modes and the mode transition mechanism. As seen, the red solid curve indicates the maximum power transfer ability of Mode 1 and the boundary curve between Mode 1 and Mode 2 under different gains. The blue curve represents the maximum power transfer ability of Mode 3 and the boundary curve between Mode 3 and Mode 4 under different gains. Six dotted curves are the required instantaneous transfer power by the resonant semi-DAB DC-DC converter at different voltage conversion gains as the load power is 1000 W, 800 W, 600 W, 400 W, 200 W and 100 W respectively. The situation of $M > 1$ is taken as an example, when the dotted curves are below than the red solid curve, the maximum power transfer capability of Mode 1 can meet the required instantaneous power, so Mode 1 is used, otherwise, Mode 2 is used. The principle of the situation of $M < 1$ is similar, so it is not analyzed for simplicity.

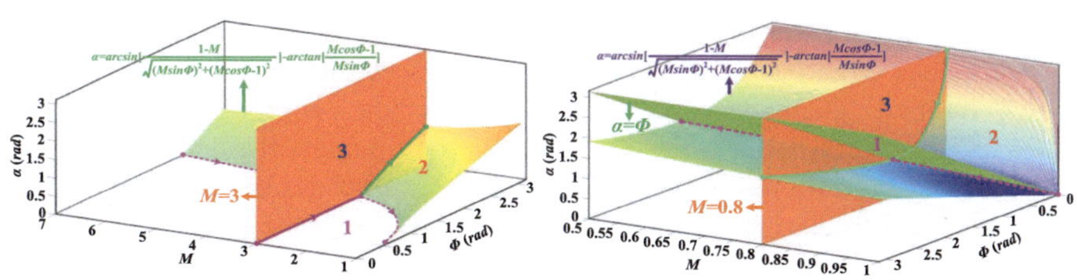

(a)M>1 (b)M<1

Fig. 11.8 The three-dimensional control trajectories under $M > 1$ and $M < 1$ of the proposed four novel decoupled resonant inductor current modulation modes

Table 11.4 System parameters

Symbol	Value	Symbol	Value
v_{ac}	220V_{rms}	C_{oh}	20 μF
f_{ac}	50 Hz	C_{ol}	10,000 μF
v_o	270~400 V	Q_1~Q_6	C3M0045065K
C_{in}	2 μF	D_{o1}, D_{o2}	RURG80100
L_r	45 μH	D_{r1}, D_{r2}, D_{r3}, D_{r4}	KBPC2510
C_r	110 nF	f_s	100 kHz
L_m	210 μH	t_{dead}	200 ns
N	1: 1	P_{rated}	1000 W

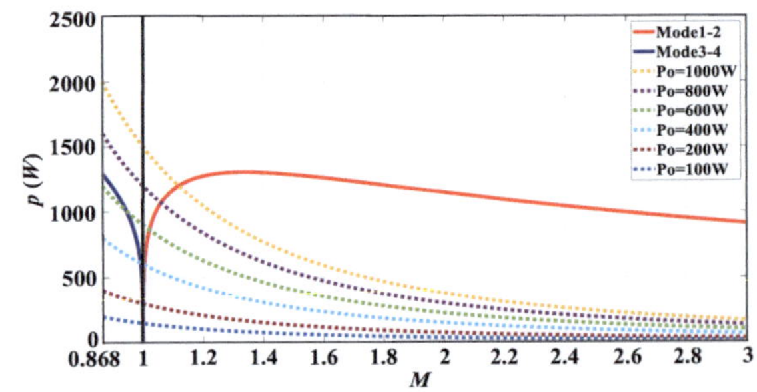

Fig. 11.9 The power transfer capability of different operating modes

Figure 11.10 shows the closed-loop control diagram. It should be noted it is composed of double closed loops including voltage and current control. The output voltage is controlled by the outer loop and its output works the amplitude of the reference grid current I_{max}, the phase and frequency of the power grid voltage can be obtained through the phase-locked loop (PLL) module. The instantaneous value of the reference grid current i_{ac_ref} can be obtained according to I_{max} and the phase angle of the grid voltage by a multiplier. Then the inner current control loop is used for current tracking to implement PFC. The output of the current loop is a middle control variable q, and the conversion gain M at different grid phase angles can be calculated. q and M are injected to the modulation scheme unit, the corresponding mode under different grid phase angles is selected according to the mode transition mechanism and criterion as the Table 11.5. Then the corresponding control variables α and Φ can be calculated according to the corresponding computed expressions as shown in Fig. 11.10. Finally, the PWM module unit outputs the driving signal for six switches according to the previous calculated control variables α and Φ. High power factor and the required wide range output voltage can be obtained by using the proposed closed-loop control strategy.

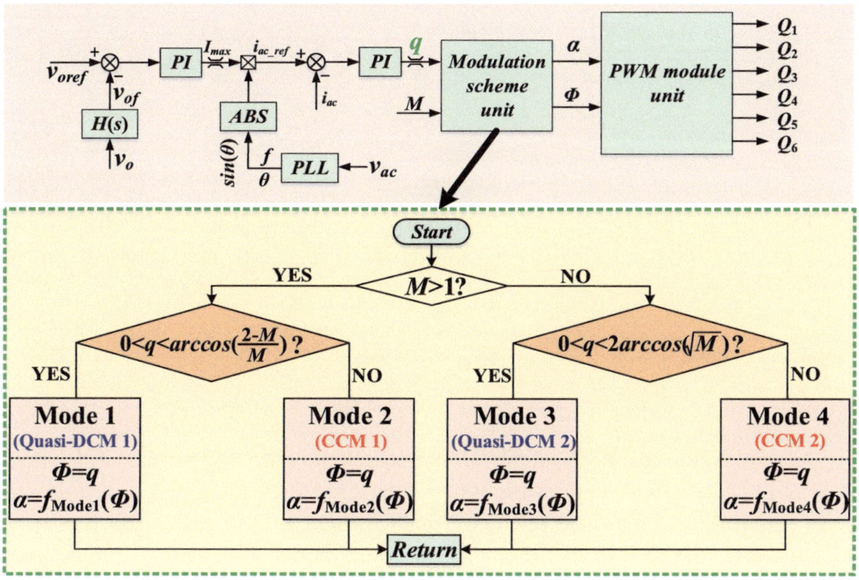

Fig. 11.10 The closed loop control diagram of the proposed control strategy

Table 11.5 The mode transition mechanism and criterion

$M > 1$	$\begin{cases} 0 < q < 2\arccos\left(\frac{2-M}{M}\right) \text{ (Mode 1)} \\ 2\arccos\left(\frac{2-M}{M}\right) < q < \pi \text{ (Mode 2)} \end{cases}$
$M < 1$	$\begin{cases} 0 < q < 2\arccos\left(\sqrt{M}\right) \text{ (Mode 3)} \\ 2\arccos\left(\sqrt{M}\right) < q < \pi \text{ (Mode 4)} \end{cases}$

11.2 Converter Resonant Tank Design Consideration

The resonant tank components affect the performance of the converter. So the resonant tank components need to be designed optimally. The design procedure of resonant tank components needs to consider the following three principles,

(1) The resonant tank components can satisfy the requirement of the maximum instantaneous transfer power within the entire operating range;

The resonant tank components determine the maximum instantaneous power transfer capability. According to previous theoretical analysis, the maximum instantaneous transfer power within the entire complete ac line period lies in mode 4 at the maximum absolute value for the gird voltage, the rated load power and $V_o = 270$ V. The transfer power expression of Mode 4 is given as the following.

$$p(\alpha) = \frac{4\sqrt{2}n^2V_o^2}{\pi^2 MZ_B\sqrt{F - \frac{1}{F}}} \cdot \cos\left(\frac{\alpha}{2}\right) \cdot \sqrt{\sin(\alpha) + 2\cos\left(\frac{\alpha}{2}\right) \cdot \sqrt{M - \cos^2\left(\frac{\alpha}{2}\right)}} \tag{11.3}$$

$$F = \frac{f_s}{f_r} \tag{11.4}$$

$$f_r = \frac{1}{2\pi\sqrt{L_r C_r}} \tag{11.5}$$

$$Z_B = \sqrt{\frac{L_r}{C_r}} \tag{11.6}$$

(2) The peak value and RMS value of the resonant inductor and primary side current should be as small as possible under the premise of meeting the requirement of maximum transfer power, and which influence the conduction loss of the switches, copper loss of the magnetic components, core loss of the magnetic components and conversion efficiency. The RMS value of the resonant inductor and primary side current within half complete ac line cycle under different F and Z_r for $V_o = 270$ V and $P_o = 1000$ W are given in Fig. 11.11a, b, the maximum peak value of resonant inductor and primary side current within complete half ac line cycle under different F and Z_r as $V_o = 270$ V and $P_o = 1000$ W are given in Fig. 11.11c, d; and the maximum turn-off current of $Q_{1/2}$ within half ac line cycle under different F and Z_r for $V_o = 270$ V and $P_o = 1000$ W are shown in the following Fig. 11.11e.

(3) The voltage ripple across the resonant capacitor Δv_{cr} should not be too large, which affects the stable operation. The maximum voltage stress of resonant capacitor within half a ac line cycle under different F and Z_r for $V_o = 270$ V and $P_o = 1000$ W is given in Fig. 11.12.

Though the above comprehensive analysis, the resonant tank components are optimally designed as $L_r = 45$ µH and $C_r = 110$ nF.

11.3 Comparisons Among the Proposed Modulation Modes and Other Similar Modulation Modes

In the previous work [9], the decoupled inductor current modulation strategy is proposed for the non-resonant semi-DAB-based single-stage high frequency isolated AC-DC converter as shown in Fig. 11.4, which could reduce the THD of the input ac current, improve power quality and reduce losses, and increase conversion efficiency. Similarly, the corresponding resonant inductor current modulation strategy is proposed for resonant semi-DAB-based single-stage AC-DC converter as shown

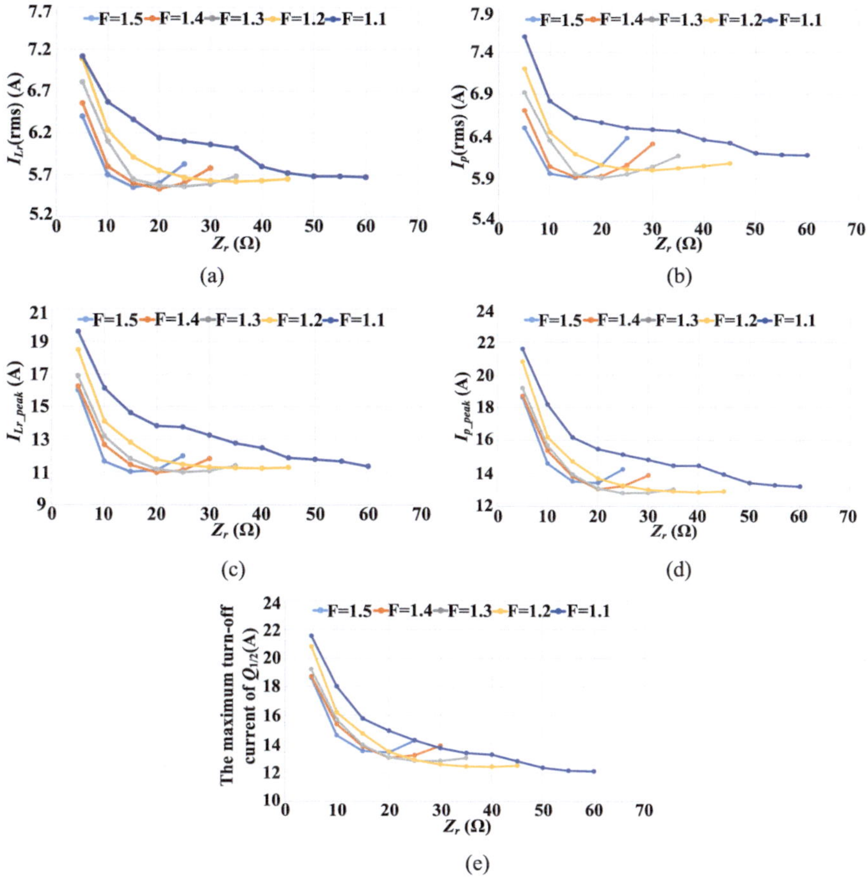

Fig. 11.11 The various current values within half a complete ac line cycle under different F and Zr for $V_o = 270$ V and $P_o = 1000$ W. **a** The RMS value of resonant inductor current. **b** The RMS value of primary side current. **c** The maximum peak value of resonant inductor current. **d** The maximum peak value of primary side current. **e** The maximum turn-off current of $Q_{1/2}$

Fig. 11.12 The maximum voltage stress of resonant capacitor within half a ac line cycle under different F and Zr for $V_o = 270$ V and $P_o = 1000$ W

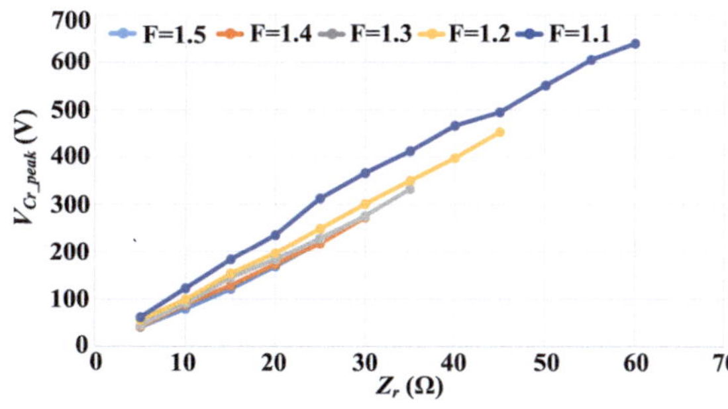

in Fig. 11.5, which could reduce the turn-off current of the switches and the electromagnetic interference. While the common shortcomings of the above two modulation strategies are that ZVS range of the primary side switches are not wide due to the existence of the bias current in Mode 1 and Mode 3 with the fixed switching frequency control strategy.

The most basic and simple modulation and control strategy for the resonant semi-DAB-based single-stage AC-DC converter is the SPS modulation strategy as shown in the following Fig. 11.13. It is found that it has the following shortcomings,

(1) The turn-off current of Q_3, Q_4 are larger for Mode 2, Mode 3 and Mode 4, and the turn-off current of Q_5, Q_6 are larger for Mode 3.
(2) The pulse width of v_{ab} is wider for Mode 2, Mode 3 and Mode 4, which increases the transformer core loss.

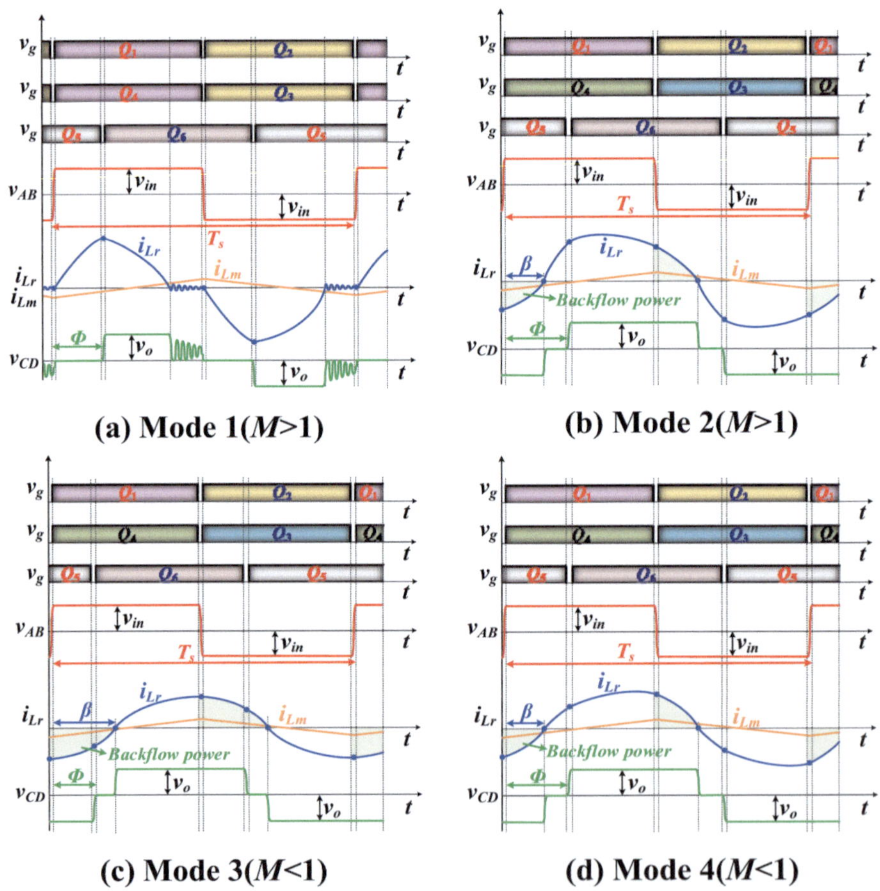

Fig. 11.13 The SPS modulation modes for resonant semi-DAB-based single-stage AC-DC converter

(3) Larger backflow power for Mode 2, Mode 3 and Mode 4.
(4) The lower limit of power transfer for Mode 3 exists when the voltage conversion gain is the minimum value $M_{\min}$, so the converter cannot operate under the middle, light load under the SPS modulation strategy for $V_o = 270$ V (The converter can operate within the full load range for $V_o = 400$ V).

It is worth noting that he proposed modulation modes with the fixed switching frequency control strategy could overcome the aforementioned disadvantages. The performance of three modulation strategies will be compared and analyzed. For fair comparison, the same hardware conditions for three control methods are listed in Table 6.4, the only difference is listed in the following Table 11.6.

The turn off currents comparison of different modulations at different grid phase angles under $V_o = 270$ V and $P_o = 1000$ W are given in Fig. 11.14. It can be seen that compared with other two modulations, the proposed novel modulation strategy could reduce the turn off current of switches. A large backflow power exists in the resonant SPS modulation, while the proposed modulation can avoid large backflow power.

Table 11.6 The only difference of parameters

Decoupled inductor current modulation for non-resonant semi-DAB-based AC-DC converter	Resonant single-phase- shift (SPS) modulation for resonant semi-DAB-based AC-DC converter	Proposed modulation for resonant semi-DAB-based AC-DC converter
$L_r = 34$ μH	$L_r = 45$ μH	$L_r = 45$ μH
	$C_r = 110$ nF	$C_r = 110$ nF

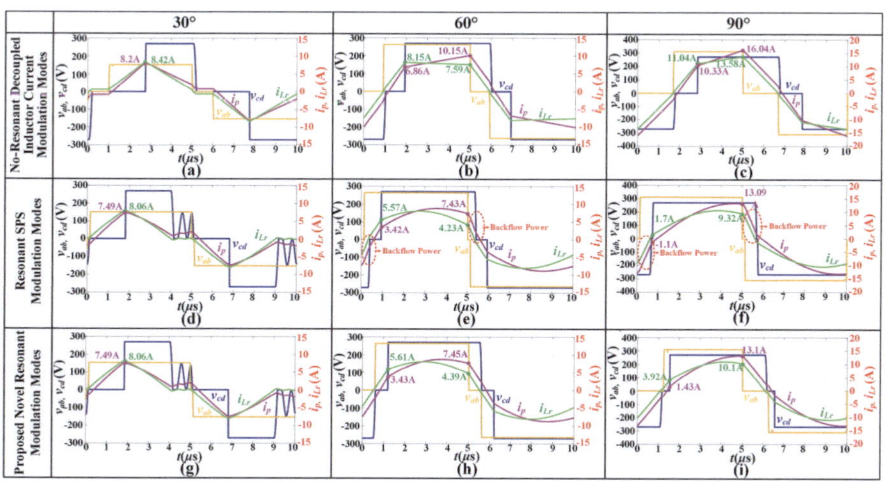

Fig. 11.14 The switching current comparison among different modulation strategies ($V_o = 270$ V and $P_o = 1000$ W)

The RMS value of the resonant inductor current and the primary side current within half a ac line cycle under different output voltages and different load powers by using three modulation and control strategies are shown in Figs. 11.15 and 11.16. As seen, the converter has the lower RMS value of I_{Lr}(rms) and I_p(rms) by using the proposed modulation strategy regardless of the output voltage and load power, which affects the conduction loss of the primary side and the secondary side switches, the copper loss of the magnetic components including a high frequency transformer and resonant inductor.

The maximum peak value of the resonant inductor current and the primary side current under different output voltages and load powers by using three modulation and control methods are shown in Figs. 11.17 and 11.18. As seen, the maximum peak value of I_{Lr_peak} and I_{p_peak} are the lowest by adopting the proposed modulation method under any output voltage and load power, which affects the current stress of the primary side and secondary side switches and the core loss of the resonant inductor.

The maximum turn-off current of Q_1~Q_6 under different load powers and different output voltages by using three modulation and control methods are shown in

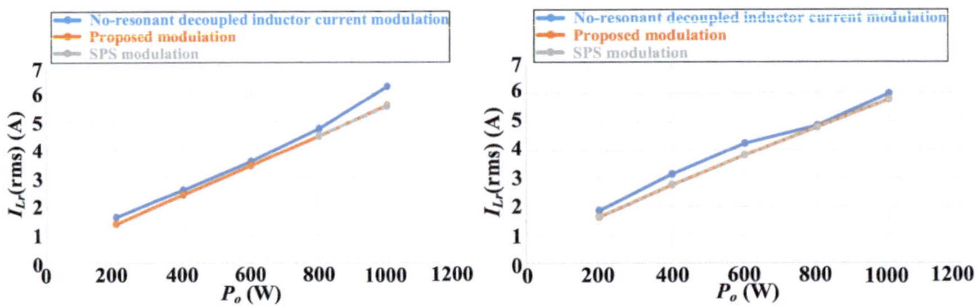

(a)V_o=270V (b)V_o=400V

Fig. 11.15 The rms value of resonant inductor current within half a ac line period under different load powers and different output voltages

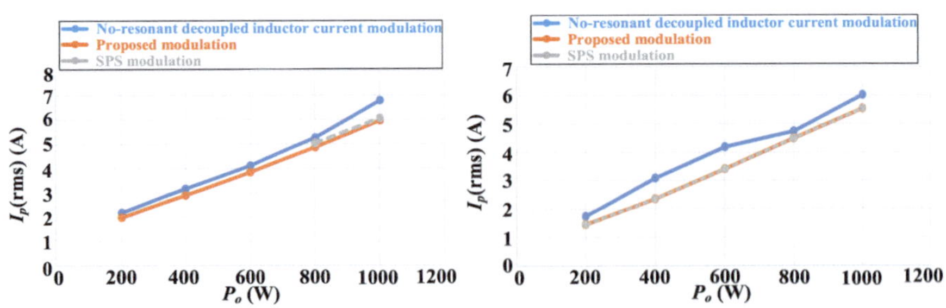

(a)V_o=270V (b)V_o=400V

Fig. 11.16 The rms value of primary current within half ac line period under different situations

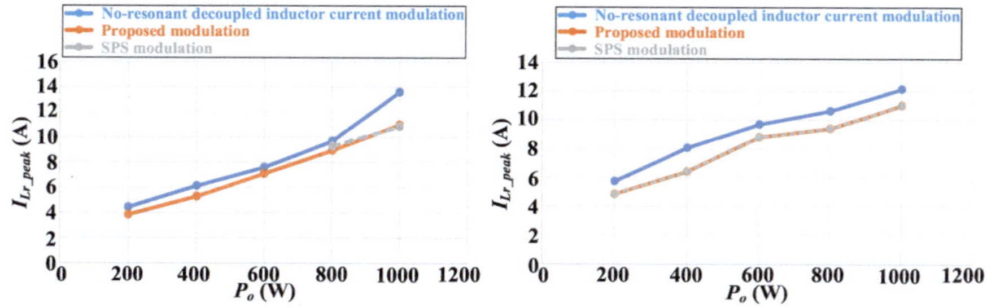

(a)V_o=270V (b)V_o=400V

Fig. 11.17 The maximum value of peak resonant inductor current under different load powers and different output voltages

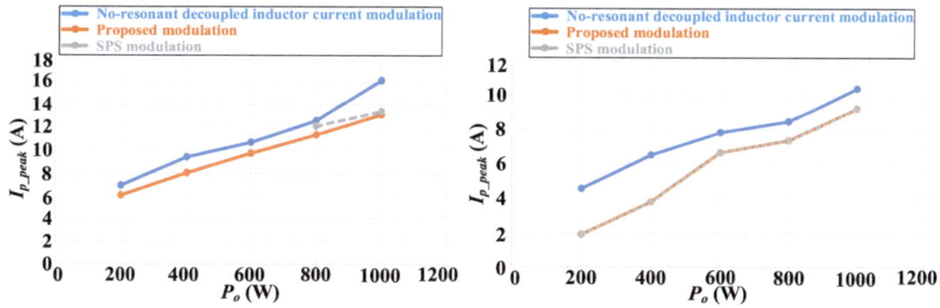

(a)V_o=270V (b)V_o=400V

Fig. 11.18 The maximum value of peak primary side current under different load powers and different output voltages

Figs. 11.19, 11.20 and 11.21 respectively. As seen, the converter has lower maximum turn-off current of the switches by using the proposed modulation strategy under most operating conditions, the maximum turn-off current of Q_3~Q_4 under SPS modulation strategy is so large compared with the other two methods, which affects the turn-off loss of the switches. It should be noted that there is a lower limit of power transfer for Mode 3 when $M < 1$ under SPS modulation strategy, so these current values under middle and light loads for $V_o = 270$ V are not given in the Figs. 11.15, 11.16, 11.17, 11.18, 11.19, 11.20 and 11.21.

The following Figs. 11.22 and 11.23 show the comparison of the ZVS situation around the zero-crossing-point of v_{ac}. It can be seen that the primary side switches cannot achieve ZVS around the zero-crossing-point of v_{ac} for the non-resonant semi-DAB AC-DC converter, while the primary side switches can achieve ZVS around even at the zero-crossing-point of v_{ac} for the proposed resonant semi-DAB AC-DC converter and the multi-mode modulation strategy.

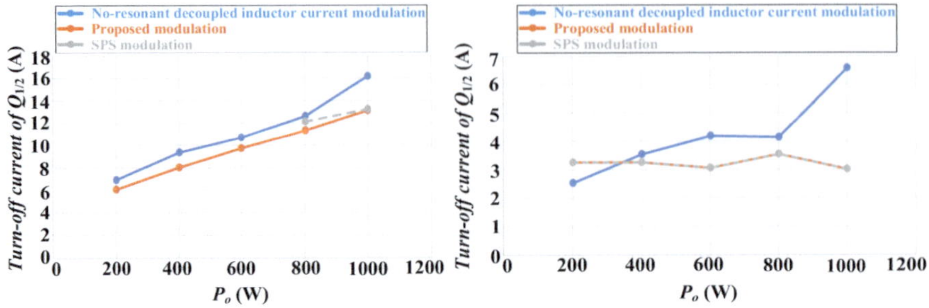

(a) V_o=270V (b) V_o=400V

Fig. 11.19 The maximum turn-off current of $Q_{1/2}$ under different load powers and different output voltages

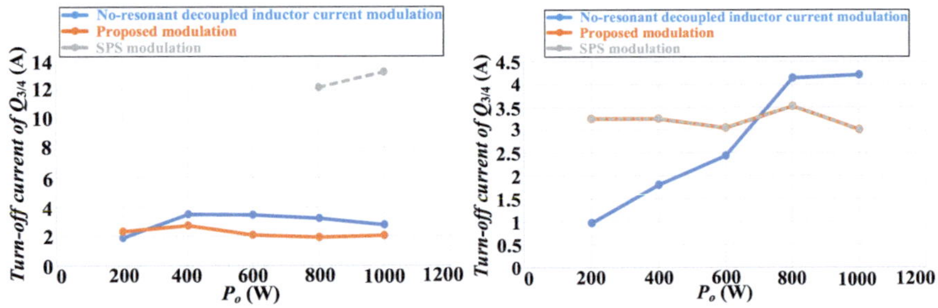

(a) V_o=270V (b) V_o=400V

Fig. 11.20 The maximum turn-off current of $Q_{3/4}$ under different load powers and different output voltages

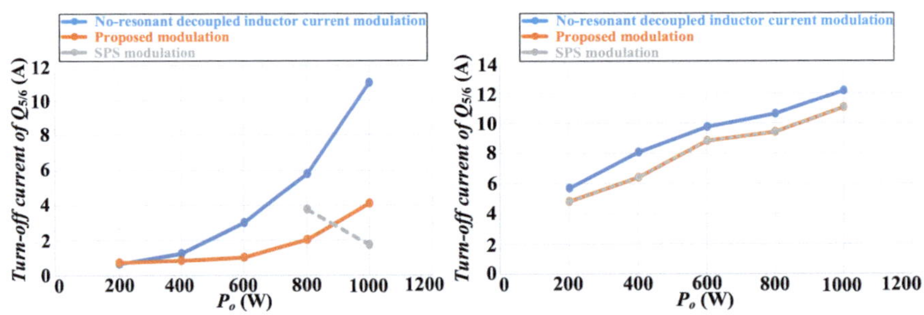

(a) V_o=270V (b) V_o=400V

Fig. 11.21 The maximum turn-off current of $Q_{5/6}$ under different load powers and different output voltages

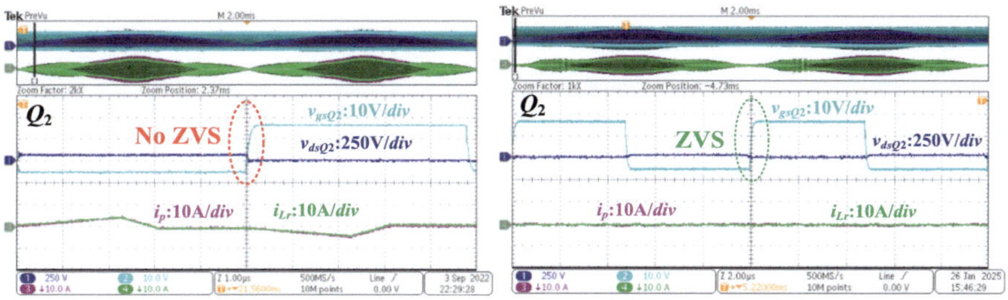

(a)no-resonant AC-DC converter (b)resonant AC-DC converter

Fig. 11.22 The ZVS achievement situation comparison for different topologies and control strategies under Vo = 270 V, Po = 1000 W

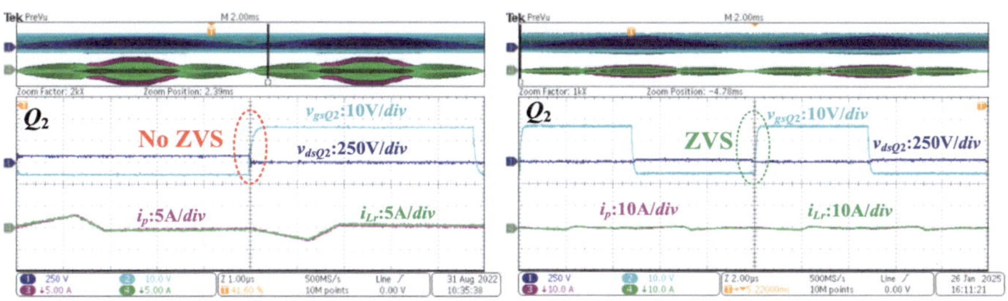

(a)no-resonant AC-DC converter (b)resonant AC-DC converter

Fig. 11.23 The ZVS achievement situation comparison for different topologies and control strategies under Vo = 270 V, Po = 400 W

11.4 Experimental Results

11.4.1 Experimental Parameters

A 1000W prototype was made to verify the effectiveness of the control strategy, the system parameters are listed in Table 11.4. The control and modulation strategy are implemented by TI DSP TMS320F28335.

11.4.2 Steady-State Operating Waveforms

Figure 11.24a–d show the steady state waveforms of grid voltage and current under the rated power and different output voltage respectively. As seen, high power quality can be obtained by using the proposed control strategy.

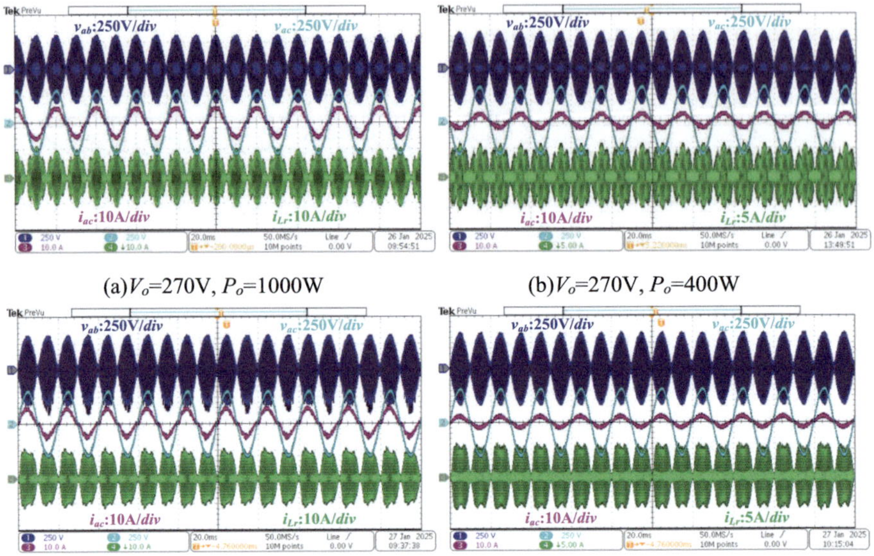

(a)V_o=270V, P_o=1000W (b)V_o=270V, P_o=400W

(c)V_o=400V, P_o=1000W (d)V_o=400V, P_o=400W

Fig. 11.24 The steady state grid side input voltage and current waveforms under the rated load power and different output voltages

11.4.3 Mode Operating Situations

Figure 11.25 show the waveforms of v_{ab}, v_{cd}, i_{Lr} and i_p within the entire ac line cycle and different operation modes under different working conditions respectively. Figure 11.25a is the result for the rated power and V_o = 270 V, there are three operating modes (Mode 1, Mode 2 and Mode 4) within the entire ac line period. Figure 11.25b is for V_o = 270 V and P_o = 400 W, there are four operating modes (Mode 1, Mode 2, Mode 3 and Mode 4) within the entire ac line period. Figure 11.25c is the result for the rated power and V_o = 400 V and Fig. 11.25d is the result for V_o = 400 V and P_o = 400 W, there is only one mode within the entire ac line period. The experiment results agree well with the previous theoretical analysis.

11.4.4 Mode ZVS Soft Switching Situations

Figure 11.26 shows the ZVS on situations of different switches at different voltage conversion gains under different load powers and different output voltages respectively. Only the situations of Q_2, Q_6 are given due to the symmetry between the upper switch and bottom switch in a half bridge. As illustrated in Fig. 11.26, the experiment results of the ZVS for switches agree well with the previous theoretical analysis.

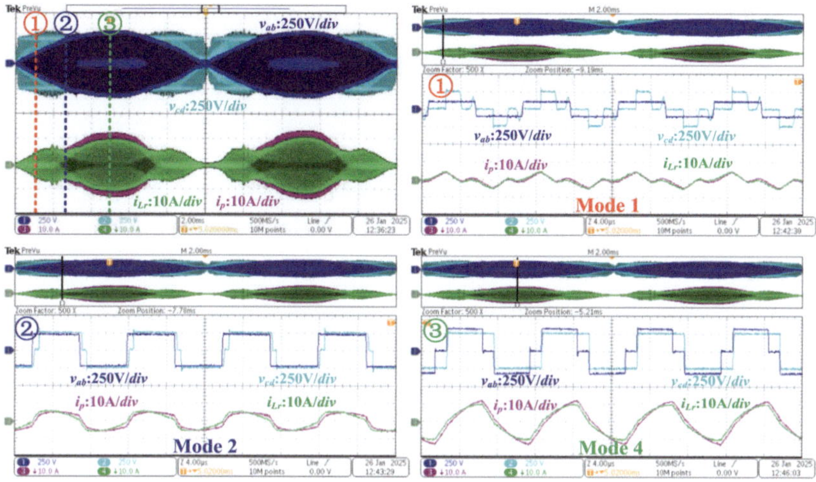

(a)V_o=270V, P_o1000W

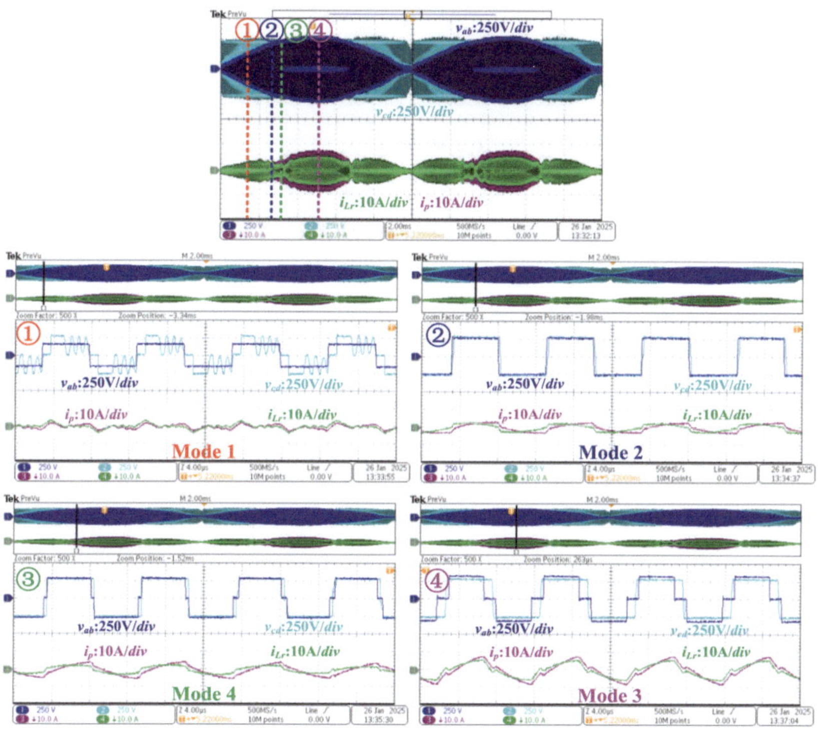

(b)V_o=270V, P_o=400W

Fig. 11.25 The mode operating situations under different working conditions

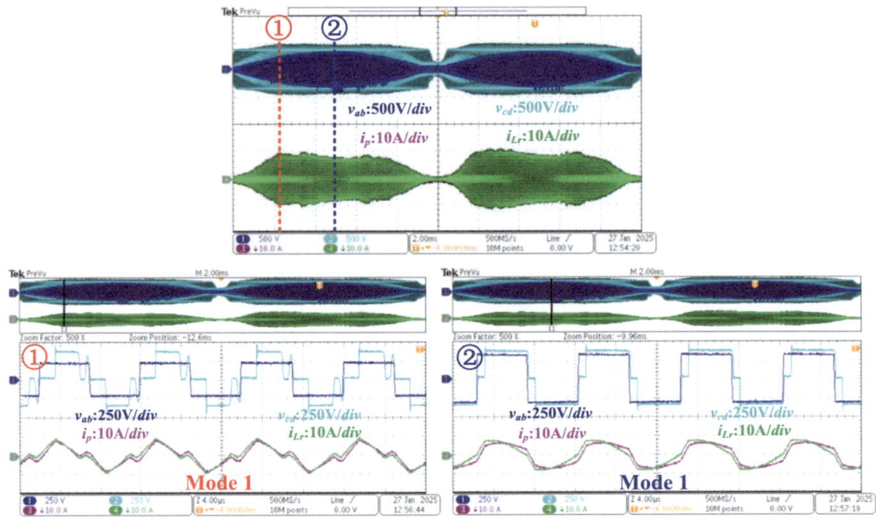

(c)V_o=400V, P_o=1000W

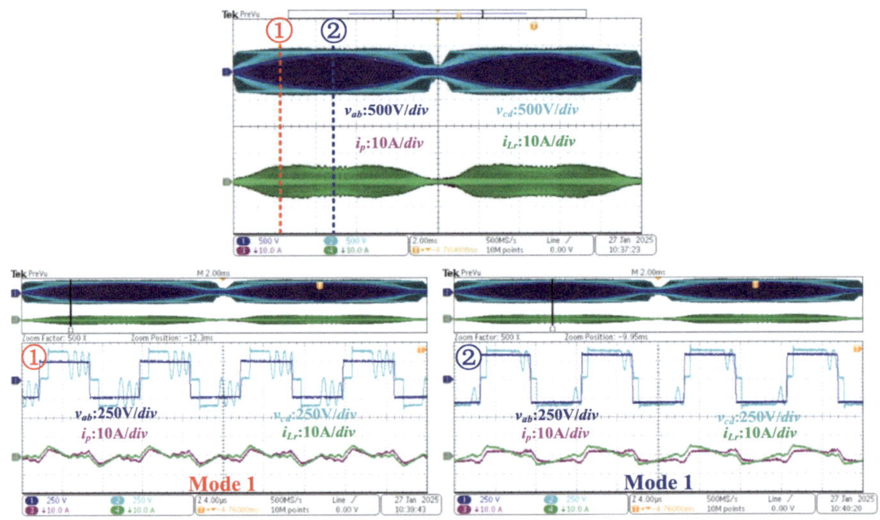

(d)V_o=400V, P_o=400W

Fig. 11.25 (continued)

11.4.5 Dynamic Experimental and Conversion Efficiency

Figure 11.27 shows the conversion efficiency by using different control strategies and output voltage respectively, it can be seen that the peak value of the conversion efficiency is 96.47% by using the proposed decoupled resonant inductor current

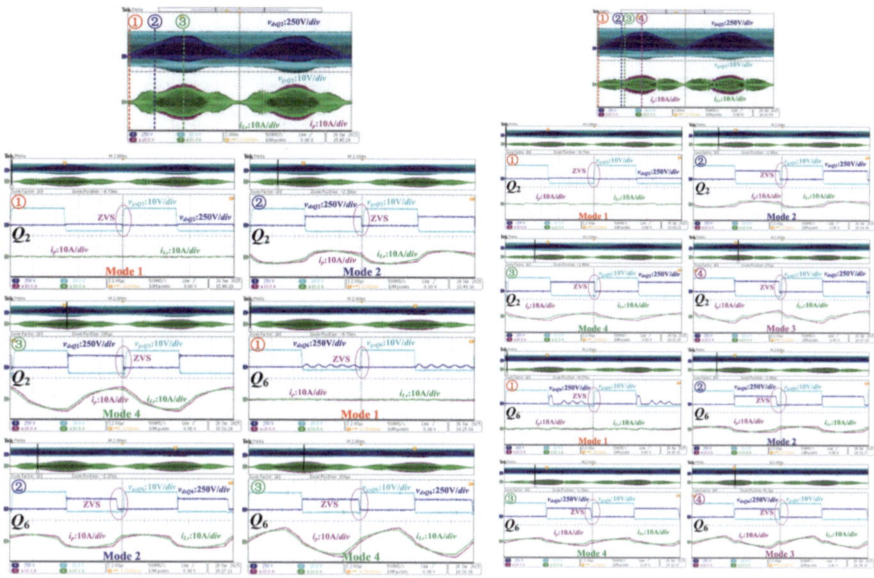

(a)V_o=270V, P_o=1000W (b)V_o=270V, P_o=400W

Fig. 11.26 ZVS achievement situations under the different working conditions

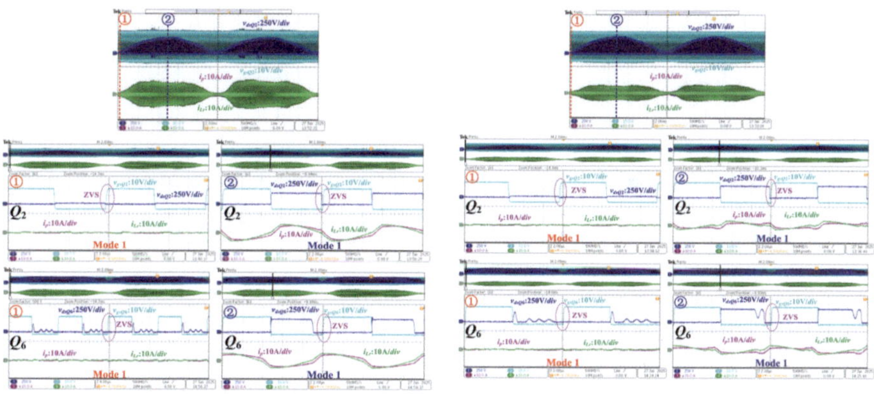

(c)V_o=400V, P_o=1000W (d)V_o=400V, P_o=400W

Fig. 11.26 (continued)

control strategy because of the lower turn-off current, the lower RMS value of the resonant inductor current, the lower peak current of the resonant inductor within each switching period and extremely wide ZVS on range.

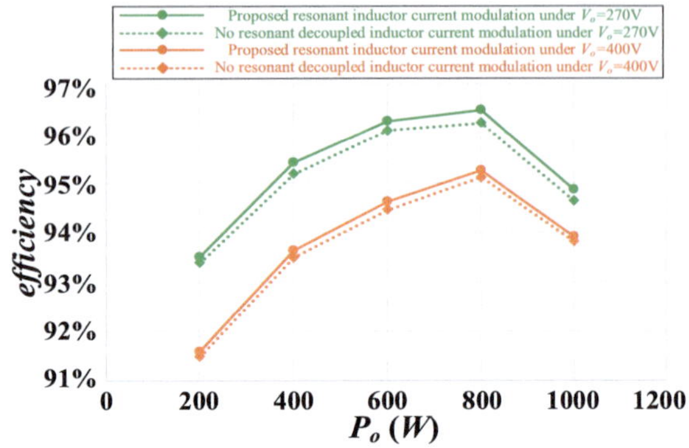

Fig. 11.27 Efficiency curve

11.5 Comparisons with Other Works

Table 11.7 shows the comparisons of the proposed topology and control strategy with some previous similar works. In order to make all switches achieve ZVS on within the full operating range for the semi-DAB single-stage AC-DC converter, an improved modulation and variable frequency control strategy is proposed in [10], while the control method is too complex. An optimal peak current control strategy is proposed in [11], while it is only suitable for the situation of $M > 1$. An DCM modulation mode is used within the entire AC line cycle for the resonant AC-DC converter in [12], which can reduce the turn-off current of switches, while the peak current is relative high due to the only DCM mode. A TPS modulation strategy is adopted within the entire AC line cycle for the LCL-DAB based single-stage AC-DC converter in [13], while the control with only one mode is not flexible for any electrical angle. Compared with the prior works, a resonant multi-mode modulation and control strategy is proposed in this paper, which could reduce the turn-off current for any switching period and the switches can achieve ZVS within the entire AC line cycle, which can improve the conversion efficiency.

Table 11.7 The comparisons of different topologies and controls for single-single AC-DC converter

	Semi-DAB AC/DC [10]	DAB AC/DC [11]	Resonant AC/DC [12]	LCL-DAB AC/DC [13]	Proposed one
Inductor current	Segmental straight line	Segmental straight line	Segmental curve	Segmental curve	Segmental curve
Rated power (W)	250	1000	1000	1000	1000
Control strategy	Variable-switching frequency	Constant-switching frequency	Constant-switching frequency	Constant- switching frequency	Constant-switching frequency
Switching frequency (kHz)	20~100	100	50	50	100
Operating modes	Seven modes	Two modes	One mode	One mode	Four modes
Current loop	/	/	√	/	√
Control complexity	Complex	Medium	Simple	Medium	Simple
ZVS range	Full range	Narrow	Wide	Wide	Extremely wide
Turn-off current	High	High	Middle	Low	Low
Maximum efficiency (%)	96.08	95.75	95.5	96.35	96.47

11.6 Conclusion

A novel single-stage high frequency AC-DC converter based on resonant semi-DAB and multi-mode control strategy with smooth transition is proposed firstly. Four novel resonant inductor current modulation modes are used and each mode has two control variables. The resonant tank is designed optimally for higher performance. The converter can make the primary side switch achieve ZVS on within full operating range using the proposed simple constant switching frequency control strategy and can obtain higher power quality. Compared with other similar modulation and control methods, the proposed modulation strategy can reduce the RMS value of the resonant inductor and the primary side current within half a complete ac line period, the maximum current stress of the primary side and secondary side, the turn-off current of switches under the most operating conditions. The power transfer capability, control principle and mode transition mechanism, parameter design are given. The effectiveness of the proposed circuit topology and the control method is verified by a 1000 W experimental prototype.

References

1. Jo S-W, Kim N-G, Kim M-H, Kim M-J, Kim M (2022) Highly efficient full-bridge resonant AC/DC converter using T-type voltage doubler. IEEE Trans Ind Electron 69(11):11076–11087
2. Uddin W, Wagaye TA, Kim M (2022) Quasi-single-stage current-fed resonant AC-DC converter having improved heat distribution. IEEE Trans Power Electron 37(11):13499–113512
3. Ma H, Lai J-S, Zheng C, Sun P (2016) A high-efficiency quasi-single-stage bridge less electrolytic capacitor-free high-power AC-DC driver for supplying multiple LED strings in parallel. IEEE Trans Power Electron 31(8):5825–5836
4. Forouzesh M, Liu Y-F, Sen PC (2023) A line cycle synchronous rectification strategy based on time-domain analysis for single-stage AC-DC LLC converters. IEEE Trans Power Electron 38(4):5077–5091
5. Luo H, Zang T, Zhao C, Chen S, Zhou Y, Qiu Y, Wang X, Zhou X (2023) Power boundary controlled single-stage LLC power factor correction converter and its optimal parameter design. IEEE Trans Ind Electron 70(12):12219–112232
6. Zhou B, Zhao C, Luo H, Qiu Y, Chen S, Zang T, Zhou Y, Zhou X, Zhao H (2024) Hybrid control strategy for single-stage interleaved totem-pole LLC AC-DC converter with input and output power decoupling. IEEE Trans Power Electron (Early Access)
7. Chen T, Yu R, Huang AQ (2022) A bidirectional isolated dual-phase-shift variable-frequency series resonant dual-active-bridge GaN AC–DC converter. IEEE Trans Ind Electron 70(4):3315–3325
8. Zhou S, Li X, Wang H, Huang P (2022) Modulation of a semi dual-active-bridge resonant converter with soft-switching and zero backflow power. IET Power Electron 15(6):515–530
9. Ma P, Sha D (2023) A single-stage AC–DC converter based on semi dual-active-bridge with decoupled inductor current modulation strategy. IEEE Trans Power Electron 38(8):10170–10182
10. Ma P, Sha D, Zhang D (2024) High efficiency and full range ZVS single-stage semi-DAB AC-DC converter with improved modulation and variable frequency control. IEEE Trans Power Electron 39(5):6003–6016
11. Sha D, Zhang D, Zhang J (2021) A single stage dual-active-bridge AC–DC converter employing mode transition based on real-time calculation. IEEE Trans Power Electron 36(9):10081–10088

12. Kim M-H, Park J-W, Kim M (2023) Bridgeless filterless neutral-point-clamped resonant AC/DC converter. IEEE Trans Ind Electron 70(9):8895–8906
13. Guo Z, Zhan W, Chen Z (2024) Single-stage isolated AC–DC converter based on LCL resonant DAB converters with single-loop control. IEEE Trans Power Electron 39(8):9305–9317

MIX
Papier aus verantwortungsvollen Quellen
Paper from responsible sources
FSC® C105338

If you have any concerns about our products,
you can contact us on
ProductSafety@springernature.com

In case Publisher is established outside the EU,
the EU authorized representative is:
Springer Nature Customer Service Center GmbH
Europaplatz 3, 69115 Heidelberg, Germany

Printed by Libri Plureos GmbH
in Hamburg, Germany